'A splendid new book ... all rational argument seems to be on his side.'
—Professor Tim Flannery, *The Guardian*

'A wonderful book, visionary, illuminating and fascinating.'
—George Monbiot

'It has shaped how I think about my farm, and the choices we make about our land ...'
—James Rebanks

'A visionary yet practical book.'
—John Burnside, *New Statesman*

'An exposé, a plea, and a vision of a better future.'
—Simon Reeve

'Ben Macdonald has an impressive track record as a field naturalist, wildlife film-maker and writer, and this passionate, authoritative, up-to-date and, ultimately, optimistic book is a worthy comparison to such seminal works as George Monbiot's *Feral* and Mark Cocker's *Our Place*.'
—Jonathan Elphick, *BBC Wildlife*

'Having read a number of the recent books about rewilding, I was tempted to think "Oh blimey, not another one!" I am now tempted to say "they left the best till last ..."'
—Bill Oddie

'*Rebirding* is beautifully written, based on deep, personal experience and a genuine love of the subject. You may not have come across Ben Macdonald before now; but believe me, you will hear a lot more from him in the future.'
—Stephen Moss

'With Monbiot's *Feral* and Isabella Tree's *Wilding*, *Rebirding* sits separate from both and is in fact an essential third book to read if you've enjoyed the others. In short, it's a captivating, fascinating and inspiring read.'
—Ed Stubbs, *Birdwatch* magazine

'This is the best book on nature, conservation and rewilding I read in 2019 – perhaps one of the best I've ever read. I finished reading it with a real sense of hope for the future.'
—Alex Roddie, *Great Outdoors* magazine

'A book about a key subject at a key time, passionate and deeply thought-through. Anyone concerned with the future of the natural world in Britain will want to read it.'
—Mike McCarthy, author of *The Moth Snowstorm*

'A beautifully written, thoughtful and yes, provocative book.'
—Dr Martin Harper, Conservation Director, RSPB

'This is a stimulating and important book, beautifully written and well researched … It provides a compelling vision for the future.'
—Dr Carl Jones, Durrell Wildlife Conservation Trust

'A must read and a good read … the type of book that grabs and keeps my attention. You should read it and I think you may well enjoy it a lot.'
—Dr Mark Avery

'I thoroughly recommend the book and applaud its breadth and detail … Macdonald's book has really surprised me. I have learned much I did not know about Britain's early bird faunas, and even the history of its mammals … The level of treatment and scholarly references are on a par with conservation science books.'
—Peter Taylor, *ECOS* magazine

'This has to be the one book you read this year if you read no other, as its messages are myriad and its import undeniable. This is most definitely my book of the year and possibly the whole decade!'
—Bo Beolens, *Fatbirder* website

REBIRDING

REBIRDING

Restoring Britain's Wildlife

BENEDICT MACDONALD

PELAGIC PUBLISHING

First published by Pelagic Publishing in 2019.

This revised and updated paperback edition published in 2020 by Pelagic Publishing.

Reprinted 2020.

www.pelagicpublishing.com
PO Box 874, Exeter EX3 9BR, UK

978-1-78427-219-7

British Library Cataloguing in Publication Data
A catalogue record for this book is available from the British Library.

Cover image © 2018 Graham Carter
www.graham-carter.co.uk

Contents

To my grandfather, Frederick Thomas Irving Giltinan (1915–2004). You were a true inspiration, a wise and wonderful guide, and are a sorely missed friend. May you rest in peace.

Foreword

Stephen Moss

Those of us who are passionate about Britain's birds and wildlife are aware of a paradox. Over the past few decades, we have witnessed a rise in enthusiasm for the natural world, beyond the wildest dreams of anyone who, like me, came of age in the 1960s or 1970s. From a niche hobby pursued mainly by men, which most of us were rather ashamed to admit to, birding has entered the mainstream. A love of birds – and indeed all wildlife – is now as readily accepted in society as, say, a passion for football, art or music. TV programmes such as *Springwatch*, and presenters like Chris Packham, Bill Oddie and of course Sir David Attenborough, have made what was once dubbed 'organic trainspotting' not just respectable, but desirable.

The paradox is that, during the very same period, we have witnessed a catastrophic decline in Britain's birdlife. Since 1966, when England famously won the football World Cup at Wembley, we have lost an estimated 44 million breeding birds. That includes the astonishing total of 20 million house sparrows – that's 50 individual sparrows every single hour, for the past half-century. Once-familiar farmland birds such as the grey partridge, linnet and yellowhammer are in sharp decline, and we face the very real possibility that the turtle dove will soon follow the wryneck and red-backed shrike into extinction as a British breeding bird.

It's not just birds: we have also witnessed a catastrophic decline of flying insects: the 'moth snowstorm' that used to occur on warm summer nights is now just a distant memory, and the sight of wild flowers and butterflies, or the sound of buzzing insects, is becoming increasingly unusual in much of rural Britain. Many mammals, including favourites such as hedgehogs as well as less glamorous species (notably bats), are also in freefall.

There have been some good-news stories as well: the return of birds of prey such as the buzzard, osprey and marsh harrier, the reintroduction of the red kite and crane, and the astonishing arrival of many species of long-legged waterbirds on my home patch, the Somerset Levels, should all be welcomed.

Likewise, the restoration and reconstruction of habitats – from the Avalon marshes in Somerset to the Caledonian pine forest in Scotland, are to be applauded as a step in the right direction. But they cannot make up for what we have lost.

Despite the claims from organisations such as the National Farmers' Union and the Countryside Alliance that rural Britain is in safe hands, these self-appointed 'custodians of the countryside' have, through their support of intensive farming, presided over the devastation of much of Britain's wildlife.

In response, the past decade has seen a subtle shift in the bestselling genre known as 'new nature writing'. Authors have moved away from simple descriptions of wild creatures, or personal reflections on place and nature, towards a more militant approach. Fired up by the concept of the 'Anthropocene' – a newly coined term to describe the devastating effects humans are having on the planet – nature writers are now showing their deep concern at what has happened on our watch.

Led by the outspoken conservationist Mark Avery, with his 2012 book *Fighting for Birds*, some authors finally began to get off the fence and say what they thought. The following year, George Monbiot weighed in with *Feral*, which offered an alternative solution to the prevailing interventionist approach to conservation. In 2016, I published *Wild Kingdom*, which examined the plight of our habitats and their wildlife, with specific examples of how people are working to bring back some of our special wild creatures and restore the places where they live. Recently, one of the most respected authors in the genre, Mark Cocker, also nailed his colours to the mast with his 2018 book *Our Place*. Subtitled *Can we save Britain's wildlife before it is too late?*, this is a devastating and deeply personal account of the very palpable loss of species the author has witnessed since his 1960s childhood.

Mark Avery, Mark Cocker and I are all of the same generation: we cut our teeth as birders in the heady days of the 1970s, when naturalists seemed to care little or nothing about the wider countryside. Ben Macdonald is of a different generation – the same age as my oldest son. He is one of a group known as 'millennials', the children of the post-war baby-boomers. Millennials are often defined (mostly by older journalists) as digitally aware, obsessed with their electronic devices, and far less likely to have a concern for the real world – let alone nature – than us older, simpler folk. That, I suggest, is utter nonsense.

There may well be millennials who spend their lives in darkened rooms staring at screens; but there are also many more who engage with the wider world – including nature – at least as deeply as we did. Indeed, I would argue that because of the rise in popularity of pastimes such as birding, there are now far more young people out and about in the countryside than there were back in our day.

Organisations set up and run by young people, such as A Focus on Nature and Next Generation Birders, have used social media to form links, connections and networks. Today – as their older members reach their thirties – they are starting to enter the mainstream of conservation organisations, politics and the media. What also marks out this generation is that they are fully aware – in a way we never were – of the catastrophic plight of Britain's wildlife. But what also sets them apart is that they are determined to do something to reverse the trend: bringing back our natural heritage. Hence this book.

What makes *Rebirding* so different from what has gone before is that it is not written from the point of view of an older author lamenting what we have lost. Instead, it comes from the refreshing, energetic viewpoint of a younger writer who knows that unless we do something – and soon – he will witness the virtual disappearance of most wildlife from the British countryside during his own lifetime.

What marks out *Rebirding* from most other nature writing is that it is not just about declines, but also provides pragmatic and workable solutions. Ben Macdonald combines a hard-nosed understanding of the political, social and economic issues faced by wildlife – and people – with an infectious positivism. Ben has worked out what we need to do to bring back Britain's birdlife, and sets out his proposals in a clear and realistic way. This is a radical manifesto for change: refuting the notion that we are running out of land, by showing that in reality this supposedly crowded island has all the space we need for nature. Ben suggests that we need to reform the way we use the uplands – allowing blanket forestry plantations and sheep-grazed landscapes to rewild – and reform shooting and hunting estates for the benefits of wildlife. Perhaps the most surprising part of this book is his well-reasoned point that we would not need to impinge on essential food production to get our wildlife back.

Ben has also backed up his argument with facts and figures that show first, how easy it would be, and second, the benefits it would bring – not least to the significant minority of Britons who, like me, live in rural areas. And he realises that instead of antagonising those 'on the other side', including many farmers and landowners, we all need to make people aware of how these radical changes could help them and their communities thrive. Put simply, if implemented, this manifesto would solve rural depopulation issues, and the decline of many countryside communities, at a stroke, by offering real prosperity to the majority of people, not just riches for the few. Nature makes money. It doesn't threaten rural jobs, as is often claimed, but creates them.

Rebirding will not please everybody; indeed, it will make uncomfortable reading for those for whom intervention, and the direct management of habitat, is still the prevailing philosophy. And while habitat management is of course sometimes necessary, Ben makes a very persuasive argument for setting large areas aside and letting nature – along with the strategic reintroduction of keystone species – do its own thing.

What I really love about this book is that it doesn't just go over the same tired and familiar issues. Instead, it offers a revolutionary new approach. This not only could work, but needs to succeed, if we are to stop fiddling at the margins and do something that actually restores Britain's birds and wildlife to their rightful place at the centre of our nation, culture and society.

Rebirding also offers a compelling vision: of a richer and healthier Britain, with flocks of pelicans drifting over our wetlands, wildcats and capercaillies returning to our increasingly wooded hunting estates, and wildlife safaris to watch lynx and golden eagles in Snowdonia – all within a generation.

Not least, *Rebirding* is beautifully written, based on deep, personal experience and a genuine love of the subject. You may not have come across Ben Macdonald before now; but believe me, you will hear a lot more from him in the future.

Mark, Somerset
June 2018

Preface to the Paperback Edition

Since the publication of *Rebirding* just over a year ago, the grassroots support and feedback has been humbling and inspiring for both author and publishers. From upland farmers to large landowners, from NGOs to individual conservationists, many people have fed back to us a shared wish to see an extraordinary recovery in Britain's vanishing wildlife.

As a result, this edition has not only been updated but now includes further insights into a number of topics. These include the pivotal role of beavers in creating and shaping landscapes, the historical place of fire in Britain's grasslands, the parts played by animals such as wolves and elk, and more about the Atlantic rainforests that would have covered our steepest hillsides. In addition, in later chapters, more attention is paid to the collapse of the ecosystems within our soil, and the importance of mixed upland farming for biodiversity in agricultural landscapes.

In addition, the book has been brought up to date with a number of new highlights in the world of British ecological restoration – from the return of white-tailed eagles to the Isle of Wight in 2019 to the hatching of the first white storks since the 1500s, on the Knepp Estate in Sussex, in May 2020.

Benedict Macdonald
June 2020

With Thanks

Hundreds of inspirational naturalists, producers, writers, conservationists, guides and friends have, in different ways, helped me write this book, both over the past four years and in the previous twenty years of my growing fascination with British nature. Listed below are just a few of those who have been most helpful and most kind.

I would like to extend particular thanks to Professor Ian Newton, for his thorough review of the science behind Chapters 3 to 6; Isabella Tree, for her assistance with the mechanics of rewilding; Nicholas Gates and Ruth Peacey, for their feedback on the manuscript. Most of all, to my parents, Liz and Ian Macdonald, for their love and generosity throughout the fledging process.

Matthew Aeberhard, John Aitchison, David Attenborough, Gary Atterton, Mark Avery, Dawn Balmer, Carl Barimore, Jez Blackburn, Mark Blake, Marek Borkowski, Adam Bradbury, Catherine Brain, Hugh Brazier, Victoria Bromley, Neil Burke, Charles Burrell, John Calladine, Geoffrey Carr, Kenneth Carruthers, Graham Carter, Peter Cairns, Peter and Richard Castell, Adam Chapman, Elisabeth Charman, Timothy Chiles, Peter Conrad, Dave Cooke, Huw Cordey, Hywel Davies, Roy Dennis, Graeme Dickson, John Dries, Alastair Driver, Shelagh Fagan, Terry Fenton, Jolyon Firth, Alastair Fothergill, Nicholas Gates, David Gibbons, Edward Gilbert, Frederick and Betty Giltinan, Terry and Elizabeth Giltinan, Ben Goldsmith, Derek Gow, Mervyn Greening, Joanne Harvey, Katherine Hindley, Ben Hoare, Mark Holling, Stephen Hollinrake, Simon Holloway, Nathanael Hornby, John Hudson, Jonnie Hughes, Robin Husbands, Cressida Inglis, Felicity Jones, Rob de Jong, Martin Kelsey, Johnny Kingdom, Gordon Kirk, Waldemar Krasowski, Werner Kunz, Alan Law, Dave Leech, Roger Lovegrove, Nigel Massen, Matthew Merritt, Catherine Miller, George Monbiot, Stephen Moss, Ash Murray, Julia and Lisa Newth, Ian Newton, Janos Olah, Chris Packham, Ruth Peacey, Bernard Pleasance, Anders Povlsen, Margaret Power, Matthew Price, Matt Prior, Ella and William Quincy, Connor Rand, Stephen Roberts, Allan Rustell, Ashley Saunders, Keith Scholey, Anna Scrivenger, Bob Sheppard, Christina Shewell, Guy Shorrock, Hector Skevington, David Slater, Ken and

Linda Smith, Adrienne Stratford, Adriana Suarez, John Swallow, Julian Sykes, Caroline Thomson, Isabella Tree, Mark and Jack Vaughan, Frans Vera, Mark Ward, Alan Watson Featherstone, Nick Williams, Rebecca Wrigley, Derek Yalden.

Introduction

The environmental movement up to now has necessarily been reactive. We have been clear about what we don't like. But we also need to say what we would like. We need to show where hope lies. Ecological restoration is a work of hope.

—Alan Watson Featherstone[1]

The sun strikes the wooded river; orange and blue. The hawthorns beside you blast with nightingales. A cuckoo bubbles away, darting between bushes in search of an unwary bird's nest into which to pop her imposter's egg. Far out on the green lawns of the river's winter wake, brick-red godwits, green-suited lapwings and dozens of ruff glow, luminous, below the early-morning sun. The electric shiver of pirouetting snipe and the rising kettle sound of curlews fill the morning air.

As the sun heats the marsh, turtle doves start purring from a stand of bushes. An elk, hiding in the beaver-coppiced willows, bursts away with a disgruntled yell. A family of ravens croak at something in the woodland far beyond. It could be the local lynx, with a freshly caught roe deer, but you will not see it this morning. You've not seen it in ten years.

The air is layered thick with skylark song. Red-backed shrikes pop up on bush tops. High on their menu, a dung beetle saunters past, looking for a place to dig a tunnel, in which to hide his precious cargo. A wild cattle bull, quietly browsing the hawthorn, gazes at you with placid lack of interest.

Beyond the valley, a wavy world of birches, willows and the iron frame of English oaks slowly shed their mist. Something is complaining; it sounds as if someone's put a pillow over an angry kestrel. It's a wryneck, calling harshly for its mate. In an open stand of trees, you pick out the boomerang circuit of a bird on an invisible string. Spotted flycatchers are seeking the first of the morning's butterfly clouds.

Giants take to the skies. Bugling cranes fan out from the valley's reedy heart, nudging one another as they float over your head. The lumbering rectangle of a white-tailed eagle rises from its willow nest. Soon there are several in the air, circling the marshes to fish. Then, something of the past: something too large and extraordinary to still exist. A Dalmatian pelican. On three-metre wings, the curly-headed giant crosses the valley without a single flap.

How wonderful to have all this in Britain. How amazing that against all the odds, we have places where our natural heritage runs free. Who would have thought, fifty years ago, that such things would be possible? How proud we must be of our own wild places; how pleased we do not need a passport to enjoy exceptional nature.

Except – as things stand, there is little prospect of such a future. In 2020, Britain has no such places. As the current consensus stands, it never will. Your grandchildren will never hear the purr of a turtle dove or the drone of a hundred bees. They will never learn the song of a nightingale or understand the meaning of 'like moths to a flame', let alone smile at the prospect of a pelican. They will walk in factory landscapes even more silent than those of today.

We are now approaching what some scientists term the sixth mass extinction. Since 1970, there has been a 58% decline in the number of fish, mammals, reptiles and birds worldwide.[2] With many British wildlife species accelerating in their decline, who is to say quite how much poorer our grandchildren's world will become?

In Britain, we have been removing fauna from our island for millennia. Now, as the insect food chain collapses around us and the populations of many fragile birds become isolated and vanish, turtle doves are set to be extinct in under ten years' time. Wood warblers, nightingales, cuckoos, curlews, willow tits and many others free-fall to extinction. Forty-four million individual British birds have vanished since 1966.[3] Wildlife bleeds from our countryside and from our daily lives.

We might blame climate change, or migration patterns, but such declines cannot be seen in the traditional farmlands of many other parts of Europe. In the last two hundred years, Britain has driven more species to the brink than any other European country. With models of conservation management having reached their limits, unable to save landscapes or rebuild our broken food chain, it's time for a new plan.

We need to restore the huge areas in our country where nature can look after itself, and many of the native mammals that once took care of our wildlife. Britain has all the space it needs for nature. Over 82% of British people live in urban areas.[4] Just 6% of our island is built upon.[5]

Snowdonia is larger than Kenya's famous Maasai Mara.[6] The Cairngorms is still half the size of Yellowstone.[7]

Birds are not dying out beside us, swallowed by new housing, but vanishing from our rural deserts – places where we have all the space needed to save them. Such areas, with failing or damaged economies, many of them funded by the taxpayer alone, await the return not just of our wild heritage, but of thousands of new jobs – and billions in income.

With everything to play for, let's take the initiative. Let's be the first generation since we colonised Britain to leave our children better off for wildlife – the first to restore the landscapes that rightfully belong to our country. It's time, at last, for Britain's diverse and mighty nature lobby to say what they *would* like.

This book takes you back in time, to when humans first set foot on these islands; revealing the fantastic wildlife that we once had and how this has changed over time. It explains why British wildlife is vanishing – and how it can be saved. It sets out ideas for what we could enjoy in the future, how that is possible, and why this would benefit not only our wildlife but our economy and culture. In the words of the great conservationist Alan Watson Featherstone, 'ecological restoration is a work of hope'. This book aims to show where that hope lies.

CHAPTER 1

Taming Britain

The Retreat of the Giants

Everywhere on earth, living systems have been radically altered by the loss of great beasts.
—George Monbiot[1]

The travellers had staggered for days across a sea of rocks. Ahead, at last, lay the prospect of new life. A sea of grasses rose, by inches alone, from the wound-eating water and the jagged coastal spikes. The new sun was raw on their backs. The dawn light washed the land ahead with an orange glow. A plain. Filled with possibility. A new life.

A wide-eyed harem of horses jumped from the water's edge as the travellers waded, foot-wrecked, ashore. The frost blinked, deep, in the grass. Strange trunk-bearing antelopes with curly pink horns burst from the travellers with nasal grumbles. Waifs in the frost, demoiselle cranes bugled with alarm, skipping away across the glinting steppe. But the giants simply stood. Tusks arched beside their heads, southern mammoths, lords of the polar Serengeti, had come to the coast to feed. Woolly rhinoceros, armed with two horns – one huge, one giant – twinkled, indifferent; the frost wedged deep in their coats.

Our travellers watched this parade of formidable wool and towering mountains of meat. All the giants saw, though, were small bedraggled mammals, walking on just one pair of their legs. An inconsequence, a hairy wreck. These sodden intruders were no cave lions.

Our travellers, in turn, may have felt wonder at what they saw. They may have admired the beauty of the giants: their granitic massivity, the sculptured shine of their horns – the tallest shapes for miles around on Norfolk's coast. But the travellers would also have seen opportunity. It would soon be time to hunt, but now, it was time to move.

Scimitar cats, the size of small horses, would soon be following the mammoths, but our smaller travellers would be far easier prey. Deep-set eyes and heavy brows set hard against the odds, the nomads continued ashore. That night, home would be hard won. Who knew what lurked in those coastal caves? Had these early hunter–gatherers shared plans, shared fear, shared excitement at the bounty of food before their eyes, it would not have been in words we understand today. This was almost 900,000 years ago. And these intrepid nomads were the first known colonists of the British Isles.

The story of changing Britain, of taming ecosystems, of wildlife decline, began even in these early hours. In this chapter, we'll explore how Britain's original ecosystems shaped the evolution of its wildlife. We'll travel through time to the Industrial Revolution, learning how we tamed Britain and made it our own. The changing fortunes of British wildlife begin with the story of us. To restore the beauty of natural Britain, we must recall what 'natural' looked like. We must learn how 'natural' changed over time. And to restore Britain's wildlife, we need to take a longer, wider, wilder view.

A Land of Giants

In 2013, on the shores of Happisburgh in Norfolk, rough seas eroded the sandy beach. Scientists scrambled to photograph something amazing: the earliest hominid footprints outside of Africa, perfectly preserved in the mud. These footprints, almost 900,000 years old, revealed that an early hominid, *Homo antecessor*, had set foot in Britain far earlier than anyone had thought.[2] Hours later, the footprints were washed away forever.

The period in which our ancestors arrived was the Pleistocene.[3] In a time-frame of hundreds of thousands of years, the Pleistocene was an age of extremes – of shifting warm and cold, of giant beasts. Glacial and inter-glacial periods would transform the character of the British landscape several times, before we arrived at our modern temperate climate.

Our pioneers had walked across land to an island yet to be: an island connected to Europe. Norfolk's climate, at that time, had warmer summers but much harsher winters than today. Our ancestors arrived as dwarfs in a land of giants. Woolly rhinoceros, giant elk and southern mammoth, casting their shadows over saiga antelope and wild horses, would have been the least of their concerns.[4]

Male sabre-toothed cats weighed up to 400 kilograms.[5] Cave lions, an extinct, larger subspecies of the African lion,[6] are known to have crunched their way through less-than-cuddly cave bear cubs.[7] The fossil record suggests these giant cats, like our ancestors, loved a good cave to call home. The early human inhabitants of Britain, no doubt, had regular tenancy issues on their hairy hands.

Early on, our ancestors would get to work hunting down the giants around them. Further excavations at Happisburgh reveal that flint hunting weapons were crafted as early as 830,000 years ago.[8] At least 400,000 years ago, early humans learned to hunt and kill straight-tusked elephants with wooden spears.[9]

Pristine habitats are shaped by giant animals, as can be seen in the wilderness of Tanzania, Botswana and Zambia to this day. Removing giant stewards profoundly changes the richness of a habitat. So habitat change in Britain has been going on for a very long time.

Early Birds

In addition to the giant animals that once wandered our shores, Britain's caves provide fascinating opportunities to uncover the birds that once called our island home. Thanks to decades of work by ornithologists and palaeontologists, we have some idea of the early birds that sang as our ancestors struggled to survive. *The History of British Birds*, by the late Derek Yalden and Umberto Albarella, provides a brilliant account of our winged fossil record.[10]

Pre-dating hominid arrival was an English albatross, *Diomedea anglica*, gracing our oceans 3 million years ago. It is one of the earliest modern bird fossils, unearthed from three sites on the Suffolk coast. An extinct relation of the short-tailed albatross, a graceful wanderer of today's Pacific, these giants once set out on five-year flights from Britain – bidding farewell to lifelong mates on a coastline still prized for its birdlife today.

Overseas, in Tanzania's Olduvai Gorge, lie fossils of birds such as corncrakes and whimbrels, which spend the winter in Africa but breed in Eurasia. These date back 1.9 million years. This suggests that the Palearctic migration, whereby birds fly from Africa to breed in Europe, was already under way at this time. The inspiring journey of travellers like swallows, seeking the comfort of the British summer, goes back a long time indeed.[11]

As archaeologists have dug backwards through time, fossils have revealed to us more about our forgotten polar Serengeti. These records include brown bears, wolverine, reindeer and steppe lemming, and a now-extinct western partridge that scuttled across our polar steppe up to 125,000 years ago. In the Creswell Caves of Derbyshire lie the bones of demoiselle cranes. We might imagine these elegant but feisty birds chasing off a nosy mammoth calf.

Our interglacial periods, less famous in schoolbooks than our ice ages, were enormous in duration. Fossils of Cory's shearwaters, now a Mediterranean seabird, have been found in caves on the Gower peninsula of south Wales, hinting at the warmth of these periods when the ice cap was far to the north. In these warmer times, freezing steppes gave way to fertile wooded grasslands, perhaps most similar to those of the Serengeti, Okavango floodplain or Luangwa river valley today.

Enormous cave hyenas, now extinct, became our commonest large predators.[12] Straight-tusked elephants and hippopotamuses grazed the fertile wooded plains of the Thames valley.[13] Beds of elephant bones from human hunts, uncovered in Essex, take us back to a time when we were harvesting the giants around us. Yet the fossils show that alongside our elephants dabbled humble gadwall and other ducks. We often forget that our birds evolved in the wake of landscape managers far larger than those of today.

The warm-era fossil record of Port Eynon cave, in Gower, is filled not only with the bones of cave hyenas, but familiar Welsh birds – skylarks, swallows, starlings and red kites. These are birds of spacious grasslands and scattered trees: habitats consistent with the action of giants. Elephants and rhinos maintain rich open grasslands with stands of trees. Skylarks and starlings, it would seem, took only much later to the human grasslands of the farm. And as we will see later in this chapter, almost all British land birds are best adapted to a mosaic of trees and open land: a mosaic that pre-dates any kind of human farmland.

Between 13,000 and 10,800 years ago, the British landscape plunged into transition, moving from polar steppe towards a warmer climate. The cave of Soldier's Hole, in Cheddar, Somerset, contains the fossils of ptarmigan, but also black grouse and hazel grouse. Such fossils reveal the changing nature of our mountain tops over time: from arctic wilderness towards a wooded world.

The trees surged back. The fossil record corroborates the suggestion of climatologists that as the climate warmed, Britain moved first towards a 'taiga' landscape, rich in willow and birch. The last records of hawk owls and pine grosbeaks, now found in similar landscapes in northern

Scandinavia, date from this time. The diminutive hazel grouse also vanished too early for hunting alone to account for its decline, which may also have been due to our taiga woodlands changing naturally in their composition over time.

The fossil record at this time reveals other exciting birds. One is the black stork, which, prior to Britain's deforestation, may well have soared over the lowland forest bogs of our country. Most fascinating of all, however, is the lost history of that goshawk-killing giant, the eagle owl. The last proven fossil of this owl comes from Demen's Dale, in Derbyshire, 10,000 years ago. But eagle owls are not tied to the taiga zone. They thrive across Europe, nesting, often, in caves. If not driven out by a changing climate, was this giant bird, aggressive around its chicks, our earliest avian adversary? Was it hunted from our cave homes – too furious to tolerate, too huge to elude our detection?

And why were so many birds found in caves at all? Some, like swallows, would have nested in them. Others may have been washed in by tides, or been brought in by ravens – or by that common giant of the skies; those white-tailed eagles that nested in our caves.

The Great Extinction

What happened around the end of the last glacial period, 13,000 to 8,000 years ago, was a shocking loss of large animals that played out across the temperate world. The Quaternary extinctions were the most extreme loss of the planet's wildlife since the disappearance of the dinosaurs. In a relatively short space of time, North America lost its four-tonne giant ground sloths, its giant armadillos and its mastodons. And similar losses occurred in Britain. Woolly mammoths held on here until 14,000 years ago.[14] Their cave lion hunters vanished at a similar time.[15] The bones of woolly rhinoceros were still being used for painting by our ancestors at Creswell 15,000 years ago, but these grazers had vanished before 10,000 years ago.[16] Their sabre-toothed predators vanished around the same time.[17] Giant elk died out in the British Isles around 9,000 years ago – on the Isle of Man.

Climate change and the 'overkill hypothesis' – the notion that we hunted too many of the giants for them to survive – have long vied as explanations for this extraordinary loss. Why did whole ecosystems vanish, in ecological terms, overnight? There are severe flaws in the theory that climate change alone drove such giants to extinction.

Our climate was certainly warming at this time, and the land was becoming more wooded. But megafauna shape the conditions in which trees grow, as surely as human foresters today. Large herbivores do not live within grasslands like the Serengeti: they create them. If our giant herds had been healthy, they might well have been able to survive the changing climate, as they had survived a changing climate many times before. But by the time the last ice age came to an end, we had harvested these slow-breeding animals, in confined areas, for hundreds of thousands of years.

Straight-tusked elephants vanished from the Iberian peninsula long before their cold-adapted cousins, the mammoths and woolly rhinos, around 30,000 years ago.[18] There is no climatic reason why a temperate-zone woodland elephant would go extinct – but hunting pressure is well documented in the fossil mountains of their bones found across Europe. Modern times remind us that when humans wish to slay giants – today, for ivory – those giants stand little chance.

In Eurasia, it is also significant that the largest species vanished far earlier than their peers. Mammoths survived worldwide until 4,500 years ago, on Russia's Wrangel Island. Woolly rhinoceroses were, however, better suited to the modern taiga climate. They survived long after the end of glaciation, but, with their populations fragmented and vulnerable, they perhaps could not survive us. To this day, however, smaller animals of the same habitat as the rhino – reindeer, muskox and bison – have all outlived the rhinos. In Europe, as in the Americas, the giants were the first to fall.

On visiting some of the most pristine national parks in southern Africa, it is hard not to feel a terrible sense of loss in seeing how our own continent would once have looked. There is a bittersweet thrill in walking beside improbably large animals, many of whose futures now seem every bit as certain as that of the last British mammoths. What would many of us not give to travel back in time, and teach our own ancestors to cherish the giants who once shaped a world far richer than any we can now imagine?

Island Nation

During the last ice age, sea levels were 127 metres lower than they are today.[19] Stunted oceans revealed land bridges that allowed humans to return to Britain time and again. For much of history, Britain's human colonists had, as palaeontologist Mark White puts it, 'a very short record of residency'. Extreme ice ages drove us out of Britain no fewer than eight times.[20]

At the end of glaciation, Britain was becoming isolated from Europe as sea levels rose. An estimated 5,000 hunter–gatherers had established here, having, it is thought, followed migrating herds of mammoth and reindeer back across Doggerland, the land bridge that joined northern Europe to the British Isles.[21] Then, 8,200 years ago in Norway, a tract of continental shelf the size of Iceland plunged into the sea. The largest known landslide in history, the 'Storegga Slides', triggered a series of colossal tsunamis.[22] The marshes of northeast Scotland vanished. Land bridges further south sank below the sea. Any of our ancestors puzzling over the tide, and how it worked, would have been crushed by ten-metre waves.[23] Our isolation was completed with improbable speed. Britain became an island – for good.[24] The creation of our island consolidated two factors – isolation and human activity – that shape the fortunes of British wildlife to this day.

With the giants gone, our ecosystems would have changed forever. Recent studies show that today's rhinos are apex ecosystem engineers, keeping open short grasslands.[25] Elephants trample trees and shrubs, maintaining space, yet simultaneously transport the seeds of the very largest and most valuable trees in their dung; planting as they go. This may seem academic now, yet the song of the corn bunting and the turtle dove would have evolved alongside the activities of our largest grassland herbivores.

These links between our past and the birds we see today are everywhere. The next time you watch a bird foraging in disturbed earth, call to mind what would have created that disturbance in the first place. From the wallowing of elephants and rhinos to the digging of wild boars, disturbance has shaped the ecology of Britain's wildlife as much as any other force.

A Land of Light and Trees

After the ice age, it is always said forests recolonized
Europe. In fact, trees recolonized Europe.
—Frans Vera[26]

The establishment of our temperate climate began as the glaciers vanished, so the early Holocene, dating from 12,000 years ago, has become most ecologists' benchmark of 'natural'. It is to the assemblage of animals at this time we must turn to discover how the landscape would have looked.

In recent decades, the long-standing theory that Britain was covered in dense forest – a habitat most ecologists point out is relatively species-poor – has given way to the better-supported and infinitely more logical idea that Britain was a wooded mosaic, dominated by a contest between trees and animals. This seemingly 'historical' point actually affects any kind of vision for the future of Britain's nature: what our landscapes should look like and how our birds could prosper. So it's worth pausing to take a look around at the last of Natural Britain. Detailed analysis of Britain's commonest trees has shown that since the Pleistocene these were sunlight-loving oak, birch, hazel, willow, alder and hawthorn. The record gets more interesting when you look at the insects, too. The widely accepted theory is that trees with the most associated insect species were, naturally, our commonest trees in recent times.[27] In 1960, the biologist Sir Thomas Southwood documented the number of insects dependent on different species of native tree. Topping the chart was oak – Britain's cathedral of life. It was followed by willow, birch, hawthorn and blackthorn. Then came poplar, apple, pine, alder, elm and hazel. All of these species are those best adapted to light – to wood-pastures, wetlands, marginal habitats or grasslands. Not one of them is adapted to thrive in dense forest.

These top insect trees tell us much about our natural ecosystem. Hawthorn's insect diversity attests to a strong presence in the ecosystem, and it remains vital for many scrubland birds. Crab apples are characteristic of wood-pastures in the New Forest, where free-roaming herbivory is still in play. Old cider orchards with domesticated apple trees form invaluable habitats for birds, in part because birds have adapted to live around wild wood-pasture apples for a very long time.

Most of our woodland birds are dependent on oaks; many time their breeding season around its peak abundance of caterpillars. The trees that fuel the greatest biodiversity in Britain are also those most intrinsically adapted to thriving not in dense canopy forest, but in sunlight.

The studies of Dr Keith Alexander have revealed that the commonest beetles in the early Holocene were those requiring open-grown trees.[28] Fossil records of two such beetles, still around today, are of those that require large-trunked oaks. If they are to reach the girth necessary to hold decaying hardwood over time, such trees require sunlight, too. Alexander's breakdown of the beetle fossil record shows that, in the early

Holocene, 28% of Britain's beetle fauna were grassland and scrub species, 13% were arboreal (open or closed woodland), and 47% relied on wood decay (wood-pasture and open woodland). Whilst Britain was filled with diverse trees, in various formations, the beetle record suggests that our woodlands were spacious and open.

At the University of Oxford, Dr Mike Allen, studying Stonehenge snails, was always presented with the 'fact' that Salisbury Plain must once have lain below dense woodlands.[29] His snail fossils, however, did not agree. Early Holocene snails from this area were species adapted to open grasslands, with fruit trees and scrub. This, in turn, is consistent with the sustained impact of nomadic grazing herds. Allen theorised that it was the abundance of such grazing animals that in turn had drawn early human hunters to Salisbury Plain in such numbers.

A range of other animal groups hint strongly that our island's natural biome was once a maze of shifting, broken habitats, with trees, scrublands and grasslands all in play, side by side. Almost all of Britain's 'woodland' butterflies, whether the blackthorn-specialist black hairstreak, the sallow-loving purple emperor or the many butterflies adapted to areas of broken space in our forests, from chequered skippers to wood whites, attest to the diverse mosaic in which they evolved. Hairstreaks thrive between oaks, scrub and grasslands. Many fritillaries thrive in sunlit glades. Few, if any, British butterflies, can survive in dense-canopy forest. The black-veined white (extinct in Britain since the 1920s) devours fruit trees as a caterpillar but as an adult competes fiercely over thistles in meadows. The large tortoiseshell (also lost) eats elm and willow species as a caterpillar, but the adults are drawn to dung, fallen fruit and open glades. Others, such as the large and mazarine blues, were adapted to low-intensity grazing, which alone can create the anthill-rich grasslands they require. It is also illuminating how disturbed open ground is perhaps the single most important habitat for the survival of most of Britain's wildflowers, too.

Bats reveal an equally clear picture of the early Holocene landscape. Of the eighteen British species, most are adapted to mosaics of trees and grassland. The noctule lives in ancient rotting trees but hunts dung beetles in open glades. The horseshoe bats thrive in pastures, scrub and open trees. Almost all our bat species are reliant on broken landscapes, rich in open-grown dead trees, scrublands and grasslands alive with beetles and moths.

At King's College in London, Dr Francis Rose spent four decades studying woodland lichens in places such as the New Forest.[30] He found that whilst hardly any lichen, moss or liverwort species lived within dense stands of trees, almost all of them thrived in sunlight – along the edges of glades, or growing on the forest's pasture trees. Rose also noticed that ancient fen plants, like butterwort, were vanishing in wood-pastures in Norfolk as traditional grazing was abandoned. This suggested that these plants had originally evolved alongside wild grazing animals.

The Chaos Animals

The reason for spacious grasslands and scrublands in a wooded land comes down to the simple fact that Britain's trees did not grow uncontested. A vanished array of large herbivores shaped the formation of our habitats. Whilst the grasslands may not have been as large as they were under the stewardship of the giants, trees would still have been required to fight against wild grazers for survival. These forgotten animal architects answer the question as to why almost all of Britain's wildlife thrives best in mixed mosaic habitats.

Aurochs, wild cattle, existed in Britain in ever-diminishing numbers until their extinction here over 3,000 years ago.[31] Today, you have to watch the actions of the Indian gaur cattle, in woodland, to appreciate how aurochs once shaped the land, though ancient wood-pastures in places like Romania's Letea Forest, with its primitive-breed cattle, can still give us some idea.

Cattle break and debark trees, effectively coppicing and opening wooded habitats – but at the same time, they carry thousands of plant seeds in their dung, and also on their hoofs, thereby transporting and planting diverse flora. Cattle munch on bushes such as hawthorn, which, growing to resist their nibbling, put out long thorns and become scrub fortresses – used by a huge array of insects and birds. Cattle, however, are mainly grazing animals, and they maintain open glades. The scale of such grasslands depends on grazing preference. Rich dung dropped by cattle promotes teeming invertebrates, including dung beetle communities. You have only to watch a flock of yellow wagtails follow cattle in a coastal Norfolk field, or a shrike foraging for beetles in a Polish grazing meadow, to gain some glimpse of our birds' ancestral dependence on the actions of cattle. The diverse protective responses of our trees provide further reminders. Even the shade-tolerant lime, often associated with closed-canopy woodlands, will, given a chance, put out enormous growths at its base, designed to repel marauding herbivores. Aurochs, through grazing around, debarking or smashing against young trees, would have wrought much chaos in our woodlands, ensuring that sunlight reached the woodland floor and its fauna, and preventing the formation of impenetrable shade.

Across Europe, where small cattle herds still roam free, oak trees do not grow in 'forests' but must fight their way through thorn scrub if they are to survive. This is one of the most compelling reasons why a closed-canopy forest cannot form. It can only do so, indeed, if cattle are taken out of the picture. Bison, like aurochs, were once the creators of diverse mosaic habitats. Fossils in Doggerland suggest bison certainly came very close to Britain,[32] and these were, most probably, native animals too. Our fossil record is fascinating, but it is no more than a glimpse. As is often said about British bison, absence of proof is not proof of absence. Not only do bison love wallowing, creating rich areas for plants to prosper, but they also smash through aspiring trees, debarking them as they go, creating broken or open habitats in the process. Open areas of soil left in a bison's wake then allow flowers and grasses to regenerate.

Tarpan, wild horses, were another keystone grazer. Horses serve distinct ecosystem functions, quite different from cattle or any other wild animal. Not only are they adapted to the very harshest of winter environments and punishing upland conditions, but they specialise in removing the toughest of vegetation, greatly increasing the floral richness of the grasslands they inhabit. By removing dry and dead grassland matter, horses open up new grazing habitats for cattle and deer. By no means limited to grassland, horses can also alter the richness and composition of wooded environments as well – browsing trees as 'scrub managers' and transporting a range of seeds, including acting as one of the most effective wild planters of the apple. In Mongolia, the presence of wild horses alters the lives of many species. Choughs and wheatears follow in their wake, feeding in the short lawns created by the horses and benefiting from the rich insect life recruited to their dung. Since their reintroduction in Mongolia, wild horses have turned rank, same-aged grasslands into rich mosaics, which have been recolonised by red deer. Broken woods and bushlands are often most effectively created by the presence of wild horses. By pruning back fast-growing willows, poplars and beech, yet leaving unpalatable species like elder, horses would once have helped maintain diverse scrub-grasslands not only in our lowlands but also in our hills and valley sides.

At a species level, wild horses have never gone extinct. Tarpan were the European subspecies of wild horse; the other, the takhi or Przewalski's horse, is very much alive on the steppes of Mongolia, and has since been reintroduced to grasslands in Europe. Most British tarpan, however, appear to have been hunted out by the early Holocene.[33] Horses were hunted with specialist spears at least 400,000 years ago.[34] By the time of their apparent extinction in Britain, around 8,000 years ago, the numbers of wild horses would have been tiny, isolated on our newborn island.

Dutch ecologist and historian Henri Kerkdijk-Otten has postulated that, rather than there being one distinct wild horse in Europe, horses, like zebras, would most likely have evolved into a range of different forms.[35] A steppe-dwelling horse, for example, evolves different characteristics to a forest-edge species. He argues that wild horses in Europe live on. Primitive breeds, never domesticated, he argues, share much of the lineage of wild horses. We've just labelled and contained them differently from other wild animals.

Fossils of horses have been known from Exmoor for 50,000 years, and free-roaming horses were noted there in Domesday Book. Whilst Exmoor ponies are assumed to be feral, their physiology is ancient, with eyelids adapted to withstand the harshness of driving rain. Horses, like Africa's zebras, are hugely adaptable. And there is no compelling evidence that they would not, if left alone by our ancestors, have continued to shape the British ecosystem, as the varied zebras of Africa shape its woodlands to this day.

Wild boars are disruptive ecosystem JCBs. They upturn and root around, creating areas of disturbed ground where seeds take root. Watch a house martin gathering mud for its nest, a grey partridge foraging for food, or a robin following your garden hoe, and the role of the boar becomes apparent. Boars reveal and aerate the soil. A range of British plants, insects and birds are highly evolved to thrive in disturbed ground, and with the megafauna gone, the role of boars would have become even more important. Over time, the actions of boar lead to the re-formation of entire landscapes. Scrubland thorn trees take root in disturbed soils. These, over time, propagate the next generation of oaks.

Beavers add a staggering level of complexity and diversity to the landscapes that they shape. As soon as the first beaver pond is created, life is supercharged. Dams retain warm, still ponds that come to act as crèches of zooplankton, the basis of life in the water. In these warm, calm oases, dragonfly larvae find the perfect hunting grounds. Amphibians, such as newts and frogs, attain maximal growth. Salmon and trout nurseries are formed; indeed, on the removal of beavers from some American rivers, such fisheries have collapsed. A host of wetland birds from ducks to waders accelerate the growth of their chicks by feeding in the rich waters. A host of mammals, from water voles to otters and badgers, will colonise old beaver lodges as their home. Others, such as weasels, will often move in to the landscape only after the formation of the rich hunting grounds provided by a beaver pond. During a complex felling operation, beavers act to both thin and coppice the woodlands around their rivers, greatly increasingly their diversity. In felling trees to leave stumps, they create the dead wood needed by willow tits and beetles. In coppicing others, they create the rich bushlands haunted by reed buntings and nightingales. Over time, as the beaver dam floods woodland, entire new swamplands spring to life.

The longer that beavers live within a landscape, the more diverse and varied it becomes. As they abandon old ponds and move to new ones, rich wet meadows form within wetlands and forests. Some of the hardest habitats for humans to recreate, these damp meadows, where present in Europe, host a growing diversity of life. Damp-grassland butterflies, like the large copper, wet-meadow birds, from corncrakes to ruff, and small mammals like the water shrew are all suited to the rich, varied meadowlands left in the wake of a family of beavers.

Over time, as rivers are dammed in multiple places, across many decades, the entire character of a waterway is transformed. Rivers become huddled, bundled creatures; with the water unable to run straight, shingle deposits form at their bends – the natural home of avocets. In releasing water slowly, over months rather than days, the beaver leaves its final legacy. What were once straight rivers, unproductive for many forms of life, expand to form shallow wetlands, the haunt of cranes and the fishing places of ospreys. And by delaying the release of water downstream, beavers act as nature's oldest and most proven engineers; preventing sudden, irruptive floods from damaging habitats, and life, downstream. Most of all, beavers ensure that landscapes remain diverse, open – and cyclical. By repeatedly coppicing and felling trees, beavers, even more surely than giant grazing animals, prevent the formation of dark, dense woodlands around our rivers and wetlands. Instead, all types of habitat are brought into being at once. For decades, British conservationists have, in our nature reserves, sought to achieve the same effect as that created by beavers, but with infinitely more limited results. No animal brings more life to the land, or creates more diverse habitats at once, than this modest and industrious rodent.[36]

The final set of 'gardening' animals in the early Holocene consisted of four specialist *browsing* animals. Browsers, unlike grazers, chop and trim rather than plucking: targeting young vegetation – flowers, buds and shoots. Unchecked by predators, such animals can increase in numbers and stunt vegetation. Kept in a state of fear, they never linger in one place for long. Red deer are preferentially browsers of wetlands and open woodland, thereby playing a beneficial role in maintaining open areas in our river valleys. Roe deer are specialists of denser wooded areas, nibbling out young shoots. In natural densities, and always on the move, roe deer can create glades and space for flowering plants within our woodlands.

Elk (moose) are strongly associated with aspen, willow and birch. Their current distribution, in Fenno-Scandinavia and eastern Europe, suggests that Britain's uplands and large fenlands may once have provided such forage in abundance. By relentlessly pruning back aspiring wetland-loving trees, elk, regulated by their predators, would have had diverse benefits in both our uplands and our wetlands. A host of birds, from black grouse to bluethroats, thrive best in birch or willow-clad mosaics, with open areas of bogland. The elk, alongside the beaver, is the apex predator of these fast-growing trees. This browsing giant would have ensured that wetlands or bogs did not become overgrown; a powerful guardian of curlews and cranes. Elk were hunted out around 3,000 years ago.[37] Reindeer, often said to have vanished from Britain as the temperatures rose, were still being hunted in the boglands of the Flow Country, in Caithness, as late as the ninth century, but their role in our ecological history remains considerably more opaque.[38]

In opposition to these many grazers, diggers, choppers and browsers, which in their different ways compete against tree formation, Britain had three apex predators in the early Holocene: lynx, wolves and brown bears. These would, in turn, have regulated many of the meso-predators now overly common in our countryside, especially young foxes and badgers, which are regularly killed by wolves, and whose dens are often raided by Eurasian lynx.

Lynx are specialist hunters of roe deer, and the only predator tied to denser woodland, the habitat of their prey. Looking at some of the last areas lynx were present in Britain, such as the Lake District, gives us some insight into the places where our densest woodlands may have grown. The steep slopes of Britain, particularly in the west, would, logically, have hosted its densest woodlands: enchanted rainforests of dripping limes and tufty-eared lynx in the shade. Steep uplands create conditions where trees, rather than large herbivores, are most likely to flourish, and thus form closed-canopy woodlands over time. The very early disappearance of the lynx, therefore, is a powerful reminder of how early on we deforested large sections of our island. Far less adaptable than the wolf, Eurasian lynx, even today, can only be found in some of Europe's largest areas of continuous woodland. As lynx vanished, a crucial gamekeeper role, of regulating and culling shy woodland deer, and protecting the growth of new trees and flowers, would have been lost forever.

Unlike the lynx, our two other lost predators, often ascribed to Britain's 'forest' past – wolves and brown bears – are not actually forest animals, but are, again, adapted to thrive in broken wooded landscapes.

Brown bears require huge volumes of berries and fruits, which grow on sunlight-loving trees, and thrive around wetlands rich in fish. A bear's ecosystem role involves a lot of tree planting, as the seeds of the fruit it has eaten pass through its gut. By bringing large quantities of fish such as salmon onto the land, bears also play a lesser-known role in transforming the nutrient content of our soils. Recent camera-trapping from Belarus has shown that bears also 'manage' a range of other animals, including juicy wild boar piglets.

Wolves hunt semi-open landscapes for prey. They effectively plant trees by scattering nervous herds of deer and elk. With the 'fear factor' in play, browsing animals do not linger long enough to browse down vegetation. In moving around a landscape, deer and elk defecate seeds and plant trees. Since the return of wolves to Yellowstone, millions of new aspens have sprung up. Like the sparrowhawks in our gardens, the effect of wolves is primarily felt not in the number of animals they kill, but in their effect on how entire communities of herbivores behave. By constantly shifting prey around a landscape, wolves would have countered the 'nibbling' effect of our abundant herbivores with a tree-planting effect; ensuring that landscapes could not become denuded of a rich diversity of vegetation, by consistently keeping fearful ungulates on the move.[39]

Those keen for the return of predators understandably place emphasis on these animals as the prime architects of a landscape. Predators like wolves, however, are limited to their dens for much of the year. Herbivores roam. Serengeti studies show that only a quarter of wildebeest are killed by predators such as lions: the rest, by starvation. Wolves help an ecosystem greatly, yet it is the guild of herbivores that are the foremost agents of diversity. Our largest animals would once have had the largest impact on our landscapes.

Slow-growing trees cannot 'swallow' fast-moving animals – but dense forests *do* grow once pristine herds diminish, as has been seen in Africa when elephant herds are poached out of existence. In other areas, the reason for forest formation would have been more natural: even in the Serengeti, steep slopes are shunned by large grazing herds and steep hillside woodlands are invariably denser as a result. The ravines of Exmoor, the craggy slopes of Cheddar Gorge, and many of the steep valleys of western Wales, the Lake District and Scotland, to name but a few, may always have tended towards closed canopies. Taking a boat down the River Wye, there seems every chance that the rich green sea of oak, hornbeam and lime flanking its precipitous slopes might always have looked this way. Indeed, certain native trees, like many of the British *Sorbus* species, some of which are now only found growing in single gorges, hint at the specialised richness of a long-lost Atlantic rainforest, where valley tree communities, relatively isolated from one another, may, over time, have engendered speciation.[40] Yet that is a very different matter to the idea of Britain, as a whole, being covered in continuous dense forest.

For decades, until recently, the incorrect baseline of canopy forest has come to the fore because only one of two key forces in nature – *succession* – has been considered. The argument that trees grow over time, swallowing grasslands, then shrublands, to form dense shade, is, of course, correct. What it ignores is the other force of nature: the opening actions of animals. Over time, as we harvested Britain of its grazers and diggers, its beavers and boar, the impact of these animals would fade – but the trees continued to grow. The more players you take out of the game, the poorer your landscape becomes. The more stewards you remove, the less diverse the picture gets.

Natural Britain: the Bird's-Eye View

*Conservation should be based on practical
observation rather than unstable theory.*
—Oliver Rackham[41]

Britain's birds are the final piece of the puzzle in unravelling how wild Britain once looked. They corroborate the story of the plants, the trees, the bats and the butterflies, the bears and the wolves, the grazing animals, the browsing animals, the fungi and the lichens, the beetles and the snails, in revealing a mosaic world of infinite variety – a wilderness of woodland, scrubland, grassland and floodplains. Virtually all of our birds evolved in disrupted, chaotic, dynamic mosaics for maybe a million years or more, under the regimes of keystone herbivores. And to this day they remain true to such a broken, jumbled and chaotic world.

Of around twenty species adapted best to truly open grasslands,[42] some – such as the skylark, great bustard or quail – have no requirement for trees at all. Other species are best adapted to floodplain grasslands that grow across the season, concealing vulnerable young birds as they do so. In this category, the corncrake, redshank, black-tailed godwit, ruff, yellow wagtail and greylag goose can all be found to thrive where these original habitats remain in Europe. Other species appear adapted to very short, tussocky pastures in harsher, rockier areas – perhaps once stewarded by wild horses. The chough and wheatear, both foraging in our coastal lawns, are two such species.

Around fifteen of our native breeding birds specialise in scrub-grasslands, where thorny bushes meet rich open sward.[43] A vanishing habitat in most of Britain, this is what we might term a classic 'Serengeti', with isolated stands of thorny scrub. Red-backed shrikes, skewering dung beetles in grassland but nesting in dense thorns, are characteristic of such places, as are grassland-breeding cuckoos, watching their pipit hosts from stands of bushes. Where dense thorn stands meet disturbed soils, a range of what are now called 'farmland' birds thrive – turtle doves, tree sparrows, linnets and grey partridges. Familiar blackbirds, song thrushes and dunnocks are all, in fact, birds that thrive where scrub joins open pasture. As with our long-lost river valleys, we cannot find true scrub-grasslands growing as once they did. And the ancient loss of free-roaming herbivores in our country has led us to forget such places, and consign so many scrub-grassland species to dependence on the hedgerow – and the farm.

Dense thorn scrubland, often tidied and removed from our countryside today, accounts for at least twenty specialist birds.[44] At each stage of succession, new specialists emerge. Most of these birds are small passerines, which rely heavily on song to communicate in a complex world of layered shade. Where scrub bursts through grassland, you have whitethroats; where young trees burst through scrub, grasshopper warblers. Dense thorn castles, admitting little light, are the haunt of lesser whitethroats and nightingales. Thick bushes like willow or hazel, aspiring towards tree height, hold willow warblers, garden warblers and blackcaps. At the most advanced, rotten end of the scrubland spectrum, hazel-led habitats with dense webs of branches are the haunt of marsh tits. The most rotten of elder, birch and willow scrub, often associated with standing water, becomes the paradise of the willow tit. Scrubland is often misunderstood. We are told that it naturally gives way to woodland unless pruned back. But with aurochs, tarpan, elk and beavers in the ecological game, it would have been pruned relentlessly, with far more ferocity than a forester today. This scrub-animal contest would have created a rich mosaic – brimming with subterfuge and song.

We also find a great array of birds best adapted to open wooded grasslands.[45] Here, in fact, we find most of our birds of prey, just as you might in the ecosystems of southern Africa. Our cornerstone scavengers – white-tailed eagles, red kites and ravens – find most food in grassland but nest in large trees or the recesses in cliffs. Likewise, a whole suite of smaller predators, not suited to competition with woodland goshawks, thrive by nesting in mature tree stands but hunting open habitats – barn owl, hobby, kestrel, sparrowhawk and buzzard. In rougher grassland habitats, hawthorn or pine stands provide nest sites for the elusive long-eared owl; in Scandinavia, wherever such places are present, they form the preferred nest site of the merlin. Two species, rook and stock dove, specialise in finding food in disturbed ground but nest in older trees, whilst two more tree-nesters, jackdaw and carrion crow, often find food, and nesting material, on the backs of herbivores in open lands.

Some of the richest habitats for birds are those where mere islands of trees lie in open habitats. These habitats are not 'degraded', and indeed can be seen in their wild form across Scandinavia, where elk browse. For example, windswept bogs or grasslands with stands of birch are the prime haunt of the black grouse, a bird clearly evolved in a world where wind, water and wild ungulates contested the growth of trees. Birches and rowans provide the black grouse with food in the winter, but a rich variety of grasses and flowers provide food for its chicks, as well as nesting cover. Woodlarks thrive best where heavily disturbed grasslands meet stands of trees, where these birds sing. All of these are birds of landscape chaos – where disturbance would have kept the trees in check.

The habits of a range of wooded grassland species recall another wild process, which would have reformed our scrub-grasslands and drier open habitats in warm, dry summers: *fire*. It has been widely and corrected observed that British deciduous woodlands, in their climax form, burn very poorly, if at all. Our uplands can, today, be burned, as all that is left on them is heather, but upland fires in Scandinavia are unheard of – wet bogs, birch woods and beaver dams would all have protected the uplands from fire. One habitat that would most certainly have fallen under flame, however, was swathes of open land: areas such as the hawthorn-studded grasslands of Salisbury Plain. Even natural wetlands, once covering another 20% of Britain, can, as reeds and grasses dry out in late summer, become extremely susceptible to fire.

Wild fire, in natural environments, enhances long-term diversity by resetting large tracts of the countryside. Many of our birds, including nightjars, which preferentially nest close to burned areas of heath, and tree pipits, which thrive in earthy clearings with isolated trees, would have been two likely colonists of landscapes in the wake of a blaze. Others, like the stone-curlew, which has a now incomprehensible love of scarified soil, would also have been early beneficiaries in the wake of wild fire.

Fire, being phenomenon, not animal, is far less discriminate in how it shapes a landscape. Tinder-dry grasslands have no defence against it, nor do dryland bushes. Yet fire is one of nature's great consistencies. Across the replete ecosystems of the world, large herbivores are, periodically, wiped out or greatly reduced by diseases such as anthrax. It is at these times that elephant-battered trees can be given years, if not decades, of respite. At the same time, however, rank grasslands can recover large areas of habitat. In the short term, these benefit species such as the ground-hunting owls. But over time, those habitats grow ever less productive. Fire acts as the great leveller – resetting areas of the grassland ecosystem, removing dead matter, and creating the basis for the formation of fresh green shoots. What is now termed heathland, kept in a state of arrest by modern conservation, is in fact the relict expression of a landscape after fire: a rich jumble of low vegetation, scarred, regrowing and adder-alive.[46]

Of all wooded habitats, wood-pastures – spacious deciduous woodlands with glades – are by far the richest for birds. At least thirty of our birds are best adapted to thrive in oak-dominated wood-pastures,[47] yet only a handful of these migrate, so to speak, into denser canopy forests. Six are pasture-feeders, nesting in mature trees, often in tree hollows, but finding food on the ground: wryneck, green woodpecker, mistle thrush, starling, redstart and robin. All of these attest to the power of herbivory in maintaining open lawns around ancient trees. Four of our birds are specifically oak-evolved, leading lives based around the bounty of these trees: blue tit, great tit, nuthatch and jay – the latter capable of planting entire oak woodlands through dropping acorns. Favouring glades and open space, where it can intercept flying insects, comes the spotted flycatcher, whilst the tawny owl favours old-growth cavities for nesting but rarely hunts in dense woodland. Two species, goldcrest and coal tit, are more adapted to a life around pine and yew, both conifers that would have grown naturally within our woodlands. Decay in our trees adds three further species – the great spotted woodpecker, the deadwood-specialist lesser spotted woodpecker, and our long-forgotten old-growth tree cavity nester: the swift. Then there are species that use wooded mosaics at a much wider scale. Goshawks and honey-buzzards nest in dense stands of trees but find much of their food in clearings, particularly honey-buzzards, which hunt wet glades rich in frogs. Woodcock nest below trees but 'rode', or display, over large clearings like strange squeaking bats. For earthworms, their main prey, to thrive, woodcock woods require sunlit soils. Colonies of hawfinches favour extensive woodlands with plenty of light-loving fruit trees like cherry, but come autumn, shift their attention to different woodlands, with more hornbeam or beech mast.

In northern Britain, where pine and birch woodlands would once have been more dominant, broken wood-pastures, not dense forests, remain the richest habitat for birds. Pine, birch and aspen pastures are a prime habitat for redpolls, siskins and crossbills. Very mature, spacious Scots pine forests are preferred by the parrot crossbill, whilst the capercaillie broods in bilberry but the males lek in open clearings with a preference for stands of dead trees. Crested tits, too, prefer the oldest and more open stands of pine. Green sandpipers, perhaps once more common when the Caledonian woodlands were larger, use old thrush nests in trees to lay their eggs in, yet spend much of their time foraging in open wetland bogs. Of all our birds, only three actively favour closed canopies. The wood warbler prefers a dense canopy, in which it sings, above a browsed forest floor with some brambles, in which it nests. Golden orioles spend entire lives in dense treetops, favouring, in western Europe, poplar, and in eastern Europe, oak. Pied flycatchers, too, appear to shun the woodland edge. But these are the exceptions to the rule, and all remain heavily dependent on oak – a tree that grows best under sunlight.

Such tenacious attention to what is 'natural' might seem a little dull. But when we consider how best to rewild our wonderful island, remembering what we have lost could not be more important. Today, the wild stewards gone, we painstakingly ascribe many of our endangered species to dozens of man-made habitats, none of which, in truth, they evolved in or require. But even now, mammals and moths, butterflies and bats, and most of all our birds, point us to the wilder world where they once lived – and remind us of the variety of landscapes once crafted by wild animals, whose numbers, size and majesty we have entirely forgotten today.

Very British Pelicans

In the early Holocene, distinct wetlands existed in huge areas, where the most influential agents in the landscape, alongside beavers, were rivers themselves. It is estimated that wetlands of one sort and another covered 20% of the British land mass.[48] Yet in spite of this, even more birds show signs of being adapted to our dynamic wetlands than to our once-dominant wood-pastures.[49]

The largest of our wetlands lay in the Great Fens of Cambridgeshire, the Avalon marshes of Somerset, the Humberland marshes of Yorkshire and Lincolnshire, and the mosslands of Cheshire and Lancashire. In Britain, we can hardly imagine the scale of these places or the exuberant chaos of areas where rivers are in charge. The Danube Delta in Romania, the Biebrza Marshes in Poland and the Pripyat valley in Belarus are the best examples left in Europe of the kind of vast wet wilderness that once existed on our island too.

Fossils preserve well in peat, and so the bird fossil record for our wetlands is a good one. We know that our two largest wetlands, Avalon and the Great Fens, were the haunt of Dalmatian pelicans, which then bred across northwest Europe. In Glastonbury, their bones date from 700 BCE, and as late as Roman times these curly-headed fish hoovers still floated over Somerset's marshes. Excavations at Glastonbury reveal that cranes and white-tailed eagles were also widespread, as we find them in Polish river valleys today. Wild mute swans graced our waterways prior to their domestication later.[50] Fenland once covered 8%, or 8,400 square kilometres, of Britain.[51]

Looking at pristine river valleys in far eastern Europe, it is likely that our fenlands were once a system of gradients, where different volumes of water, at different times of year, delivered different habitats and birds. Within one untamed river system in eastern Poland, for example, you can still find shallow grazing marshes with yellow wagtails, herb meadows with corncrakes, shallow water with black terns, sedge fens with spotted crakes, damp grasslands with ruff, beaver-coppiced woods with willow tits, pockets of reeds with bitterns, and oxbows with colonies of grebes. Every configuration of 'wet' thrives side by side. Beavers play a keystone role in ensuring no one type of wetland dominates the land. This, in all probability, is what our fenlands looked like in their original state. They were vast and unwieldy – a freestyled world of beavers and pelicans: self-governing, soggy and wild.

The Softly Wooded Hills

The height of Britain's natural treeline is a contested subject, but the best-acknowledged example can be found at Creag Fiaclach, in the Cairngorms. Here, at 640 metres, or 2,100 feet, stunted Scots pine and juniper fade away, at last, into windswept heather. But the height of a treeline is no constant matter. Orkney's, for example, lies at zero metres. Here, the wind hammers the island's vegetation and 'above the treeline' birds, such as curlews, begin at sea level instead. If we set fixed rules for our ecology, we underestimate the range of natural forces in play – such as wind, flood, grazing animals, beavers or the gradient of the land – and the variety these factors would have created in our landscape.

It is likely that our uplands were soft-gradient habitats, whose degree of woodland was shaped by many factors. Elk in Scandinavia contest the growth of willows, keeping the open moors we are sometimes told only gamekeepers provide. Wild horses in Mongolia prove themselves capable stewards of upland birch woods. Beavers create wetlands, then meadow systems, within the densest of woods. Where such forces would have lessened, trees would have surged back, with juniper a characteristic species of the treeline: the last tree to give in.

It is such a gradient, and the processes that create it, that we have lost in our uplands for thousands of years. And the restoration of our lifeless hills, recreating natural ecosystems in place of the farming of grouse, deer, sheep and spruce, will be explored at length later in this book.

Clearing the Land

The Neolithic, 6,000 to 4,500 years ago, was when we settled down. Hunter nomads, who had followed mobile prey into Britain, settled into communities whose lives were based around static livestock and crops.[52] This was perhaps the biggest single change in human history.

Decreases in the elm pollen record suggest that our prehistoric ancestors began clearing trees very early on in the Neolithic.[53] Five thousand years ago, stone axes were being chiselled out of flint: axe industries formed at places like Langdale in the Lake District.[54] Four and a half thousand years ago, we took up the bronze axe – and, by Roman times, our native 60% of woodland had been reduced to 15%,[55] and our aurochs and elk were long forgotten. In reality, then, a complete ecosystem has not been seen in the British Isles for at least 3,000 years – including horses, even longer again. And that is, in itself, quite a thought.

With our woodlands vanishing, this paved the way for the isolation and hunting of our predators in the more treeless centuries to come. Four thousand years ago, brown bears wandered on Dartmoor. One, we know, was turned into a pelt and buried with a princess.[56] The princess belonged to a local tribe of the time. The bear, in the wrong place at the wrong time, was less able to thrive in a world without trees. A bear wandering a wooded valley is a quiet prospect. A bear wandering a Celtic field is a source of attention.

The haunting expanse of northern Dartmoor seems as wild and ancient as it gets. Yet our ancestors, long ago, cleared the upper reaches of its woodlands with fire. The peat in Dartmoor's wild bogs is the result of acidic soils, which in turn arose from humans burning away the trees.[57] Dartmoor may haunt us with its wildness, but it is as tamed as anywhere in Britain: a windswept brownfield site.

By 1000 CE, brown bears vanished from Scotland. Manuscripts suggest lynx may have persisted in the Lake District until 700 CE, with fossils from North Yorkshire around the year 600.[58] The last wolf officially fell dead in Perthshire in 1680, but wolves have become world experts at eluding humankind, and were reported from Scotland until well into the eighteenth century. We are often told that each of these animals was killed, but for lynx in particular this would seem improbable, and there are few records of this in folklore. Harvest a fragmented woodland of its roe deer and you effectively starve a lynx population. On the other hand, there are accounts that entire forests were burned down in Scotland to remove just a handful of wolves. Beavers, which were hunted for their fur, skins and glandular oil, vanished from England as early as 1308, and from Scotland by 1684.

Bronze Age clearances, continuing into Roman times and beyond, would however have led to an new increase in grassland birds – the corn buntings and skylarks whose fields we worry about today. Bird fossils from the Roman period abound with the remains of grey partridges, house sparrows and barn owls, but are for the most part devoid of woodland birds.

In removing most of the woodlands of Britain, we shifted and affected many of our birds. But if you disrupt a grazing mosaic, dependent on disruption, with new domestic animals in tiny herds, you can, accidentally, continue to support quite a lot of the original fauna. And the Bronze Age put in place an agricultural life-support system that would sustain a lot of British species for thousands of years.

The Windmill's Tale

Some 3,300 years ago, people settled in the Fens. Remains at Flag Fen, near Peterborough, reveal how they built walkways through the reeds. Boardwalks, which now take us to bird hides, once helped our ancestors expand their mastery of Britain, float above the water, and hunt effectively.

By 43 CE, Dalmatian pelicans are thought to have deserted Britain. Some may have abandoned their colonies as humans moved in, but cut marks on bones prove that our pelicans were eaten too.[59] Fenland as a whole remained the last area of Britain to be conquered by William, after his arrival in 1066. It wasn't until the 1100s that the monks of Sawtry began to cut draining dykes, in a time when eels were still so common they were used as currency. By the 1200s, the monks of Woodwalton Fen were draining land to graze sheep in summer.

And then, in 1630, the 'gentleman adventurers' arrived. 'Adventurer' at the time meant investor. These investors were funding the drainage of the fens. Over the next seventy years, engineers working for the Earl of Bedford constructed the Bedford Rivers, which drained huge areas of fenland north into the Wash.[60] From 1685, an unfamiliar silhouette loomed over the marshes. It struck fear into the hearts of local people, who had lived off the fens by wildfowling and fishing for many centuries. The windmill wrought as much destruction in a century as 10,000 bronze axes. Cranes had been deemed the 'noblest quarry' for centuries. Anecdotes tell of King John killing up to nine cranes in a day with his gyrfalcon, in Lincolnshire, in 1212. In 1465, the Archbishop of York excelled himself, serving 204 cranes at a banquet he hosted.[61] It is believed cranes vanished from England by 1542. Harvesting for sport would have reduced the abundance of cranes, but their demise came as they were also drained out of Britain. Other species such as the night-heron (or 'brewe'), as well as little egrets (both served regularly amongst the London game dealers of the sixteenth century) appear to have vanished by the seventeenth century too.

Hunted Out

White storks are famous for building huge nests on village houses and chimneys across the older countryside of Europe, and feeding in the meadows around. In Sussex, the village of Storrington was known, in 1086, as Estorchestone and, by 1185, as Storketon, meaning a 'homestead with storks', whilst towns as far north as Storkhill, in Yorkshire, first recorded as a settlement in 1086, act as further cultural records of the stork's once-familiar presence.[62] Various old tavern names around the country also suggest the presence of storks within villages, depicting them unmistakably on their signs. Surprisingly little trace can be found of the charismatic white stork's departure from our lives, but as late as 1507, long after the last documented pair famously nested on Edinburgh's St Giles' cathedral, storks were still fetching big money – up to 48 old pence per bird – in the London game markets. In an age before refrigerators, such a market would not have been possible were storks not being regularly harvested from the landscape around the city. In addition, a number of strange factors may have hastened the stork's demise as, after the restoration of the monarchy, they were regarded as the symbols of Republicanism and may, like the Cornish chough, have been wiped out not for what they were, but for what they represented.[63] Unlike in many other countries, white storks would soon vanish from British rural life. Being so accessible, upon our houses or churches, they may have provided far easier targets than the more elusive cranes or bitterns of our wetlands. This was the time when cultural attitudes began to change. The birds and animals around us, formerly revered in myth, folklore or religion, became, in our minds, the enemy.

The archaeological record reveals that in Roman times white-tailed eagles were the dominant giant of lowland skies: soaring from Southwark in London, then a vast marsh, to Avalon in Somerset, the fens of Cambridgeshire and the Humber estuary.[64] In all, 1,500 years ago, up to 1,400 pairs of white-tailed eagles and 1,500 pairs of golden eagles are thought to have darkened Britain's skies.[65] In millennia before, when vast woodlands and bogs covered lowland Britain, the range of the golden eagle would have been far greater again, spanning not only the hills of Dartmoor and the Mendips, and those of northern England and Wales but, as in Estonia or Latvia, our larger lowland woodlands and bogs.[66] Whilst it is believed that some pairs of white-tailed eagle may have been persecuted early on, by monks protecting fishponds, it would now become a widespread national enemy. What would follow over the coming centuries was as extraordinary as it was wasteful.

In 1532, the Preservation of Grain Act provided the ultimate 'shoot to kill' licence for an assault on our nation's wildlife. Bounties were put on the heads of a range of species, very few of which were actually damaging to local stocks of grain. Wildcat bounties in England and Wales were fixed at one penny. In the seventeenth century alone, 5,000 bounties were paid out for decapitated cats. The Scottish wildcat was not Scottish once, but lithely hunted all the parishes of our island.[67] Hedgehogs were prized at four pennies, in the belief they sucked milk from cows at night. In the seventeenth and eighteenth centuries, over half a million hedgehog bounties were paid out. Pennies were paid for the heads of ravens and red kites. Polecats, pine martens and badgers were hunted down. Other birds, from shags to choughs, which had centuries earlier been celebrated on coats of arms, were killed for purely superstitious reasons.

As late as the seventeenth century, golden eagles were still nesting, at least, in Derbyshire's Derwent valley and in Snowdonia, and in North

Yorkshire they hung on until the 1790s.[68] By 1800, eagles had vanished from these last strongholds too.

White-tailed eagles had once been deeply ingrained in our folklore. Hundreds of places in lowland England bear the place-name 'earn'. The tombs of Isbister, on Orkney, contain hundreds of eagle bones, which must have held significance for the Iron Age people of the time. With their ability to carry off sickly lambs, these eagles would now become public enemy number one. By 1625, bounties on Orkney rewarded anyone who killed an erne with eight pence. By 1774, this bounty had risen to half a crown. The Lake District became the last stronghold of the eagle in southern Britain. Here, meticulous accounts, written by Crossthwaite's church wardens, reveal the organised level of their destruction. In 1713, John Jackson took 'an old eagle'; Edward Birket, 'a young eagle'. In 1719, John Jackson took another 'old eagle': clearly he'd found his niche. By 1794, the last breeding pair of white-tailed eagles in England was gone.

Few today are aware of this extraordinarily destructive period in our history – and for anyone with the stomach for more, *Silent Fields*, by Roger Lovegrove, is a fascinating read.[69] This was a purge few other countries carried out to quite such a degree. Indeed, by the time Victorian hunters started killing mammals and birds of prey on their shooting estates, they were merely hunting the relics of what were already severely depleted populations.

Almost all of this destruction was *cultural* in its ferocity – and worst of all, we so nearly took a different route. In earlier centuries, kites and ravens were respected for providing what we might now call ecosystem services. Goshawks were revered for their hunting prowess. Choughs were admired by the Cornish, with a legend told that King Arthur turned into a chough after he died. But then something changed in us. We turned these species into enemies – and wiped most of them out.

A New Settlement

With Britain tamed by the early Middle Ages and our avian giants driven into pockets of Scotland or exterminated completely, did the rest of our birds vanish as well? Strangely, they did not. This, in part, is because the mammal herbivores had been replaced with another, equally disruptive mammal – ourselves.

In place of aurochs-grazed pastures, we put to work tiny groups of cattle and pigs as woodland stewards. In place of wild wood-pastures, we planted orchards. An organic orchard holds twice as many earthworms per square metre as dense woodland.[70] If you were a blackbird or any other pasture-feeding species, you were winning. Birds that evolved in scrub-grasslands, whether red-backed shrikes or turtle doves, thrive under extensive grazing. If you were a shrike on lightly grazed common land, you were winning. Coppicing woodlands, as the late Oliver Rackham observed, recreates the tree-breaking actions of bison – or the zealous activity of beavers. If you were a nightingale or willow tit, you were winning.

In place of wild fires in our grasslands, men burning charcoal made clearings in our woodlands. If you were a nightjar or woodlark, you were winning. Hay meadows grow, over decades, to hold up to 400 species of wild flower, and insect densities that can rival those of natural grasslands, and traditionally these were rarely drained. If you were a species of our lost floodplains or wet meadows, like the corncrake – you were winning.

As our lightly farmed grasslands gave many British birds a new lease of life, our stewardship of the pasture-woodlands continued as well. In 1066, William I, having conquered, wanted somewhere to hunt. He saw a large tract of agriculturally poor land west of Southampton and pounced. Here, in exchange for his new subjects' right to graze livestock, William secured the right to hunt. As is typical of many of the best conservation successes, our last true woodland mosaic, the New Forest, was safeguarded entirely by accident. Across Britain there came to be many hunting areas, 'forests' and parkland estates, that influential landowners removed from agricultural circulation, deploying grazing animals to preserve some of the lost richness of our original landscapes.

For centuries, by creating an earthy grazing mosaic of their own, with a scattering of light-loving trees, early farmers accidentally ensured that the majority of birds did not face extinction. Many species would rely on these fragile, proxy ecosystems for centuries to come. Then – we would take them away. And that brings us to the story of John Clare.

CHAPTER 2

The Anthropocene

The Killing of the Countryside

Earth might not yet be irreparably transformed in a short period of time but handled with care, so it could be handed down to others. This gives the Anthropocene a meaning not only for scientists – but for everyone.

—Christian Schwägerl[1]

It's spring. Poet and self-taught naturalist John Clare heads out into the countryside around his village of Helpston in Northamptonshire.[2] He revels in the life that the birds, the trees, the communal countryside bring to him – and the rural people whom he meets on his walks. By the standards of today, John's bird notes err on the side of the romantic, but he also has an eye for detail – a remarkable eye.

John spots the nests of nightingales 'lost in a wilderness of listening leaves' in tousled hedges near his home. He observes the patience of a female yellowhammer, leaving her nest as local boys raids her eggs, only to return and re-lay the day after: 'the yellowhammer never makes a noise, but flies in silence from the noisy boys.' He angers a wryneck in a green woodpecker hole, as 'the sitting bird looks up with jetty eye, and waves her head in terror to and fro.' He observes how the cuckoo's song tails off as the breeding season goes on: 'when summer from the forest starts, its melody with silence lies.' By night, he explores a 'furze-crowded heath' alive with nightjars and relishes the 'fern owl's cry'.

John was an extraordinary naturalist. His poems describe 145 species of bird and provide the most comprehensive catalogue of the life that persisted in Britain's pastoral farmlands in the early nineteenth century. Now our giant elms, hay meadows, wrynecks and common corncrakes live on only in his poetry – and the shared farmlands of eastern Europe.

By the time John is writing his later poems, in the 1840s, however, we realise that the story of vanishing birds, and ever-quieter fields, is not a new story at all. Something chilling, destructive, is ripping the life from John's countryside. Birds and flowers are vanishing. He is shocked at the senseless felling of an ancient elm:

> *Self-interest saw thee stand in freedom's ways*
> *So thy old shadow must a tyrant be.*

John Clare lived from 1793 to 1864, long before the Common Agricultural Policy. But what he documents is the start of a bird-removal scheme even more destructive – and long forgotten. Enclosure was on its way. John feels the pain of the landscapes dying around him, the trees being ripped out, the people being fenced out of shared lands by landowners sealing off the countryside for an ever smaller number of farmers:

> *Now this sweet vision of my boyish hours*
> *Free as spring clouds and wild as summer flowers*
> *Is faded all – a hope that blossomed free,*
> *And hath been once, no more shall be.*

Animals John respected, such as moles, are found hanged in their droves under the new farmland system. Nightjars are killed for sucking the teats of goats – but John knows from observation that nightjars feed on insects. John's rural paradise, his local birds, and his own mental state will be slowly eroded as the century goes on. John ends his life in an asylum. Over time, his mind wanders more and more. Yet even in his latest works, he never loses sight of the greatest sadness of it all: the derangement of the countryside itself.

The sharing arrangement between most birds, and most people, was about to end. It was the advent of industry. The countryside was becoming a commodity. It was the start of a new era in world history too.

The Dominant Ape

The latest geological epoch, the Anthropocene, marks a time in which the most dominant changes to the planet have been brought about not by nature but by people. Many scientists recognise the start of the Industrial Revolution, around 1760, as the time when human action began to dominate the planet.[3]

In Britain, everything from our canalised rivers to contemporary climate change has been driven, largely, by human action. Since the

Industrial Revolution, nature itself has come almost entirely under our control. From the 1760s on, a series of small-scale landscapes would give way to a country uniquely dominated by vast crops of cereals, cattle, sheep, timber, grouse and deer. This is the story of how Britain became a factory, unique in the world in terms of its intensity.

This is also, therefore, the most depressing chapter in this book. With each passing decade, you will see how much we had – and lost. But the purpose could not be more different. To restore our wildlife, we must remember what was ours. Only by doing so can we hope to bring it back – in our more educated and enlightened age.

1760: Last of the Common Lands

Until the eighteenth century, much of Britain functioned under a system of open fields. A large volume of labourers worked small sections of land, side by side, under their landowners. These landowners were paid rent in terms of a share of the harvest, or tithes, and were entitled to a share of their tenants' labour – but they didn't sell on much of their crop. In open fields, crops were, as they had been since the Neolithic, designed to feed local communities.

You can still find open-field farming in Estonia, Poland, Romania and other eastern European countries, where a larger rural work force helps feed the local population. Small fallows, small meadows, small areas of corn, small areas of open soil, orchards and copses jumble together, in a small area, to provide a rich mosaic for birds. This paradise of variety is human-led, but recreates some of the richness of our original habitats.

We know from European studies that when the balance of these fields shifts, birds shift too. For example, studies of strip-farming in Siedlce, Poland, from 1999 to 2003 showed that red-backed shrikes were doing well in meadows – but, with a decrease in fallows, whinchats and corn buntings were temporarily in decline.[4] Nothing around Siedlce, however, was vanishing for good. Everything was shown to cycle up and down, alongside the changing crops.

The open-field pattern is why, if you visit a strip-farming system in eastern Europe, you will see configurations of birds you would not consider 'normal' today. Lapwings and turtle doves, red-backed shrikes and whinchats – all living side by side. Landscape variety – and food – is the key to their success. This interplay of habitats, of small-scale grazing, of disturbed wooded grasslands, explains why a Romanian farm still holds more bird species, in greater abundance, than many British nature reserves.

Enclosure would remove such a varied landscape from Britain forever. By the early years of the eighteenth century, most of the common lands in western and southeastern England had already been enclosed.[5] And then, starting in 1760, several Enclosure Acts were passed. These limited the amount of common land nationwide, for the first time. Land, once collectively owned and farmed individually, came under the control of an ever-decreasing number of landowners, with ever-greater estates.

In the older farming system, a decision affected a tiny strip of land – and a few pairs of birds. With light grazing, earth and trees, you had wrynecks. By putting a fallow there instead, you replaced them with corn buntings. Under the new system, one decision could affect whole populations. By planting one field where there had been twenty habitats before, landowners could turn a mosaic into a monoculture – in a very short space of time. And that narrative of simplification, a disaster for wildlife evolved over millions of years in diverse habitats, is what continues to this day.

At the same time, an even greater ecological change was about to take place. For the first time ever, free-roaming animals such as cattle and horses, which for centuries had taken the crucial roles once played by wild grazers and browsers, would become an ever scarcer sight across our countryside. The mosaic of grasslands, scrub and trees created by grazing common land would never be replicated by the farming methods of enclosure. Birds which had called our grazed farmlands home for a very long time, from beetle-loving shrikes and nightjars to thorn-nesting nightingales, were soon to be forgotten as widespread 'farmland' birds.

1790: Hedgerow Housing

In spite of the damages of enclosure, for both communal farming and many of Britain's small-scale habitats, we have nonetheless inherited a few better legacies as well. One was the spread of the hedgerow. Hedgerows are the bushy expressions of enclosure: thorny barriers for keeping livestock from crops, and of marking where one large farm ends and another begins. Even before 1800, around 12,000 square kilometres – an area eight times the size of today's Greater London – had been enclosed with hedges.[6] Grey partridges, linnets and turtle doves all nest in the dense cover of traditional hedgerows, feeding on seeds and insects in disturbed soils close by. Victorian cereal farming, yet to decimate its insects and flowers, became a new kind of haven for such birds. Our native scrub mosaic, where seed-eating birds foraged in disturbed grasslands, and nested in dense bushes, was reincarnated in the hedgerow.

We now mourn the loss of hedgerows and their wildlife, yet these are a recent invention in the great scheme of things. This life-support system briefly expanded the scrubland empire of birds, whose declines have come back to haunt us as, two hundred years later, over half of our hedgerows have been removed once more.

1800: Draining Britain

In the Fens of the early 1800s, lekking ruffs glowed on soggy pastures, black-tailed godwits yodelled over damp fields and, until the 1850s, the nests of now-vanished floodplain birds such as Baillon's crakes were being found around the Isle of Ely.

Between 1829 and 1835, the brick-red gleam of breeding black-tailed godwits faded from the Fens. Ruff vanished from Ely before 1840. In 1769, the naturalist Thomas Pennant had been 'deafened' by hordes of nesting black terns in Lincolnshire, but by 1850 these too had fallen silent. In spite of brief resurgences of some of these birds, we have never regained the freestyling beauty of our vanished flooded grasslands. To find the teeming terns, ruffs and godwits that we've lost, you must now travel to areas like Poland's Biebrza, where rivers still shape the land, and birds move, and thrive, with the changing state of water across the season's course.

In Britain, draining waders would soon become a national business. In the early nineteenth century, Richmond Park was greened not with parakeets but with lapwings. These had vanished by 1830. The same fate, across Britain, befell the snipe and, particularly, the redshank – species whose livelihoods depend on feeding in damp soil. Today, you must travel to places like the lovely island of South Uist, in the Outer Hebrides, to find damp meadows where redshank, snipe and lapwing are the sky-dancing afterthoughts of tiny farms. Here, the late cutting of tiny meadows allows the wader chicks to survive, and the damp, insect-rich grasslands allow them to find food throughout the season. But how often now do we swerve to avoid lapwings as they pirouette across the road? Our desiccation of Britain has been terrifyingly efficient. Yet even two centuries later, the drainage of waders from Britain continues as you read this book.

1820: The Vanishing Earth

In paintings of eastern England by Constable or Gainsborough, you can see scruffy wood-pastures with bare earth and a large amount of *nothing* going on.[7] There are small sheep flocks, herds of five or six cattle, and disruptive village pigs. This oak, willow and apple-dotted earthscape, forgotten to us today, preserved a part of the grazing mosaic vital to one charismatic summer visitor: the wryneck.

Wrynecks were once so common they engendered several strange myths. Calling just a few days before the first cuckoos, wrynecks earned themselves the title of 'cuckoo's mate'. In Gloucestershire, they were referred to as the 'cuckoo's footmen'. The Welsh name *gwas-y-gog*, meaning 'cuckoo's servant', came from the idea that wrynecks built the nest and hatched the young of the cuckoo. Across Europe, wrynecks were believed to hold powers of sexual magic – their head-turning able to turn the heads of wives back to 'cuckolded' husbands.

One thing, however, is guaranteed to remove wrynecks – and that's removing anthills. Between 1750 and 1850, Britain's human population tripled. More crops had to be sown, which meant there was less exposed soil. Lightly grazed wood-pastures gave way to more productive fields of corn. In 1820, the wryneck had begun to decline – and 150 years later it was effectively extinct as a British breeding bird.

The wryneck's was a world where little was done with the land – a land of trees and scattered nibbling animals. The moment the earthy wooded jumble of southern Britain, with its little herds of grazers and diggers, gave way to a sea of crops, the wryneck's farmed Serengeti would vanish. To this day, not one pair has returned to its original English haunts and you must again travel far east in Europe to find places where they call commonly in villages and apple-studded farmyards every summer.

1832: The Very Open Land

The great bustard, the world's heaviest flying bird, appears to have been expanding its range as late as the fifteenth century. It was seldom recorded from earlier feasts, a good indicator of a tasty bird's abundance, but was served at the Salisbury Assize by 1600. By the fifteenth century, Britain's southern plains, on which the bustard roamed, were no longer under the custody of wild herbivores, but extensively grazed by sheep.

Much of the bustard's habitat at this time was shaped by a forgotten agricultural method known as sheep-grain farming. Flocks of sheep, grazing on nutrient-poor heaths by day, deposited their dung on poor-quality arable land by night. These nutrient-poor grasslands, in a time long before artificial fertilisers, were quiet lands of an emptiness we might now scarcely imagine. This stony tundra rolled across the Brecklands of East Anglia, the southern downlands, and the Yorkshire Wolds – the low chalk hills east of York.[8] By all accounts, these were featureless expanses, with no settlements or trees for many miles. The Wolds were described by one naturalist as 'a desolate, grassy and stony sheep-walk'. That a bird as large as the bustard expanded its range on our island so late in our history attests to the thinly populated nature of southern England at this time.

Three hundred years ago, you could have walked for many miles across these wastes with only birds for company: skylarks and wheatears, stonechats on outposts of gorse, whinchats or black grouse near clumps of birch, and stone-curlews scuttling across the broken soil. These odd plains preserved treeless grasslands, perhaps unrivalled in scale since the elephant days. This was, after all, the low point for tree cover in Britain's history.

The bustard's fate, however, was sealed as soon as 'sheep-wastes' became the haunt of people. In Yorkshire, it was hunted to oblivion. In Norfolk, conifer shelter-belts planted in Breckland fragmented its open plains. Some landowners tried valiantly to protect the species, but in ever more isolated pockets, the birds were hunted down more often than not. Like the cranes of the 1600s, bustards were now deemed the 'largest, noblest and most highly prized of birds'. As a result, they vanished soon after. In 1832, one of our last British bustards was killed on Salisbury Plain, though they have, inspiringly, been reintroduced there in the last decade.

Conspicuous species, targeted by hunters, like bustards and black grouse, benefit more than most from an absence of intensive human pressure. Today, even the southern wild they once haunted is gone from our minds. The Yorkshire Wolds, for example, now lie below rolling crops. We have found ways to fertilise our land – and colonise its soil. So our stony tundra plains have become as distant to our memory as hay meadows.

1840: The Last British Penguins

One thousand years ago, Viking warriors, who had pillaged their way across Britain, trembled below a mountain. Something wailed in the moonlit rocks above. In reality, the source of their fear was not an ogre, but a burrow-nesting seabird. For centuries, long before the Vikings arrived, the eerie calls of Manx shearwaters had haunted the mountain by night. The Vikings, for all their military rapacity, feared the sound of the shearwaters, because they thought the birds were trolls. That mountain, to this day, retains the name they gave it: Trollaval. It lies on the Isle of Rum, off Scotland's west coast. Even now, Britain's pocket albatross, the fluffy terror of the Vikings, remains the mountain's dominant life force, long after its wolves, and Vikings, have gone.

Britain's seabird cities are the jewel in our wildlife crown, and have survived amazingly well when compared to our other vanished animals and landscapes. As far as we know, only one species did not make the journey to the present – and to our loss, it was a pretty good one.

Birds generally evade capture by means of flight. Flightless birds, therefore, have an outstanding track record of extinction. Great auks, it seems, were conscious of their less-than-aerial tendencies. They selected specific sites in the north Atlantic. Whilst smaller auks jump and fly, great auks would have jumped and flopped. So they chose islets, free from predators, with ramps where they could slide into the waves.

Being very oily, very tasty and very flightless, the great auk probably suffered a reduced distribution very early on in human history. Pleistocene fossils from southern English caves suggest it may once have been far more widespread. Six thousand years ago, there were perhaps millions of great auks across the north Atlantic: a 4,000-year-old burial site in Newfoundland contains two hundred great auk bones. The largest great auk colony of recent times, Funk Island, in Newfoundland, was once so thick with birds that sailors could barely put their feet on the ground.[9] By the 1780s, however, the traveller George Cartwright watched men spend months on this island, harvesting thousands of birds for their feathers. He wrote, with prescience, 'The whole breed will soon be diminished to almost nothing.'

What is remarkable is that we know almost nothing about the life of great auks on our own shores – only how they were wiped out. Just seven great auk colonies were known in recent times. Britain's was on the formidable rock of Stac-an-Armin, St Kilda, with perhaps another on Papa

Westray in Orkney, but it is clear these were depleted very early on. In 1697, the naturalist Martin Martin, arriving on St Kilda, was told the birds arrived in early May, leaving by mid-June, suggesting, tantalisingly, that great auks may have spent around fifty days on British land each year.[10]

In 1775, the Nova Scotian government asked the British government to ban the hunting of the auks, which were being harvested worldwide at a terrifying rate. Our government agreed – but there was, of course, a clause. Fishermen were still allowed to kill the birds and use their meat as bait. In 1840, three Kildan sailors from the main island alighted on Armin and took a great auk, tying its legs together; perhaps hoping it was a valuable prize. They kept it for three days, but in terrible storms, grew fearful of it. They condemned it as a witch – and stamped it to death. It was Britain's last. Four years later, the great auk went globally extinct, the only European bird confirmed to have done so in historical times.

It is remarkable to think that such acts of ignorance, and fear of witchcraft, took place in the same era as the invention of the steam locomotive. Yet if our removal of our heritage can seem senseless, our ingenuity in bringing wildlife back can also be inspiring. The story of Britain's penguin may, even now, be incomplete.

In 2016, the American research body Revive & Restore met with global ornithologists to discuss rewriting history.[11] By editing the genetic data of great auks, sequenced from fossil specimens, into the embryos of razorbills, their nearest living relative, and planting those within another bird, capable of laying a great-auk-sized egg, it may now be possible, theoretically at least, to bring such a species back to life. History has proven that far-fetched proposals, like moon landings, can come to pass if both the method and the funding are sound. The Farne Islands have been mooted as one sanctuary for great auks, in the event of such a venture succeeding. Maybe, just maybe, riding a boat on a choppy sea, the guano cliffs looming up ahead, our grandchildren may once again espy Britain's penguins on our shores – flopping ungracefully into the waves.

1845: The Empty Highlands

Before 1755, over half of Scotland's people lived in the Highlands. By the mid-1840s, huge numbers of destitute tenants were 'cleared' by landlords who felt no longer able to support them. Many of their homes were taken down, often making way for sheep pasture. By 1851, at least 85,400 Highlanders had been displaced from their homes.

The depopulation of the Highlands consolidated one of the largest 'empty' areas in Europe. Drive from Aviemore to Ullapool today and you pass through areas of a relict human landscape, where cuckoos hawk in the rough grasslands that have swallowed villages. Nature has bounced back at great human cost. But this, one of the most depopulated areas in western Europe, also sets up remarkable possibilities for wildlife and ecotourism, which we will return to in Chapter 8.

1850: Industrial Killing

In 1465, a visitor to London remarked on its clouds of kites and ravens, noting that both were heavily protected by law as invaluable scavengers of insanitary waste.[12] Red kites once bred in almost every county in Britain, having been, until the Middle Ages, extraordinarily common. By the middle of the nineteenth century, kites would be relentlessly hunted down, as they often snatched poultry from farms. Having outgrown their social use, the last English kites nested at Ludlow, leaving just twelve pairs in Wales by 1900.

On the Volga Delta, in Russia, white-tailed eagles, which can tolerate one another as collective scavengers, can still be seen in their hundreds. Like bald eagles, they fulfil, in temperate wetlands, the function of vultures: a communal scavenger and predator that fits perfectly into the temperate European ecosystem. Even in the early nineteenth century, the white-tailed eagle remained numerous across Scotland, where it flew not in tens but in hundreds. By the 1820s, however, those Tudor vermin laws were given an industrial twist. One estate in Caithness removed 295 eagles, plus 60 eggs and young, in just a few years. In the 1860s, bounty-hunters on the Isle of Skye killed an estimated 57 white-tailed eagles on one sheep-farming estate.[13] On a Sutherland estate, 171 adult and 63 young eagles were killed in just three years.

Large, defensive of their nests and susceptible to poisoned carrion, white-tailed eagles proved devastatingly easy to eradicate. The last nest was destroyed by 1916, on an island, Skye, whose skies had, just decades before, been filled with eagles. So often, we now look back on such events with some misplaced idea that all of this was, in some way, inevitable. Yet, just five years later in Poland, a 1921 statute was raised to protect white-tailed eagles,[14] and Polish eagles now number 1,900 pairs. This is a reminder, perhaps, that not every country went to such extremes – to wipe out what is, preferentially, a fish-eating bird with a penchant for scavenging meat.

Golden eagles fared little better at this time. You can visit old hotels in Scotland that still hold paintings of golden eagles sweeping lambs, sheep and occasionally children off the hills. These master predators are in fact specialised hunters of mountain hares and grouse. By 1850, under sustained persecution from shepherds, golden eagles had vanished from much of southern Scotland. William MacGillivray reported the destruction of 'vast numbers' on Iona and Mull in the 1830s. By the 1870s, just 80 eyries were known – compared to over 500 today.

The Domesday Book's records suggest that 24 goshawk eyries were well known in Cheshire in 1086, whilst medieval Scotland had a ready supply for hunting, harvested from the wild.[15] As the gun replaced the hawk as a means of killing game, goshawks went from admired weapons to dangerous competitors. The last known English female was shot out of Westerdale, Yorkshire, in 1893. It would be a century before goshawks would return.

One shooting estate's infamous record reiterates the sheer industrial scale of the killing. On Glengarry Estate, in the Highlands, between 1837 and 1840 alone, 27 white-tailed eagles, 15 golden eagles, 18 ospreys, 275 red kites, 63 goshawks, 462 kestrels, 285 buzzards, 63 hen harriers and 198 wildcats were killed – the records meticulously noted by the person who killed them.[16] This estate was perhaps not exceptional, just exceptionally well documented, in the extent to which it removed our national heritage of wild animals. Glengarry's records give, ironically, the final glimpse of how rich the Highlands once were for wildlife – and just how much we have lost. This was a heritage stolen from everyone, by just a handful of people, without any permission at all.

1870: The New Factories

Five thousand years ago, pollen records suggest the Scottish Highlands were a maze of native trees. The eighteenth century saw a tipping point for the remnants of these magnificent woodlands, which had themselves been cleared over centuries.[17]

Around the 1830s, a new fashion, deer-stalking, arose. Having cleared people to make way for sheep, a growing number of estates now cleared sheep to make way for deer. One observer succinctly wrote that 'all one had to do was buy a large piece of hill land, build a suitable house on it, clear the ground of sheep and wait for the numbers of deer to build up.'[18] In 1811, just a few 'deer forests' were managed for stalking. By 1873, that number had risen to 79. By 1900, 2.5 million acres, over 10,000 square kilometres, an area larger than Yellowstone National Park, had been converted to 150 deer estates.[19]

For over a century, meanwhile, 'walked-up' grouse shooting had quietly catered for a couple of expert marksmen stalking a moor, dogs at their feet, guns at the ready. Paintings of grouse hunting from the eighteenth century show black grouse fleeing hunters through a jumbled landscape of birch, quite similar to the uplands of Norway. But this was the Industrial Revolution. Grouse were now an industry too.

To feed red grouse to hunters en masse, the owners of grouse moors made a globally unique decision in the world of hunting. They created grouse farms. What is now known as 'traditional' burning, the muirburn, invented less than two centuries ago, was brought about to rejuvenate the heather fed on by red grouse. And now beaters, or grouse chasers, would drive grouse towards the waiting 'guns'. Without the slog of having to work the moors, this kind of shooting appealed to those with cash to spare, but without the skill or inclination to master the stalk. Under such conditions, one Lord Walsingham personally shot over 1,000 grouse in a day's shooting, including 94 in just 21 minutes.[20]

Burned treeless uplands developed. The purge of predators, threatening the prize animal on the farm, the red grouse, rose to new extremes. Before the 1830s, hen harriers ghosted across the rough grasslands of southern England, in places like Raventhorpe Common in Lincolnshire. They were also common across south Wales – but truly abundant on the English uplands and in Scotland. Yet again, the reminders of abundance come to us through the scale of the killing. Between 1850 and 1854, as grouse production intensified, 350 harriers were wiped out in Ayrshire alone.[21] So effective was the harrier cull that the species was soon banished from mainland Britain entirely, and would take decades to return from its refuges on Orkney.

By the 1870s, huge areas of Britain's land were being preserved for a specific lifestyle choice embraced by a fraction of the population – the killing, in a particular manner, of farmed deer and canned grouse.

1880: Rise of the Mowers

Until the 1880s, the rasping of corncrakes in our hay meadows was a quintessential summer sound. The birds were so familiar that they featured in Mrs Beeton's cookery book. The first corncrake declines took place in the agriculturally wealthy south and east of England. Here, the horse-drawn mower was born. Meadows were now cut rapidly, from the outside in, killing the corncrake's flightless chicks. The tradition of spotting nests on foot, whilst hand-cutting, was forgotten, and as meadows were now cut in a fraction of the time, escape routes for young birds in meadows were increasingly cut off. But the mower heralded the start of something else – the gradual decline of the hay meadow itself.

Hay meadows are human creations where nature then takes hold. Year on year, they yield ever more flowers and insects. They are cut late each summer, then browsed gently by livestock, before rebuilding their full glory the following year. A phoenix, the meadow is never truly dead.

The hay meadow was a place where we perhaps created something even more diverse than nature had intended. In Romania's Carpathian hills, hand-cut hay meadows, for example, hold eight times more butterflies than adjacent grazed pastures, and can grow to hold 400 species of flower. Red-backed shrikes, long vanished from our own farmlands, remain one of the commonest birds in such places, often feeding on beetles exposed as the meadow is slowly cut in late summer. The fields are thronged with corn buntings, turtle doves and many other species vanishing elsewhere in Europe.

The horse-drawn mower and then, in more recent times, tractor-drawn mowing machines would slowly remove the hay meadow from the British landscape almost entirely. Since the Second World War alone, Britain has lost 97% of its hay meadows,[22] but the narrative goes back more than a century. Only on the islands of the Uists, and the Inner Hebridean farmlands of Iona, Tiree and Coll, can we still enjoy the full bounty of hay meadows – and the common grating call of the corncrake. Away from the Hebrides, however, British hay meadows are now little more than ghosts: places of a richness we can scarcely imagine without travelling to the most ancient farmlands of Europe.

1890: The Woolly Empire

Whinchats, dapper songbirds that migrate annually from Africa to Europe, do not require sheep, and did not evolve beside them. Conservation management plans, though, often advise how many sheep whinchats require. This is really odd. Whinchats have graced our grasslands for many thousands of years. Sheep have not. So how did a migratory songbird from Africa become dependent on non-migratory sheep, from the Middle East?

Whinchats benefit from the presence of sheep because what is left, after sheep roam a hillside, is bracken, grassy ground and well-established trees. If you're walking in the Brecon Beacons in summer, on the ferny hillsides known as the *fridd*, you may glimpse a whinchat watching you, its mate hidden away in a bracken tussock. By the 1890s, whinchats were shifting their distribution patterns alongside sheep, so dominant had this animal become.

Having increased since medieval times, British sheep numbers had reached over 15 million animals before 1900. That number has since risen to 22 million, by 2012.[23] In 1801, however, the majority of Wales's 587,000 people were rural and still working in agriculture. There was, at this time, a semi-nomadic practice, now largely forgotten, called transhumance. Shepherds would move with their sheep and cattle, and the seasons, from a low-lying valley farm, a *hendre*, in winter, to an upland farmstead above the valley, a *hafod*, in summer. The same happened in Scotland, too. By endlessly shifting animals, shepherds would have created an upland landscape more diverse than that of today; one where flowers and bushlands had more chance to flourish and survive, as can be seen in the hills of the Alps or Carpathians. By the end of the nineteenth century, this ancient tradition would change. Enclosed farms were established, rendering landowners independent, with ever-growing flock sizes and an ever-diminishing workforce. Only a small proportion of the Welsh population, even by this time, owed their survival to the land.

As early as the 1860s, labourers were moving to the cities, beginning a 160-year narrative of rural evacuation.[24] Before 1900, the nomadic sheep-farming traditions had been replaced with new, stationary sheep farms. As sheep numbers increased, entire landscapes, once varied, gave way to the green lawns of grass, brown lawns of heather and single oaks that we have grown to accept as the uplands of Wales and western England. Yet these are some of the most denuded and nature-starved areas perhaps not only in our country, but on our planet – not only for wildlife, but for many communities as well.

1900: A Century of Change

In 1801, when black grouse bubbled on Bodmin Moor, the English population was 10.5 million. By the next census, in 1841, it had grown to 15.9 million. By 1901, it had almost doubled to 30.5 million people – compared to 55 million living in England today.[25] The full effects of our growing demand for food, however, had yet to impact something special: the little lives of the grass.

In the early twentieth century, spotted flycatchers were one of our commonest migrants. Whinchats and nightjars bred around the scruffy edges of our villages. Wrynecks raided anthills in Kentish gardens. Colonies of marsh warblers thrived in withy-beds, from Sussex to the Severn. Cirl buntings chattered on Wimbledon Common. The birds of weeds and seeds, grey partridges, turtle doves and tree sparrows, could be found in their droves. Cuckoos were noted on their arrival in newspapers, across almost all of our villages. Nightingales blasted out their songs across southern England. Britain was a tamed land but fizzed with insects. In the first decades of the twentieth century, insectivorous birds were still to be found everywhere: the pretty afterthoughts of rural life. Butterflies, we know, still flew in their thousands. Many birds, adaptable to our every change, sang beside us in numbers no one alive can now remember. We had yet to tidy up the insects, flowers and weeds – the bits in between. In the coming century, the final act of taming would begin.

1914–1918: Birds of the Great War

Between 1914 and 1918, Britain lost just over 700,000 young people on the battlefield,[26] and by the end of the war the country was close to financial collapse.[27] Birds, however, have a curious habit of capitalising on human disaster. The wartime government, realising that food supplies were running short, called for wood-pastures to be ploughed up and replaced with new cereals, but there were fewer hands on deck and the wartime countryside lapsed into a degree of depopulated scruffiness not seen for decades. Marshy farmland came back: birds like snipe, suppressed by centuries of drainage, drummed over a silent England. Buzzards, relieved of direct persecution, began to recover in numbers, and spread to eventually recolonise the whole of rural Britain.

1919: Swift Conversions

Deep in the heart of Abernethy Forest, in the Scottish Highlands, stand some very special trees. These dead pines have woodpecker holes drilled into their trunks. But if you wait long enough, you'll realise these holes are no longer home to woodpeckers. If you are very lucky, you will see a far stranger sight.

Once or twice a day a bird comes down from the sky. It has flown over gorillas in the Congo and camels in the Sahara and now, for just a few months, it flies at dizzying heights over our own heads, outpacing peregrines in level flight. It is a relic of our ancient, largest trees: the swift.

These Abernethy trees are the last reminder we have in Britain of how swifts lived before there were houses.[28] Swifts, evolved to nest in trees, and crawl downwards into holes, have since adapted to nest in buildings, and crawl upwards into nooks. They have become masters of architectural yoga.

Swifts had declined across the nineteenth century,[29] but conservationists pin their recent downfall to one year, 1919. The Housing and Town Planning Act of 1919 would see over 200,000 homes built in just a few years.[30] The order of the day was quantity. From this time on, haphazard old houses, with their loose stone, slate or brick roof tiles that, for swifts, act as man-made trees, would not be seen again. It is believed that 10% of all homes built before 1919 are able to harbour swifts. For houses built after the Second World War, however, that figure falls to just 1%.[31] And the number of swifts in Britain continues to fall each year.

1929: Sprucing up Britain

In the wake of the First World War, the Forestry Commission was born. Faced with a new low in British woodland, covering just 5% of the country after the war, the Commission was founded to increase the production of timber. By 1929 it managed 600,000 acres (243,000 ha), planting 138,000 of these (56,000 ha) with new trees.[32] The effects of this planting are still being felt, preventing life in the deciduous wooded landscapes where we should have most life of all. In the 1920s, the order of the day was bulk. Softwoods that grew fast were desired, and so non-native, fast-growing Sitka spruce, suited to unfertile, acid soils, heaths and bogs, became the favourite.

In their quest to turn our woodlands into timber factories, the government imperative of the time, the Forestry Commission would not only colonise huge areas of peat bogs and heaths with plantations, but destroy 38% of Britain's ancient woodlands, to replant them with conifers.[33] Whilst ecosystems from Brazil to Indonesia have been wrecked by felling trees, Britain's wildlife has often been smothered by planting them. In our largest afforested area, Galloway Forest Park, covering 770 square kilometres, we find just 6% of native trees. In Kielder Forest, an area of 610 square kilometres, the figure is 7%. The list goes on and on across our forest parks.[34]

Whilst the big march of forestry happened after the First World War and led to the picture described here, the now-forgotten desecration of many of our finest woodlands had taken place long before this time. In the first half of the nineteenth century, an enchanted national monument stood in the New Forest. It was called Old Sloden Wood. One writer described it beautifully:

> Hollies, yews, and whitebeam of the largest growth
> stood singly or in small groups at intervals, for the
> full appreciation of their form and colour, and for
> glimpses of distant landscape. Here and there a
> shapely oak or beech overhung the evergreen clumps,
> and aged birches or hawthorns studded the open
> spaces.[35]

A few decades later, the *Journal of Forestry and Estates Management* described it once more:

> By an Act passed in August, 1851, the Commissioners
> were empowered to remove the deer, and to plant trees
> other than oaks ... This power has been exercised in the
> most barbarous and destructive manner imaginable ...
> The Commissioners pounced upon the richest and most
> picturesque parts of the forest, cut down the ancient
> trees – the living mementoes of bygone centuries – tore
> up with the plough the rich greensward that had
> existed for ages, and reduced some of the loveliest
> bits of landscape scenery to gloomy and monotonous
> plantations of black fir.

It's worth noting that the forestry practices considered normal by the standards of today were, by the standards of the 1850s, a disgrace to the foresters of that era. Whereas our ancestors could marvel at Sloden's cathedral yews, we must walk instead through Sloden Inclosure. Growing in straight, planted lines, the New Forest's plantations have swallowed something infinitely richer – and this story has played out, unchecked, across our country.

Such planted forests have radically altered the woodland landscape of the British Isles. Whilst goshawks, crossbills and siskins have all been able to recolonise Britain as the crops have matured, and such forests are not entirely lifeless, these dark silent places, alien ecosystems for much of our native wildlife, are one of the greatest deserts in our country.

By the 1940s, the Forestry Commission was the largest landowner in Britain, but nowhere in the annals of wildlife recording do we find a resurgence in most of our native woodland birds. Today Britain faces the fastest woodland bird declines, and the most imminent woodland bird extinctions, of any European country.

1930: The End of an Affair

The basic system of crop rotation was developed 8,000 years ago in the Middle East, long before soil chemistry was understood.[36] Until the 1930s, rotation was trundling along and its premise was simple. Your soil can only provide so much. Growing the same crop in the same place depletes the soil's nutrients.

All crops draw on nitrogen, potassium and phosphates for their growth. Traditionally, to put those nutrients back in, farmyard manure was spread before the planting of a nutrient-hungry crop, such as corn. In addition, planting the same crop in the same field over two consecutive years risked, in a time before pesticides, the crop falling prey to disease and pests. To get around these problems, pre-war farms would often rotate their crops, sometimes on a four-to-five-year basis. Crops such as peas, beans or lucerne were of great use, sucking nitrogen back into the soil, and so improving conditions for the next crop planted on the land. At a season's end, the remnants of these crops were ploughed into the soil, further improving its productivity. This was the only way, traditionally, that nutrients could be restored to the soil.

The side effect of crop rotation was promoting bird diversity. Birds benefited from a range of food and nest sites, because whilst one field was doing one thing, another was doing something else. This increased the variety of the menu on offer across the year, and promoted the small-scale mosaic of habitats in which many of our scrubland birds evolved in the first place.

From the 1880s to the 1930s, it was boom time for arable birds. This may have been the best time in history to be a grey partridge or tree sparrow. Infinite stubble. Infinite hedgerows. Infinite insects. Infinite weeds. These weedy seedy birds thrived beside us in cereal farms, in part because Britain's fields 'rotated' side by side. If you live in Britain year-round, you need food year-round. For what are now farmland birds, winter is about stubble, and stubble is about seeds. Summer is about insects – building proteins to lay eggs, feeding your chicks and your chicks learning to feed themselves. Provided you have seeds, insects, and a safe place to nest, life is good. By having fields in different 'states' all at once, the varied seasonal diet of farmland birds is far more likely to be provided for.

From the 1930s onwards, a word now indelibly associated with farming – fertiliser – became popular. Inorganic fertiliser rigs the productivity of soil. With fertiliser, you no longer need to rotate your crops, because you can restore the nutrients artificially. Pesticides, meanwhile, allow you to head off the problems of disease. This means you can plant the same crop, in the same field, year after year. But while you can rig productivity for farming, you can't rig a food chain for birds.

Fertiliser leads to your starting over each season – not just with crops but with plants. The buggy and weedy complexity of a field is replaced with a blank slate. The sterility of this field only increases over time. Each year leads to fewer flowers, fewer insects – and fewer birds.

In combination, fertilisers and pesticides compromised a cycle of insects and seeds that had worked in sync with British birds for tens of thousands of years. They infiltrated landscapes, and still do so today, promoting the rapid development of species-poor grassland, of the kind you often see at the edge of our roadsides. From the late 1930s, grey partridges and tree sparrows started to decline. Cirl buntings began a retreat at the edges of their range. For the first time in over 830,000 years, we had the tools to turn on the plants and insects that had kept so many of our birds alive, against the odds, for so long.

1939–1945: Wartime Birds

The Second World War ravaged Britain, but as huge numbers of people left for the front line, more birds quietly drifted back in. In 1939, hen harriers floated back into the Scottish Highlands from their refuges on Orkney, and continued to expand until, by the end of the war, they had reached southern Scotland. With the conversion of sheep estates to deer forest, golden eagle killing had reduced in Scotland by 1900. Across the war years, eagles recolonised many more of their Highland homes.

If there are any ecological lessons to learn from this harrowing time in our own history, it's how quickly wildlife can bounce back, how quickly landscapes can be retaken by nature, and how, in an age where conservation has defined itself by management, we've forgotten quite how many birds thrive best on peace and quiet.

1946: Ripping Out the Roads

Aerial photos from the late 1940s show a rambling Britain, filled with non-woodland oaks, old willow lines and oodles of hedge. Now hedgerows, the relics of enclosure, would themselves become obstructions to ever-larger farms.

A hedgerow made of bushes is a habitat. A hedgerow with trees is a highway. From a bird's perspective, an ancient hedgerow is the continuation of a wood. Ancient hedgerows with oaks don't just increase feeding and nesting sites. They connect woodlands to one another, allowing birds to move across landscapes. If you rip wooded hedgerows from a landscape, you're destroying not just homes – but highways. Even in a countryside devoid of large woodlands, small copses, connected by hedgerows, can function as big woods. Many birds, including marsh tits, do not seem to notice that their hazel woodland has become synthetic, long and thin. Nor do Britain's dormice. Then again, they were probably asleep.

On average, Britain has lost 50% of its hedgerows since the war, and many arable counties, especially in eastern England, a great deal more.[37] In the absence of our wooded grasslands, hedgerows kept our remaining trees connected and alive. This is why ancient hedgerow removal is so devastating – and one reason why isolation is such a crushing problem for our woodland birds. From the 1950s, the highways of the countryside were closed. Countless lines of open-grown oaks and cuckoo-haunted willows were ripped out forever.[38]

1960: The Vanishing Clouds

The post-war years were probably the last time people could take for granted teeming multitudes of grasshoppers and butterflies, and hordes of moths on a summer's night. For many of us, our grandparents were the last generation who could scatter grasshopper clouds from below their feet – or doze under a drone of bees. Younger people may smile indulgently when an elderly relative describes clouds of butterflies, yet records and collections show that these were real.

In the 1950s, cuckoos, specialising in hairy moth caterpillars, began declining in eastern England. Butterfly- and bee-specialist spotted flycatchers were monitored from 1965 by the British Trust for Ornithology (BTO) Common Birds Census. Between 1965 and 1976 alone they declined by 50%. Familiar birds of our open wooded lands, cuckoos and spotted flycatchers had dovetailed into villages, farms and the countryside at large until this time. But these birds are deceptive – they have specialised diets: diets that, before the 1960s, nobody would have had cause to notice at all. Chapter 3 will reveal in greater detail what has happened to our insects and our birds since this time. But for now, one fact will suffice.

Back in 2004, the RSPB organised a 'splat test' to 'see if insects were really in decline'. This survey yielded one truly chilling result. As car owners drove through our countryside, assiduously counting 'splats' on their registration plates, they recorded, on average, one insect death – for every *eight kilometres*.[39] If there is one statistic that tells you why shrikes are extinct in Britain and spotted flycatchers are forsaking your village, you need look no further. Having eliminated the top end of the food chain centuries ago, we are very close to wiping out the life at the bottom. The birds, caught in the middle, vanish around us every day.

1962: The Great Wildlife Cap

If those Tudor grain acts seem barbaric to us now, future generations will surely look back on the European Union's Common Agricultural Policy (CAP) as one of the greatest extinction policies ever created and carried out for wildlife. The CAP, conceived in 1962, aimed to squeeze every last inch of food from the European landscape, wreaking by far its greatest effects in its western countries. Reading this history, you'll have realised that agricultural practice has intensified not over decades but over centuries. The CAP, however, was its crowning glory.

The CAP incentivised farmers to produce more food by subsidising them; keeping prices artificially high. By 1999, the CAP was costing the EU taxpayer more than 40 billion euros per year: over half the total EU budget. In England, in 1998, the situation was so extreme that farming income was £2.17 billion, whereas subsidies made up an additional

£2.67 billion.[40] With a rigged market and unable to fail, farmers were paid to perfect Factory Britain. Land drainage, hedgerow removal, conversion of pastures into crops and increased agrochemicals and pesticides had all been in play before this time. Now they were wheeled out everywhere – at once.

If there is a choice within a system, there will be variety. Some farmers will maintain scruffy farms, some will practise rotation, others will leave fallows. Some will prize old orchards or veteran trees; others will not worry about taking out hedgerows or draining every field. Each farmer has, given a *choice*, a slightly different method of farming – and a highly individual tolerance of nature. But if there is one paid directive to remove all these things – they get removed. 1532: remove the kites and wildcats. Get paid for each one. 1962: cleanse the countryside for yield. Get paid for every square centimetre you put on the market. Get paid a huge part of your income to turn your land into a factory. In this regard, 'fault' for the desert state of our countryside might not, in fairness, be directed at farmers – but at policy.

The 'green revolution' of the CAP achieved what it intended to do. Against a backdrop of continual population rise, by the 1990s Britain was producing 25% more food per head than thirty years before. As of 2015, the UK is three-quarters self-sufficient in food such as homegrown crops, and 62% self-sufficient in all food. However, because of what we export, we actually only supply 54% of our own food in total. The rest, we still import.[41]

That 25% increase in food per head, however, has not wrought a 25% reduction in our wildlife. It has wrought a collapse, the worst in Europe. In the past four decades, we have witnessed approaching local and national extinctions in many birds of the countryside, which show no sign of reversing at all. Here's what we know so far.

1973: Recording Extinction

If the British Trust for Ornithology were to invade another country, it would be a ruthlessly logical affair, conducted one kilometre at a time by dark green volunteers armed with notebooks. Fortunately for other nations, the BTO has confined itself to acting as our leading research institute for birds.

In 1968, the BTO initiated the fieldwork for an atlas to record the breeding distributions of all the nation's birds, and published it in 1976. Since then, it has produced two further atlases, the last (covering 2007–2011) in 2013.[42] So over the last four decades, we know exactly how, and exactly where, our birds are returning and vanishing – in grids of every 10 square kilometres in Britain. It is partly because of the success of these atlas statistics that 1970, not 1760, has become the benchmark for modern bird decline. It is to the 1970s we look, misty-eyed, for a time with more tree sparrows – yet a time long after many birds had already lost most of their populations and abundance in our country. Since 1966 alone, Britain has lost a minimum of 44 million individual birds. Twenty million of these have been house sparrows, as their insects, seeds and bushes have been tidied out of farmland.[43] This leaves 24 million individual birds of a wide range of other species. Almost all of these are, however, the insectivores or seed-eaters of our vanished wooded grasslands or wet meadows: birds that adapted to farming over time, gambled on its food supplies – then lost.

In recent times, turtle doves committed to the weedy margins of arable fields. In the late 1960s, as many as 250,000 of these birds bred across Britain. Older Norfolk naturalists recount them smothering telegraph wires as they prepared to migrate each autumn. Fewer than 4,300 pairs remain at the time of writing – next year, there may be a thousand fewer.[44] With their weed seeds sprayed into oblivion by herbicides, turtle doves have starved, halving the amount of chicks they are able to raise. They've crashed by 96% since 1970. Within a few years, they could be gone.[45]

Since 1970, corn buntings have declined by 90%. Since the 1930s, second to house sparrows, we've lost more biomass of tree sparrows than any living bird: 97% have vanished since 1970 alone, and for every tree sparrow we see today there were 30 in the early 1970s. These birds' dual diet, of summer invertebrates and winter seeds, has been cleansed from modern farmland all at once. Since 1973, we've lost 93% of our grey partridges. Herbicides, again, have starved their chicks, which now have just a 30% chance of survival.[46] The story goes on for all of the birds that survived in our mosaic farmland – whose seeds and insects the CAP has, for four decades, paid farmers to weed out.

The last relics of our wet grassland bird communities continue to drain away. With only isolated populations left, intensely vulnerable to predators, their decline is remorseless. Since 1960, lapwings have tumbled by 80%. Since 1970, yellow wagtails have declined by 60%. And 62% of snipe vanished from our wet grasslands between just 1982 and 2002. These are the birds of damp soils, and small herds of grazing animals. Soil compaction, insecticides and herbicides destroy their prey base. The CAP, since its inception, has actively encouraged land drainage, drying out the soil in which wading birds feed – but it is the increase in cattle herd sizes since 1970 that is perhaps even more striking.

If there is one number that explains the lack of buzzing meadows and yellow wagtails in Britain, it's *142*. That was the size of your average British dairy herd in June 2015.[47] As recently as the 1970s, the average herd size was 30. In Poland today, that average herd size is still no more than *five*.[48]

British livestock farming now bears no resemblance to the way in which original grazing animals once shaped the land. The pea-green dairy fields you drive past are by and large mown and shorn of life. Their open lawns afford nowhere to hide – and nowhere to feed. Cattle in their hundreds, confined in fields, create compacted lawn. Medicated dung, the result of worming chemicals, acts to sterilise the soil and its beetle communities. The once soggy, buggy cattle field, amenable to wet-meadow birds like yellow wagtails, has become a desert. Many are too dry, hard and insect-poor for ground-feeding birds such as waders to survive.

Away from our towns, the starling's habitat is wooded, unsprayed pasture, teeming with juicy leatherjackets (crane-fly larvae) below the soil. Not only have the pastures used by starlings for feeding become compacted by static livestock herds, but glyphosate, widely sprayed onto our fields, devastates the growth of ground-dwelling invertebrates. Across Europe, RSPB scientist Dr Richard Gregory has calculated that starlings have vanished at a rate of 150 per hour since the 1980s. In Britain, 80% have been lost since 1980, and more from countryside landscapes than from our towns.

1980: The Woodland Crunch

Scrub, for farmers and conservationists alike, has often become a prime candidate for destruction in our countryside. The figures are now in on how devastating this has been. Nightingales have declined by 90% since 1970; we have lost nine of every ten. Willow tits have declined by 88% between 1970 and 2006, and now race towards extinction. They are our fastest-vanishing bird. The dank jumble of elder, used by willow tits, or the outgrown blackthorn of an old hedge, used by nightingales, are two such details that have en masse been tidied out of Britain. Scrub has been removed as hedgerows and margins, from our brownfield 'wastelands' and from our nature reserves. Many scrubland specialists are now red-listed. At the same time, our wood-pasture species – those birds adapted to the spacious, climax vegetation of the British Isles – have vanished for different reasons.

Hawfinches rely on a diversity of mature, light-loving trees now rarely seen in one place: apples, cherries and oaks in the summer, hornbeams or beeches for their seed in the autumn. Hawfinches need entire wooded landscapes to survive, using each area in different ways across the seasons. With our shady woodlands no longer opened by either cattle or coppicing, light-loving trees like apples and cherries have grown rare. Only in rich varied woodlands like the New Forest do hawfinches now find their varied menu of ancient trees.

Lesser spotted woodpeckers thrive where caterpillar-rich trees such as oaks, willows or alders are left to rot in peace. Copses, old orchards, riverside trees and outgrown hedgerows benefit these deadwood birds, which require huge areas of such trees to survive. Since its inception, the CAP has paid arable farmers to take such trees from the landscape, along with three-quarters of our lesser spotted woodpeckers. These birds have also starved in their nests, as caterpillars have vanished from our woodlands.

Our vanishing wood warblers are oak caterpillar specialists. In eastern Europe's woodlands, they remain a common sight. In Britain, they face extinction. In recent years, our native great and blue tits have moved their breeding cycle forwards with the earlier spring, taking advantage of the earlier emergence of caterpillars due to climate change. At the same time, ever-increasing armies of deer have ripped out the bramble and other food supplies that once fuelled the butterfly populations in our woodlands. Wood warblers, flying in from Africa, not only arrive late to the party, but find an ever-poorer banquet. As our woodlands starve of caterpillars, such delightful summer visitors have been the first to lose out.[49]

Britain, one of Europe's least wooded countries, its woods mostly alien conifers, or growing as isolated islands, could not be worse placed to avoid a mass extinction of its woodland birds. As each species vanishes, we can learn, if we pay due attention, about the growing desert in which we live. A lesser spotted woodpecker, for example, tells you that a landscape is rich in native trees, maturity, insects and diversity. Its disappearance tells you that those things have gone.

1990: Tipping Point

The phrase 'tipping point' refers to a series of events that may spell a point of no return for our planet's environment. Professor Tim Lenton, a leading climatologist, has outlined what such tipping points could be.[50] One example he gives is Antarctic sea-ice melt. The world's seas would rise by 7 metres if Greenland's ice cap melted. But if Antarctica melted in its entirety, it would add 70 metres to our oceans. The melting of the Arctic would be disastrous. Antarctic melt could, without exaggeration, signal the end of civilisation as we know it.

Lenton cites boreal forest die-back, and Amazon rainforest die-back, as two other tipping points for the planet. In this case, the world's lungs would cease to work. Overheated forests are less able to take in carbon dioxide or expel oxygen. In both cases, the world's natural systems are pushed past what Lenton calls a 'critical state'.

Since the 1970s, what we have seen in western Europe has been a new tipping point for our native wildlife, as the relics of lost ecosystems have finally collapsed. When you reach a tipping point, things go horribly wrong. All the problems that have been building up don't ease off. They intensify.

Carried out yearly since 1994, the BTO's Breeding Bird Survey (BBS) monitors the status of our 'countryside' birds every year. Given that since the late 1700s, our birdlife has been in rapid decline, you would have thought that by 1990 we might have reached the bottom. Yet BBS shows that between 1994 and 2011 alone, turtle doves have fallen by a further 90%, willow tits by 79%, wood warblers by 65%, whinchats by 57%, nightingales by 52%, spotted flycatchers and starlings by 50% and curlews by 48%. These are resident and migrant birds alike. These birds have not 'declined': they are declining.

In the early 2000s, my father and I would drive to Norfolk in winter, and compete on who could count the most kestrels between Wisbech and Peterborough. In 2004, we counted nineteen. In 2017, driving the same road at the same time of year, I counted two. Such personal memories are sadly supported by hard national facts. British kestrels declined by 20% between 1995 and 2008, but by a further 36% between just 2008 and 2009. Declines are speeding up in many birds, not slowing down.

In recent decades, as some of the worst effects of early CAP policies have wound down, other new forms of damage have intensified: none more so than our ceaseless war against the soil itself. Beset by chemicals from worming drugs to weed-killers, the fragile insect communities below our feet are now more imperilled than ever before. At the same time, deep-ploughing can destroy such communities entirely. Mycorrhizae, the network of subterranean fungi on which the root systems of many plants, including trees, depend, are ripped out by deep-ploughing, along with unseen worlds of crane-fly larvae, worms, annelids and beetles. As some forms of factory farming have faded out, others are coming to the fore.

The British landscape reels. Temperate ecosystems are extraordinarily resilient, but as a child in my garden, in the 1990s, I never contemplated a world without honeybees, or a garden without starlings. We have now pushed things too far. Pesticides, herbicides, monocultures, drainage, isolation, invertebrate decline, the loss of oaks; the tidying of scrublands; intolerance of the kestrel's grassland haunts. It is the collective force of sterilisation that has starved and isolated wildlife populations, which now free-fall to extinction by themselves.

2010: Bipolar Birds

In July 2010, Norfolk birder Connor Rand and I crept into the Washington Hide of Norfolk's Holkham Freshmarsh. Rumours were rife. *Something* was afoot. But the Norfolk grapevine was full of rumours that summer. Murmurs of hawk owls in Breckland. Whispers of a colony of hoopoes near Cley. In the British capital of birding, we all wanted to believe there was more life out there than any bird atlas could unearth. So we followed one more whisper, and sat quietly in the hide at Holkham on a hot summer's evening. And then we saw it. A spoonlet.

A newly fledged spoonbill chick is perhaps the most adorable thing on the planet. It's white, fluffy, doddery on its legs and born with a spoon in its mouth. It was mooching around in front of the hide. As we watched, not one, not two, but five adults came drifting in to the stand of willows above. There was something brazen about it, as if they'd been doing it for decades. To set eyes on a spoonbill colony, the first in Britain since

1620, was a moment I will never forget. That year, under the careful eye of Natural England, four pairs of spoonbills fledged six young. In 2019, twenty-eight breeding pairs, and 244 newly fledged spoons, graced the Holkham coast.

In recent decades, the story of birds returning to our wetlands is inspiring. Cranes, drifting over the derelict drainage mill at Horsey Mere, in Norfolk, must be one of the most triumphant sights in the world of British wildlife. History here has not moved in a straight line but a circle. The crumbling windmills that drained cranes from Britain have come and gone. Cranes, symbols of long life, have outlived them.

Bitterns, on the brink of extinction in the 1990s, now outnumber grey herons in the marshes of Somerset. An army of other herons has capitalised on warming global temperatures and our expanding reedbeds. In 2017, ten pairs of great white egret, five pairs of cattle egret and a family of night-herons all raised young in these marshes. Mediterranean birds unthinkable in my childhood, like glossy ibises, are moving in. Black-winged stilts are nesting in our eastern marshes.

Yet it is in the bipolar ratios in Britain, between managed wetlands, returning birds of prey and the rest – the dying organism of the countryside itself – that we find the most damning verdict on the health of our countryside. We have more goshawks in silent spruce crops than godwits in wild river valleys. We may soon have more white-tailed eagles than turtle doves, and more bitterns than functional woodland ecosystems. At current rates of decline, the next generation will see more birds of prey than caterpillar-eating cuckoos. We have more avocets, a specialist of gravel, and more peregrines, an apex predator, than hawfinches, drawn to ancient oak woodlands – our once commonest habitat. And that – is a mess.

In 2017, a bulletin from the Somerset Ornithological Society lamented that 80% of bird records submitted came from the Somerset Levels alone.[51] It is not hard to see why. Concentrated within the Avalon marshes is excitement – the thrill of life itself. Bugling cranes. The boom of a bittern in the mist. Families of otters. The emerald flash of a kingfisher, its nest hole plumbed into a bed of peat. The bubble of a female cuckoo, wickering in the reeds. Each year more life, not less, comes to makes its home in Avalon.

Outside, beyond these carefully protected wetlands, lies silence. There is so little chance of stumbling on a hidden woodland, fluting with nightingales, of discovering an Exmoor valley bubbling with curlews, that few people bother to write off their precious work-free days to disappointment. In Somerset's silent fields, nature springs not one surprise.

2020: Freefall

When faced with such rapid, devastating declines in our wildlife, the temptation can be to see Britain's problems as part of some international malaise – the inevitable consequence of a changing climate, or a growing population. Across the world, these are indeed two key drivers of wildlife decline, but they do not offer an adequate explanation for the unique desert we have created in Britain. One problem in coming to terms with the sheer sterility of our own country, national pride aside, is the syndrome of 'shifting baselines', whereby we have adapted our expectations of the countryside to the standards of just the last couple of decades. Most often, a conservation baseline will simply reflect what has been lost in the last generation. If you read a press release on insect loss, for example, it usually mentions the 1970s as the 'start' of a decline. Nothing could be further from the truth.

In reality, a huge amount of Britain's insect abundance and diversity was lost before any record-keeping began. True butterfly clouds vanished from most places before any of us were born. So did wood-pastures fizzing with enormous anthills, and wrynecks. So did mile after mile of beetle-filled meadows, and red-backed shrikes. Beetle declines began at least around the mid nineteenth century with entire species vanishing before 1900. Now, our long-forgotten natural history lives on not in any living memory, but in the living countryside still seen in large parts of eastern Europe and other areas of our continent that have yet to be turned into factories.

Many press releases tell us that Britain's bird declines are 'being seen across Europe' – but this is not entirely true. Western, industrialised, intensive Europe has seen these declines. Pre-intensive far eastern Europe, yet to feel the full destructive force of policies such as the CAP, has not. Whether it's a turtle dove from Africa, or a partridge native to a Norfolk field, the loss of weeds in farmland achieves the same result: extinction. The farmlands of eastern Poland, Romania, Latvia or Lithuania – where the CAP has yet to wreak its worst devastation – are still alive with much of what we've lost.

Across Europe, the degree of decline in many birds correlates not with where those countries are, but with quite how intensively their countryside has been managed or cleansed. In hay-meadow-rich countries like Hungary, its 64,000 or more pairs of turtle doves are considered to be stable. The 120,000 or more pairs in the ancient farmlands of Romania have been found to fluctuate – but not, as yet, to decline.[52] In countries with intensive farming but a large amount of less intensively farmed land, like Germany, turtle doves have declined by a half. In Britain, as described above, they've declined by 96%. Germany, a nation of intensive farming that feeds its own, still has 8,500 more pairs of wrynecks than Britain's average of zero.[53] Comparative studies show that as early as the 1970s, mixed farmland in Suffolk held 2.2 pairs of spotted flycatcher per square kilometre. In the same period, parkland in West Germany held 100 pairs per square kilometre.[54]

Germany is home to around 630 pairs of white-tailed eagle, but 150,000 pairs of red-backed shrike. That ratio, between a large predator and a bird happy with a spiky bush and beetles, is consistent with a modern landscape that has kept room for some of its insects. In 2014, the Netherlands, highly developed and populated, still held enough insect-rich pastures for 33,000 pairs of black-tailed godwits.[55] Britain, with more acreage of low-lying farmland, had, in the same year, just 50.

The causes for bird decline lie in the unique silence of the British landscape. We have tidied and removed life with greater zeal than our neighbours, as we did in both Tudor and Victorian times. Today, 99% of walks in the wider countryside will take you through a landscape that is not only free from wolves but free, at last, from most birds and insects too.

In addition to our collapsing insect food chain, another factor leading to our current wildlife freefall in Britain is something called 'extinction debt'. Once populations have become isolated, you don't need to do anything wrong. Such pockets of birds are already too small to survive the typical fluctuations of a normal population. With no recruitment of new birds, it only takes a few unfortunate events for each of those island populations to vanish. A bad winter, a wet summer, a persistent fox at a colony of curlews – and those birds will vanish, one island at a time. For example, 80% of our remaining turtle doves have vanished since 1994 alone. But that doesn't mean farmers started doing anything much worse after 1994 – extinction does not, after all, work in a straight line. You cannot, for example, watch birds starve in the nest unless you study them each year. But suddenly, one year, your population of these birds will vanish, like some awful magic trick, because not enough chicks have been able to find food. Resident birds such as willow tits can sing for years – then go silent. Behind the scenes, their terrible yearly survival rates guaranteed extinction: these birds were the living dead.

In reality, ecosystems and their inhabitants do not respond to damage straight away. You punch them and they often react years or decades later. Not only do chemicals take years to fully sink into food chains, but birds do not fall off their perches overnight. It is only through successive years of nesting failure that birds are left unable to replenish their populations. Extinction debt is a term used for where future extinction is *already* guaranteed, as an inevitable result of events in the past. And sadly for Britain's birds, a huge number of our bird populations are extinct – even as you watch them.

Freefall is under way, and has been for twenty years. It would be naive to think that many species in serious decline will be here at all within a human generation. At least six of these – turtle dove, wood warbler, willow tit, lesser spotted woodpecker, nightingale and curlew – are extinction-critical. And still other birds are just beginning their declines. Who would have thought we'd worry about pied wagtails, bobbing around our motorway service stations in their dapper pinstripe suits? Since 1994, they have declined by 11%. As even our chaffinches fall silent, we make a dangerous mistake if we think we've reached the bottom. But we have also the ingenuity and opportunity to turn things around.

Britain has millions of wildlife-loving minds who – acting intelligently, differently than before – can restore our wildlife to amazing heights, provide protection against climate change, and ensure a future for rural jobs. And that is what the second half of this book will be about.

CHAPTER 3

The First Imperative

Finding Food in a Starving World

Insects of all kinds literally swarmed. Butterflies were in profusion. The silver-washed fritillary was in hordes ... they were so common that, as the sun touched their overnight resting places, they dropped from the trees like an autumnal shower of falling leaves.

—Frederick William Frohawk, ca. 1880[1]

We were deep in the Leuser rainforest. The last place on earth where orangutans, tigers, rhinos and elephants survive in one place. The last island outpost of Sumatra's once endless jungle. We had come to film the island's orangutans, here in their last refuge. Yet all around us, villagers were felling giant trees to plant chillies.

Most of the time, in a rainforest, you hear a lot but see very little. Sound shivers and mutates, distorted, through the twisted skyscrapers of trees. Three-quarters of the jungle's inhabitants call unseen 60 metres above your head. But with the jungle on the brink, a cross-section opened up. The margins of one clearing exposed the fairy-tale diversity of Sumatra. One tree will always stand out in my mind. It was a giant fig.

Figs are to the rainforest what 500-year-old oaks are to Britain's woodlands: mines of life. Whilst Sumatra's jungles are home to over 1,000 tree species, figs are not only invaluable but rare in any one stretch of forest. Animals commute many miles to find them – and the parade we saw was incredible. One morning, the black spider silhouettes of siamang gibbons could be seen. A little baby was fumbling around, learning to climb 60 metres off the ground. Another dawn, a grumpy and mossy-looking binturong was seen coming to feast. Hordes of fruit doves and, in total, eighty species of bird, came to pay homage to the fig. Each morning, we learned a little more about the magic of Sumatra, all through the branches of this one single tree. But there was one family of pilgrims that I'll never forget.

Nothing can prepare you for a flying rhinoceros. How on earth has it survived? How outrageous that it still exists. That we aren't, already, having to convince people that a bird with a cartoon-red rhinoceros horn once croaked like a hoarse duck over the jungles of east Asia. Yet rhinoceros hornbills aren't, quite, history. Each morning, a pair of these Disneyesque birds would visit our fig. A cartoon assembly of their friends – wrinkled and wreathed, pied and bushy-crested – came to join them. It was as if someone had opened a field guide. There were the hornbills: all in one tree. Then, one day, the villagers cut the fig tree down.

Like anything lost in a food chain, or trophic cascade, you cannot see it once it's gone. It passes in the blink of an eye. One day, as we passed the clearing, there was simply silence. Sunbirds flitted away in the palms but there was a gaping hole in the sky. Our pair of flying rhinoceroses flew around, fig-less. They perched absurdly in palm trees. Hornbills depend on figs for their survival. Before our eyes, their jungle's lifeblood had drained out. For three days they'd repeat their pathetic circuit of the clearing. And then, they were gone.

Once the fig tree had fallen, who would ever notice that it was supposed to be there at all? Future generations of tourists, wandering through the palms and admiring the sunbirds, will not feel its absence – just as British children cannot feel the absence of a cloud of butterflies or the hum of bees, or collect the caterpillars dripping from a vanished oak. The British countryside, like that now denuded clearing in Sumatra, can often appear to us a land of plenty: a halcyon vision of green fields. Flying in to Britain from overseas, the green comes as relief to the eye – a carpet of promise. But much of this is now a cruel trick – one played upon the majority of the wildlife that shares our country.

The Absence You Cannot See

Finding food is referred to by some ecologists as the 'first imperative'. Mating is nice, but it's the second imperative. To pass on your genes and keep the species going, you first need to eat. Finding the *right* food, at the right time, is key to the survival of any species on our planet. Whilst primates like ourselves are omnivorous, most birds are not. Many have highly specialised diets evolved over hundreds of thousands of years.

Others have nuanced ways of capturing that food, like the flycatcher sallying for insects from a tree.

Our birds of prey, taking large items we can easily see, are easy to study in terms of their diet. Larger species like bitterns, with a known preference for rudd, have quickly recovered as their feeding areas have been restocked. Mostly, however, food loss is a narrative of growing silence, played out over decades and centuries, with each human generation forgetting what has been lost. Whilst habitat change can be witnessed, food loss is much harder to see – and even harder to quantify. Fortunately, one piece of the puzzle has been provided by the Rothamsted Insect Survey.

Since 1964, Rothamsted's insect traps have provided the longest-running study of changes in some of our insect populations. Their original purpose, however, had nothing to do with conservation. Aphids are the UK's number one agricultural pest. Knowing how many aphids you have is an indicator of how well agriculture is keeping on top of them. Fifteen of Rothamsted's suction traps, 12 metres tall, are sited around Britain. They operate daily in the aphid season – from April to November. These insect-hoovers, taller than an average house, have enormous reach. The data from one of these traps represent not just the aphid population of the field in which it stands – but a landscape-scale analysis of families such as moths, from an area of 100 square kilometres.

It is from this that we know, with accuracy, that two-thirds of common larger moths have declined in the last forty years. Once-common garden species, such as V-moth and garden tiger, the latter's caterpillars a vital food source for the cuckoo, have decreased by more than 90% since 1968 and now face the threat of extinction. In collaboration with Butterfly Conservation, Rothamsted has also found that the total abundance of moths in southern Britain has declined by 40%.[2] Butterfly Conservation has since confirmed that common butterflies, like the small tortoiseshell, have plummeted by 78% since the early 1970s. Sixty per cent of our butterflies are in long-term decline.

But the picture is far more extensive, far more devastating, than Rothamsted alone has been able to reveal. A mapping project of dung beetles shows species tumbling towards extinction, as more and more cattle are fed with antiparasitic drugs.[3] Avermectin, a worming chemical whose residue carries into cattle dung, inhibits the chance for dung beetle larvae to develop, reduces egg production, and increases mortality amongst young beetles. Now, three-quarters of beetles, overall, are in decline. Half of these are diminishing at a rate of 30% per decade.[4] Six bumblebees are rare in Britain and, in addition, three are now extinct. You are rarely bothered by a plague of wasps on your ice-cream now, but twenty years ago you were. The numbers are not yet in, but wasps are known to be in serious decline. Declines in managed bee colonies date back to the mid-1960s.[5] Britain lost half its honeybee colonies between 1985 and 2005 alone – double the loss in mainland Europe. But even on the continent the story of insect loss is being played out, with devastating effects.

The Krefeld Study

Krefeld is a city in western Germany, near the Dutch border, and home to the Krefeld Entomological Society. In 2013, the group returned to an insect-trapping site they'd first visited in 1989, and noticed what was surely a mistake. The mass of insects they had caught had dropped by 80%. Assuming a terrible season, they tried again in 2014: the result was the same. To put this into invertebrate weight, for 900 grams of insects caught in July 1989, just 150 grams were collected in July 2013.

In 2017, a paper written by Caspar Hallmann and colleagues pulled together the results. It analysed comparable data collected over 27 years in 63 protected areas, areas you would expect to resist the worst effects of trophic collapse.[6] It was found that since 1989, flying insect biomass had decreased by 76% over the course of a summer, but 82% during mid-summer – the critical time when most insectivorous birds are raising their young. Almost all flying insect families are involved in the Krefeld study, not only the well-known orders. In the summer of 1989, one reserve trap collected 17,291 hoverflies – in the summer of 2014, just 2,737. That's 14,000 fewer hoverflies, over a summer, for a bird like a flycatcher to eat. Overall, birds across this wide-ranging study area have had four-fifths of their summer diet removed – since 1989 alone.

In Britain, we'd be complacent to think we have fared better. We have, almost certainly, fared worse. Germany's nature reserves still contain insectivores such as red-backed shrikes, attesting, as they do, to insect populations long lost from our own shores. Indeed, insectivorous birds in Germany, including cuckoos, are seeing shallower rates of decline than in Britain.[7]

The 82% of mid-summer insect loss in the Krefeld study would, in Britain, correspond to rates of insectivorous bird loss in the same period – spotted flycatchers, cuckoos and wood warblers – even without other factors thrown in. It would account for the continued decline of whinchats, and other species dependent on the bounty of insects they have evolved to hunt over millions of years.

The reasons for such catastrophic food loss are structural, as we have seen earlier in this book, and, more recently, chemical too. For decades, pesticides have been able to directly remove insect life. But in their latest incarnation, neonicotinoids (thankfully now banned at the time of writing), they have finally achieved what we haven't been able to do since the earliest days of farming: wipe out insects systemically, from the bottom up. Applied to the seeds of crops, rather than sprayed onto vegetation, neonicotinoids are water-soluble. Studies by David Goulson at the University of Sussex show that nectar from wildflowers, in fields adjacent to the spraying, can even end up containing higher concentrations of neonicotinoids than the intended target crop.[8]

Neonicotinoids, ingested by insects such as bees, cause mental confusion, preventing them from navigating and communicating, and accelerating colony collapse. Overall, an astonishing 1,500 peer-reviewed studies now align to show that neonicotinoids are, at the very least, extremely damaging to bees.[9] In the Netherlands, Caspar Hallmann found that 95% of neonicotinoids enter the wider environment.[10] And when surface-water concentrations of the chemical exceeded 20 nanograms per litre, bird populations declined by over 3% per year.

Ironically, the intense chemical warfare directed against our nation's insects, and the chemical killing of our birds, is not only avoidable, but destructive to farming interests too. Studies at the University of Helsinki show that oilseed rape yields are decreased, not increased, by neonicotinoids.[11] It is estimated that it would cost £690 million, in Britain alone, if farmers had to pollinate our crops without the free assistance of insect pollinators – and this is now a very real prospect indeed.[12]

Overall, what the Krefeld study shows is that whilst nature reserves can act as a refuge for individual insect species, they are completely powerless to protect the abundance of insects as a whole – or, as a result, populations of insectivorous birds. The Krefeld study is emphatic that insect loss exists *regardless* of habitat characteristics – and plays out at a huge spatial level. Our tiny nature reserves, with their relict, refugee cuckoos, cannot possibly cope with this scale of national food-chain collapse. However well managed, nature reserves cannot fight against the fact that insects flow across our countryside, as does starvation for adult birds, and their chicks, when these insects vanish.

A devastating collapse in the invertebrate food chain is the greatest driver of contemporary bird decline in Britain. Yet the absence of that food chain is invisible to us – a problem we often overlook because its scale is too enormous to be comprehended. Insects fade without fanfare as they are tidied and removed from our lives. There is no drama to their departure. After a generation, nobody remembers – nobody *expects*. So let's find out quite how fast Britain's loss of invertebrates has crept up on our birds. How bad is the collapse of the food chain?

The Million-Anthill Bird

The wryneck is an open-woodland species that, at first glance, appears happy in a range of habitats consisting of woodlands and grasslands. In Britain, wrynecks often thrived in ancient orchards. Green woodpeckers, however, another ant specialist, are doing well in modern agricultural Britain, including our orchards, whereas wrynecks are extinct. The difference is simple – and lies in the ground.

Wrynecks depend not on ants, but on larvae snaffled out of anthills with their extraordinarily long tongues. In the nesting season, Swiss studies show that most of the food fed to wryneck chicks consists of ant pupae and nymphs. So how many anthills does a wryneck need to survive?

Swiss studies in orchards and wooded grazing farmland, the habitats formerly used in Britain, found that pupae are collected by wrynecks in visits to one visible ant nest per feeding foray from the nest. Over the course of a season, up to 68,000 ant nests in a territory will be raided. More than 8,000 of these nests will be visible above the vegetation. Upwards of 400,000 pupae are needed to raise a single family. And a wryneck raises two families in the course of a summer.[13]

This means that around 800,000 or more ant pupae must be available in nearby anthills over the summer. And this is just for *one* pair of wrynecks to succeed in a season. This number only takes account of the demands of the chicks during their time in the nest. In reality, well over *one million* ant pupae will be required by one pair – from the time wrynecks arrive in Europe, in late April, to their departure in late July.

As we will explore in Chapter 6, however, birds do not function in single units but in landscape populations. A small viable population of 50 wryneck pairs would, then, in the course of one summer, require closer to 50 million ant pupae to survive. That's a landscape with up to 3.4 *million* anthills. If you're wondering where such an earthy grazing mosaic, maintained by tiny herds of pigs or cattle wandering through groves of apple or oak, can be found in southern Britain – you might search for quite some time. Since the 1970s, not one single pair of wrynecks is known to have returned to England to breed. Thinking about food gets us thinking about the processes that create it – and how those food supplies have vanished over time.

The Beetle Butcher

Tumbling in numbers from 1850 onwards, red-backed shrikes are now effectively extinct in Britain. A bird of our long-lost extensive grazing pastures and hay meadows, red-backed shrikes remain one of the commonest birds in the farmland of far eastern Europe. Nesting in places from scrub stands to woodland edges, gardens and single bushes in meadows, it is the diet of shrikes, not their choice of land, that makes them so fussy – a diet that makes them garden birds in eastern Poland, yet history in Britain. In Polish farmland, 98% of the shrikes' breeding-season diet consists of invertebrates; only very few lizards and birds, the most famous prey items, are impaled on thorny bushes. Of the invertebrates, one study showed that 54% were beetles, 17% bees, wasps and flies, 16% grasshoppers and crickets, and 6% other bugs.[14] Shrikes favour large prey items – and pack as much protein into one shopping trip as they can. So how much food does a pair of red-backed shrikes need, to get through the summer?

As you examine the world of food-provisioning, the statistics become long and complex, and so, for the sake of readability, I have moved these to the endnotes. My calculations are both conservative and incomplete, but even so, for just one family of red-backed shrikes to survive the summer, they will need over 26,000 large invertebrates.[15] For a small viable population of 50 pairs to survive, you would require almost *1.33 million large invertebrates* in a single landscape. That is a landscape hopping with grasshoppers and literally crawling with dung beetles.

If you're walking in eastern Europe, you'll find extensively grazed farmlands. In the middle of these are thorny bushes, and you may well see a shrike perched on top, skewering an extremely dead mouse onto a thorn before heading off for another course of beetles. In Britain, you won't. Yet the structure of these landscapes looks similar. Almost all of the difference lies in the ground.

In eastern Europe, small organic cattle herds create pastures teeming with dung beetles. The ground is a living organism. In Britain, herds of medicated cattle create poo without profit. Herd sizes are so large that a mosaic cow pasture, hopping with grasshoppers, is a forgotten sight. The menu of red-backed shrikes vanished from our countryside not ten, but fifty years ago. Now, perhaps, the mystery of shrike extinction in Britain becomes less mysterious – once their hunger becomes known.

The Caterpillar Cleaner

Outside of the New Forest, cuckoos face the real prospect of extinction in England, though large areas of the Scottish Highlands, unsprayed and untidied, offer them a longer-term future. Best-known for their parasitic tendencies, adult cuckoos have a means of feeding evolved over hundreds of thousands of years, quite possibly a lot longer. Though fond of dragonflies on their arrival in Britain, the main skill of all cuckoo species lies in dealing with hairy caterpillars. This was noticed by the American naturalist E.H. Forbush as early as the 1940s.[16] He wrote, of black-billed cuckoos:

> *The cuckoos are of the greatest service to the*
> *farmer, by reason of their well-known fondness*
> *for caterpillars, particularly the hairy species …*
> *Wherever caterpillar outbreaks occur, we hear the call*
> *of the cuckoos. There they bring their young, and the*
> *number of caterpillars they eat is incredible.*

During one invasion of forest tent caterpillars in Massachusetts, in 1898, Forbush recounts how one observer, a Mr Frank Moser, watched a cuckoo catch and eat '36 of these insects in inside of five minutes. He saw another in Malden eat 29, rest a few minutes and then eat 14 more.'

The European cuckoo, like its American cousins, *loves* toxic, hairy caterpillars. That comes down to its gut instinct. Before eating, cuckoos slice the caterpillar and shake the insect at one end until its toxic contents are released. Once ingested, the hairy spines of the caterpillars lodge in the cuckoo's stomach, giving it a furry lining. Unfazed, the cuckoo regurgitates these spines as pellets – and carries on as before. Only cuckoos have this adaptation.

An evolutionary specialisation is no accident. It points towards something integral to an animal's existence. Cheetahs, for example, are fragile animals that have to quickly bring down fast-running prey, suffocating them rapidly to avoid injury to themselves. So they've adapted their physiology to run very fast, for short bursts of time, and bring down only particular species of antelope. Likewise, hairy caterpillars, correctly served, provide cuckoos with the large protein reserves they need to lay eggs across a protracted period of time, in a number of different birds' nests.

If our cuckoo's American relatives can, in a single feeding foray, consume around 30 caterpillars, let's allow our more restrained British cuckoos 15 caterpillars in a sitting – and 10 sittings per day. In the long daylight hours of May and June, this equates to less than one feeding spell per hour. So one adult cuckoo may be taking at least 150 caterpillars per day. This, in all probability, is a serious underestimate.

Cuckoos occupy breeding territories in Britain from mid-April to the end of June, around 75 days. That means a minimum of 11,250 hairy caterpillars, for one bird, in a season. 'Pairings' in cuckoos are complex affairs, with gender ratios and pair bonds not always well defined, but in theory one pair of cuckoos would therefore need upwards of 22,500 caterpillars to survive a summer on our shores. Females, during this time, must build the calcium reserves required to lay an average of nine eggs in the nests of their hosts, so the female's caterpillar intake will push this number higher.

Cuckoo populations require really large areas of land, because of their nomadic nature. Dartmoor, one of England's last cuckoo strongholds, is believed to hold around 100 singing males. The New Forest may hold a similar number, and here the population is thought to be stable. So a landscape may need to hold at least *2.25 million* hairy caterpillars in a season, for its cuckoo population to stand the test of time. As you wander through farmland England, it's worth looking out for what places might accommodate those caterpillars. Or you could save a lot of time: the cuckoos are showing you the way. Only landscapes free from intensification have healthy cuckoo populations. This is because 'hairy caterpillar' moth species are in terminal decline, but most of all in southern England. This, in turn, comes down to what *they* like to eat!

Each of the cuckoo's favourite caterpillars has favourite food plants of its own. Drinker moth caterpillars prefer damp habitats like fens and marshy grassland. Oak eggar moths prefer scrubby habitats, sallows and blackthorn. Garden tigers need plants like nettle, dock and burdock. The garden tiger has declined by over 90% in Britain since 1968.[17] Magpie moths, a favourite of cuckoos in the Highlands, have declined by 69%. White ermines, crashing by 70%, need nettles and docks.[18] All of these moths are the specialists of scrublands and margins, habitats that are entirely at odds with our tidy chemical farmlands. Most of all, cuckoos spend enormous amounts of time foraging in old stands of willow. These are the moth cities of the British countryside – home to caterpillars both abundant and diverse. Yet in many areas of the arable countryside, such once-iconic restaurants, and the food they harbour, have long since been grubbed out.

As our moths have vanished, our cuckoos have followed. Cautious conservationists put food on the back burner because they cannot quantify, exactly, how the catastrophic loss of insects affects each one of Britain's birds. But they do not have the time to try. Conservation is a race against time, science developed in the context of a warzone where the application of universal rules should be a weapon in the fight. And it's time for a shock. Our food calculations may be way off. Way under.

Away with the Butterflies

Spotted flycatchers, like red-backed shrikes, prefer to take larger insects that, in fewer protein parcels, accelerate the growth rate of their young. A three-county British study found that over half of flycatcher chicks are fed with adult flies, and most are fed with butterflies and moths. Three orders – butterflies and moths, true flies such as hoverflies, and sawflies, wasps and bees – make up three-quarters of flycatcher diet.[19] Almost all food is caught in the air. Research has found that only a small amount of prey is collected on the ground, generally in wet conditions, when flycatchers have to take prey such as aphids – far less nourishing to their young.

Spotted flycatchers are familiar birds: better-studied than most in terms of their diet. It is possible, then, to map out the needs of a flycatcher pair in far more detail. And the more we uncover, the more insects we realise are required. For just one pair of spotted flycatchers to raise their two broods in Britain over a summer, they will need, at the very least, 109,000 insects. A significant proportion of these will be flies, butterflies, wasps and bees – caught on the wing. And for your landscape to support a small viable population of 50 pairs, more like *5.5 million* flying insects will be needed. In this light, surprise at the flycatcher's decline, and the quest for complex reasons as to why, might give way to amazement that Britain can still feed any flycatchers at all.[20]

The demand of a flycatcher can never decrease, or compromise, as Britain becomes ever more devoid of insect life. Birds can certainly 'modernise' and adapt to the human landscape. What they can never do is go on a diet.

Thinking about those five million flying insects, you may begin to wonder why such extraordinarily complex attempts have been made to unpick bird decline. In many cases, food shortage, alone, provides sufficient reason for the rates of bird decline in our country. When did you last visit a place capable of providing five million flying insects in the course of a summer? The last time you did, you were quite possibly standing on Dartmoor's wooded fringe or on the edge of a New Forest copse, or roaming the meadows of the Inner Hebrides – hearing a cuckoo, or watching a flycatcher skewering an unhappy bee.

Off the Menu

The collapse of the food chain in Britain extends far beyond the loss of flying insects. The basis of bird survival – voles, plants, even weeds – has been effectively removed. Almost every time we examine the micro-worlds that support our birds, the picture of cleansing continues.

Field voles are best served surprised. In open grasslands with barns and scattered trees, they are top of the menu for barn owls and kestrels; in rough grasslands with pines or hawthorns, for long-eared owls, and, on grassy moors, for short-eared owls. The last three of these species are all in serious decline, with the 'eared' owls, in particular, specialising in short-tailed field voles.

It's normal for voles to fluctuate in four-year cycles, but our vanishing vole hunters point to a catastrophic decline. Whilst field voles remain widespread as a species overall, any one kestrel or owl will require many thousands in one place over the course of a season to survive. A loss of furry food pervades the countryside.

Almost all kestrels tested in the UK have been found to contain rat poison.[21] Wood mice, voles and young rats take poisoned baits, and so poison quickly makes its way into the kestrel's diet. A barn owl poisoned with rodenticides will take up to 17 days to die, after eating as few as three mice containing anticoagulant rodenticides.[22] It is not only birds but other declining wildlife that ingest these poisons. When tested, it was found 67% of hedgehogs contained at least one rodenticide.[23] In recent months, new regulations, from within the farming community, have at least realised the scale of chemical warfare waged against the very bloodstream of the countryside – and things may, slowly, begin to change.

We often forget the decline of life *below* the soil, because it is even harder to observe. Leatherjackets are the larvae of crane flies, found just below the ground in open pastures devoid of chemicals, and probed from the soil by ground-feeding birds. What an eastern Polish farm and a London park like Hampstead Heath have in common is a lot of open ground, leatherjackets – and a lot of starlings. But away from our capital's sympathetic parklands, the chemical imprint in our rural pastures, the compaction of soils by cattle, and the drainage of land, rendering the earth too hard to probe for food, have all jeopardised the starling's ability to dig for leatherjacket gold.

Medication in cattle passes pharmaceutical chemicals, via dung, deep into the soil, contaminating prey in the ground. The Dutch neonicotinoid study, referred to above, also showed that this waterborne insecticide, sinking into the soil, wipes out starlings by removing their prey, along with a range of pasture birds like yellow wagtails.[24]

A recent study from the University of York showed that concentrations of the antidepressant fluoxetine, if flushed into the environment, can enter the metabolism of earthworms.[25] This can cause starlings, which eat earthworms, to lose their sex drives and feel less incentivised to feed at the crucial times of dawn and dusk! Every study is unearthing strange new ways in which chemicals are soaking into our soils, our landscapes – and our birds.

Same Tree. Same Field. No Birds

You drive, say, past a 200-year-old oak, standing in a field in southern England. At first glance, the grassland around it looks pretty similar to how it did 30 years ago. So you could reasonably expect to see the same birds. In the 1970s, your oak may have had spotted flycatchers on it.

There may have been turtle doves purring on a dead branch or cuckoos holding forth. Today, most often, the same oak has none of these birds on it at all. The tree in a field has not changed – but the birds have gone.

In Plates 1 and 2 you'll see two green fields, each with some cattle. The first resembles a classic scene from Constable, yet is in fact Brzostowo village, in Poland. In May 2016, when I visited, this vista held breeding wrynecks and starlings in the trees and yellow wagtails amid the livestock. There were yellowhammers, red-backed shrikes, cuckoos and whinchats along the field edges. At a glance this looks like any number of places in England, yet its diversity of birds recalls a Britain of the early 1800s.

The second image shows a similar green field, with cattle and trees, in southern England. But livestock in these numbers act as a mowing force to remove plant life. Trampling compacts the ground and renders food hard to access. Plants struggle to grow even if pollinators are present. The sterility of such a landscape is not self-contained. Insects 'flow' in functional ecosystems between grassland and trees.

The difference between the two fields shown in these photographs is not down to how they look to the human eye – but to food. In the Polish field, there has been no chemical use or monoculture planting to remove the flowers that build insects. There is small-scale variety, with food sustained across pasture, meadow, old trees and fallows: a functioning wooded grassland. Anthills, beetle pastures, open and vegetated areas, and flowers, generate thousands of bees and butterflies.

Both our Polish and English photos depict a pastoral scene with cattle, fields and trees – but only one functions as a living organism. Our lowland habitats have often *appeared* unchanging, as food has vanished from within them. But appearances can be deceptive.

Our soils often become more amenable to life in northern Britain. When you start driving north towards Scotland, the green fields appear much the same as those further south. They still have cows or sheep. Yet suddenly there are lapwings and curlews beside the motorway. Often, the difference doesn't lie in what you see – it lies in what you can't see. Curlews feed on leatherjackets, but also beetles and caterpillars. Snipe, feeding at the wetter end of the field spectrum, prefer earthworms, leatherjackets and beetles. Adult lapwings feed on beetles and earthworms, whilst their chicks transition from small beetles to earthworms as they grow.

Black-tailed godwits, largely lost to Britain since the 1840s, are fussier again – feeding on beetles, grasshoppers, mayflies, caterpillars, annelid worms and molluscs. Most of their diet is surface-dwelling, in spite of their long bills. Yet again, large invertebrates are of enormous importance.[26]

Plate 3 shows the beacon of a black-tailed godwit gleaming on a cattle field in Texel, the Netherlands. These Dutch fields, with their wading birds, are those where insect removal has yet to take place. The diverse, small flowers in these meadows have grown alongside cattle for centuries. The soil isn't over-trampled and the fields aren't too dry. The number of cows in the Texel fields keeps the meadow low but not lawn – just how godwits like it. A cattle field in Texel still resembles a native floodplain with native herbivores. You can still drive past many an identical-looking buttercup-filled cattle field on the Somerset Levels. But why, as we drive to see bitterns in Somerset's reedbeds, don't we pull over to watch yodelling godwits in these grazing lowlands? The answer lies in history – and in the soil.

Often, you are looking at a landscape that has been hammered not once, but dozens of times, by successive waves of chemicals, trampling and overgrazing. So whilst the buttercup fields you are left with may be attractive to look at, the food chain below has long since collapsed. Examples like this show why apparently 'nothing has changed' – yet the birds have gone. This is why we need to trust our native species to point us towards the collapse of the food chain. Habitats cannot be read by human sight alone.

Indeed, when the Knepp Estate in Sussex began its rewilding project (see Chapter 4 for details), Tamworth pigs were set loose to dig the soil. Within one particular field lay a disused footpath, invisible to the naked eye. Unlike the surrounding land, it had never been ploughed or sprayed. The pigs ignored the sprayed fields entirely, digging only the length of the organic footpath to find food. The human eye is readily deceived, but a pig's snout is not. Pigs are remarkably intelligent animals, and a humbling reminder that after centuries of living in houses, and decades of foraging in supermarkets, our human ability to read an ecosystem is infinitely inferior to that of the wilder animals around us. It is the habitat assessments written by cuckoos and curlews that matter – not those committed to thousands of sheets of paper each year. The birds are telling us, by their absence, as much as their presence, quite what a wasteland we have created in Britain – even if the wider aspect of trees and fields appears the same. Britain's birds are pointing us to trophic collapse. We need to listen, and learn.

The Empty Shape of Oaks

In 1892, the naturalist Sidney Castle Russell visited Ranmor Enclosure, in the New Forest, at a time when almost the entire forest was deciduous and its rides filled with flowers. He recorded the natural state of a British woodland's butterflies in summer – a spectacle we have entirely forgotten. On the brambles lining forest rides, he found silver-washed fritillaries in such droves that he couldn't move his feet on the ground, together with thousands of dark green and high brown fritillaries, and white admirals. Purple emperors, their caterpillars feeding on sallows, flew in their dozens. There were thriving colonies of wood whites, whilst colonies of large tortoiseshells and black-veined whites were also on the wing.[27] One hundred and twenty years later, I visited the woodland meadows of the Bukk Hills in Hungary, shown in Plate 4. Here, thronging in genuine clouds, every ride, every bramble, every flower, was as Castle Russell had described. An orange blanket of fritillaries shivered on brambles and cow parsley. The ground was thick with heath fritillaries – you had to watch your feet. And there was the purple emperor – flying not in ones but in dozens, bounding along the edges of streams.

The temptation, on hearing about clouds of butterflies, is to dismiss it as nostalgia. It is, in fact, the normal state of a healthy woodland edge. Long gone from our own island, woodland butterfly flocks can still be seen in the woodlands of places like the Carpathian mountains. In Britain, by contrast, a nectar-free bracken wilderness now carpets most of our woodlands. The vital carpet of bramble, clearly visible in Victorian photos of the New Forest, has been grazed out over time, along with a host of woodland-edge flowers. Deer leave only bracken as they browse our woodlands. Bracken, seen in most of our forests, is ecologically of little use to many insects – providing no nectar to build a food chain from the bottom up. It indicates not wildness but desert.

However pretty they may be, bluebells, too, are shunned by most of our insects. They are the only flower left after deer browsing has ripped the nectar sources from a wood. Bluebells have no caterpillars associated with them because they are toxic, whereas brambles and nectar-rich flowers 'feed' a woodland and its birds. As nectar resources in our woodlands vanish, caterpillar specialists, from resident lesser spotted woodpeckers to migrant pied flycatchers and wood warblers, bleed from our woodlands each summer. Enter many of our cherished forests, from Sherwood to the Dean, and you enter a nectar desert. The structure at the top is the same as at the top of forests in the rest of the Europe – but the bottom is radically different. Deer in unnatural densities have created a trophic wasteland in our woodlands. Non-native muntjac deer raze down young shoots, whilst deer of all species eat flowers and so destroy the woodland food chain from the bottom up.

Food loss in the wider countryside has impacted our woodland birds even more. The richness of woodland margins, their pollinators and flowers, is irrevocably tied to the health of the landscape as a whole. Most British woodlands suffer trophic isolation. Parklands across Britain were once home to lesser spotted woodpeckers, which extract caterpillars from dead timber. If butterflies and their caterpillars fade from our grasslands, these birds will struggle to find food in the oak woodlands.[28] But as we grow accustomed to our starved trees, we have forgotten how rich they once were for our native wildlife.

Making Sense of Trees

Sessile, but most of all, English oaks, have more associated insects than any other tree. Southwood's study found 284 invertebrates associated primarily with oak.[29] Around 380 of Britain's 900 moths are wood-dependent, with 220 alone supported by deciduous oaks. In *Wilding*, Isabella Tree points out that an open-grown oak has six times the woodland cover of a forest tree. But most of all, oaks need *time*.[30] As the late Oliver Rackham pointed out, whilst 200-year oaks are quite a sight, they have nothing on 500-year oaks,[31] which constitute entire ecosystems. This is one of the main reasons why the New Forest's oldest woods are so unusually rich in life. Over time, oaks become cornerstones for birds. Caterpillar abundance in oak grows over time. The most established trees feed lesser spotted woodpeckers. Blue and great tits time the hatching of their young to coincide with the opening of insect-bearing oak leaves. Wood warblers and pied flycatchers, arriving from Africa, have founded ancient migrations around the caterpillar bounty of our remnant Atlantic oakwoods.

Our second richest wildland tree family is willow. Home to at least 266 species of insect, willow does not need to grow as old as oak to become invaluable for birds. In its lower forms, bushy willows provide for willow warblers, garden warblers, willow tits, finches and many other birds. Cuckoos feast in its branches: willow has more moth caterpillars associated with it than any other tree.[32] Ancient white willow, in particular, provides huge reserves of food, nesting cavities for owls, and treecreepers sneak under its peeling bark to make their homes. Sallows, a type of broad-leaved willow, are the food plant of the spectacular purple emperor butterfly. The more willows you have in a landscape, the richer the menu for its native wildlife.

With 229 associated insects, silver and downy birches are another vital tree. Birch in its younger form is a profitable larder for black grouse, and its clusters are perfect song-posts for tree pipits and woodlarks. Birch pastures in the Highlands, untidied and growing naturally, form some of Britain's richest woodlands. A range of warblers, finches, redstarts, woodcocks and many other birds all come to thrive around birch woods over time.

Scrubland trees come with a whole range of birds attached. Hawthorn, with 149 associated insects, and blackthorn (sloe), with 109, have lists of specialists as long as that of oak. Cirl buntings, yellowhammers, linnets, whitethroat species, bullfinches and red-backed shrikes all feed around these spiky restaurants. Shady birds, like nightingales, feed inside. Diverse thorn scrublands are vital to any healthy landscape.

A range of sunlight- or wetland-loving species makes up the rest of Britain's most important trees. Native black poplars, with 97 associated insects, are rarely seen in the landscape today, but are the favoured tree of golden orioles in western Europe, which feast on large moth caterpillars.[33]

Crab apples, with 93 species, are hugely rich in hornets and fly species, and much of this diversity is shared with domesticated apple trees too. Spotted flycatchers love old, organic orchards for this reason. Wild apple groves in the New Forest still hold robust colonies of hawfinches. Apples are rarely seen in their wild form, yet when ancient orchards have vanished, we've realised, too late, how important ancient apple groves were for birds like the wryneck and the lesser spotted woodpecker.

European pines, with 91 species of insects, are cornerstone trees, particularly on nutrient-poor, sandy soils and in northern Britain. Wetland-loving alder, with 90 associated insects, is a valuable food source throughout the year, its soft branches loved by woodpeckers and its seeds of great value in winter, attaining ever greater value as it rots. Elms (of which the native wych elm is now rare, the field elm largely gone in its full form, and the European white elm extremely rare) have all been decimated by Dutch elm disease, but were once a huge reservoir of food, its leaves feeding the caterpillars of the now-vanished large tortoiseshell butterfly.

Other tree species come into use for our wildlife at certain times of year. Beech and hornbeam are of little use to breeding birds; hornbeam has only 28 known insect species tied to its branches, but the fallen 'mast' of these shade-loving trees, in autumn, provides nutrition for seed-eating finches. Limes, being dense in aphids, fill an important summer gap in the diet of small insectivorous birds. The fruits of wetland-loving elder are eaten by at least 38 species of bird, while 37 birds eat rowan berries, 23 the fruits of birch and pine, 22 the fruits of light-loving cherry and 18 the fruits of yew.[34] Hawfinches are uniquely evolved to crush cherry kernels. For our resident birds, light-loving, fruit-bearing trees have a long-evolved importance come the autumn months.

Some trees appear to have evolved as a lure for certain specialists, wherever they are found. Marsh tits are seldom found in marshes, but strongly drawn to dense hazel, which may harbour particular insects most suited to their needs. You can often find willow tits without willows, but rarely without rotting stands of elder.[35] Elder has late-summer berries and soft wood that can be excavated, but is also noted for its communities of small moth caterpillars.

Reading the above, you may realise how rare many of these special trees have become in the landscape close to where you live. Some, like the black poplar, you may never even have even set eyes on, in their full, aged beauty. Many of the trees most important to feeding our native wildlife, such as willow, are both shunned by foresters and absent from modern farmland Britain. Others, such as full-grown elms, veteran oaks or towering ancient blackthorns, have vanished catastrophically from our countryside. And as we will explore in Chapter 9, Britain's woodland crops now bear little relation to the rich menu once provided to our native wildlife by these invaluable restaurant trees.

Swallow Clouds

*Ours is a dwarf and remnant fauna, and as its size
and abundance decline, so do our expectations.*
—George Monbiot[36]

If I were to ask you what a 'reasonable' swallow population might be in a small British village, you might say five or ten pairs. But how about 500? If you were to consider what constitutes a good house martin population on a house, five pairs would be outstanding. But what about 130?

In the small village of Josfavo, in Hungary, in July 2017, I was treated to a sight common to almost any village in the Carpathian mountains. Here, wildlife cameraman John Aitchison and I counted around 1,000 swallows and house martins, adults and young, thronging the wires after their second brood had flown the nest. The wires were black. The village was booked out. Josfavo has just 100 or so houses, but lies within a countryside thronging with meadows, insects, and untouched by either chemicals or any significant air pollution.

It would be easy to think that a village black with swallows and martins has always been a southern European sight, but that would be wrong. In 1870, the owners of a country house in Essex, wanting more house martins, became frustrated with the 'injurious' house sparrows that were booting the martins from their nests. The owners unceremoniously snuffed out the sparrows. By the early 1880s, that one Essex country house held 130 pairs of house martins. The air would have been thick with them.[37]

Because swallow numbers are relatively stable in Britain as a whole, it's easy to forget that our insect clouds, our bird clouds, vanished long before avian statistics were invented. Most of our aerial feeders have been lost over the course not of decades but of centuries. Only written evidence, and European living history, preserves such numbers now. Today we see just the relics of an aerial food chain, expressed in the single pairs of martins left on a tiny percentage of our houses. What scientists have also forgotten, as they seek complex explanations for why house martins are thriving in northern Scotland, but vanishing in southern England, is the effect of air pollution on the food chain above our heads.

European studies in areas like the Czech Republic, where air pollution is only now becoming an issue, show that in many rural areas house martin colonies dramatically collapse – as air pollution increases.[38] In Britain, we tend not to carry out such studies, and have forgotten that house martins are aerial canaries, pointing to the very cleanliness, or otherwise, of the air that we breathe. At the time of revising this book, at the height of the Covid-19 epidemic, the distribution patterns of such little birds, and their abundance, must surely come to be seen, over time, as vital indicators not only of a landscape's health – but of our own health as well.

Swifts, too, have suffered from the long-term effects of aerial food-chain collapse. Oxford, for example, still screams with swifts each summer as they hawk through the college courtyards and wedge their nests within the city's plenitude of churchy nooks. In 1971, the naturalist Chris Perrins estimated there to be four pairs of swift per *thousand* households in Oxford.[39] In Warsaw, it was recently estimated there are around four pairs of swift *per building*.[40] Warsaw lies in a landscape still dominated by woodlands, river valleys and strip-farming, where insect flow blackens your windscreen as you drive out of the city. If you visit most cities in eastern Europe, especially older ones, the evening sky is screaming with swifts in their hundreds, sometime thousands. Again, these screaming clouds point to an airspace, a landscape, alive with flying prey.

Swifts carry back to nests, on average, 5 grams of food, or 20,000 sky-snatched invertebrates *per day*. Polish studies in Poznam city estimate that swifts harvest 28 million insects in and around the city each summer, keeping down numbers that would otherwise become a 'pest'.[41] Polish people are worrying that without swifts, their insects will reach pestilence proportions – and that tells you a lot about the buzzing state of eastern Poland. Whilst the Poles are worrying about declines in thousands of swift pairs, we are worrying about tens in our own towns and cities. Airborne swifts are the living expression of airborne insects. It is ecologically impossible that the food-chain collapse over centuries has not decimated their numbers. We know swifts have declined in Britain since the early 1800s. This was long before we modernised our roofs, but long after we began the industrial pollution of our island's lungs.

We might consider two pairs of corn bunting on a hedge-line a success, but in the insect-filled grasslands of Spain the bushes are thick with thousands. Single meadows in eastern Europe are so rich in stag beetles they can hold a dozen pairs of red-backed shrike. There are enough oak-dwelling caterpillars in central Europe to sustain thousands of hawfinches in a single woodland. Swift clouds, swallow clouds, butterfly clouds, ecosystems shivering with wings, may be forgotten – but that does not stop them from being rightfully ours.

To restore the food chain, we first need to realise that in Britain we lie close to Ground Zero. Food must become, for our conservationists, the first imperative. But to restore food chains, you need to restore ecosystems. And to do that, we need to talk about stewardship.

CHAPTER 4

The Lost Stewards

Return the Stewards, Restore the Birds

*When we try to pick out anything by itself, we find it
hitched to everything else in the Universe.*
—John Muir[1]

To the British eye, there was something strange about the rolling hills. In
place of green grass or purple heather lay a soft carpet of oaks, broken by
grassland. Within those hilltop woodlands were animals lost to us for a
millennium – wolves, lynx and, from the Carpathians, a few returning
brown bears.

It was the height of summer, and I was standing with naturalist Rob
de Jong on a flowery hilltop in Hungary's Aggtelek National Park. Here
on the crown of a meadow, hill-topping swallowtail butterflies looked
delightful as they battered one another senseless for the right to mate.
The purr of turtle doves smoothed the air like a balm. I remarked to Rob
what an amazing template this would be for the largely lifeless uplands
of the UK's national parks.

Rob's response was simple. 'Still dreaming, I see …' Then he paused.
'And why would you want such forests back? Yes, you get more lynx.
But most birds do not like forests. Nor do most butterflies. Our wildlife
loves people. Most of the birds in Hungary need people. Trees swallow
the birds. Birds are adapted – to us.' He paused. 'For most birds, there is
no such thing as the wild'.

Rob described hundreds of turtle doves thronging the wires in local
villages each autumn, each raised at the edge of a hay meadow. We talked
about the Alcon blue, a butterfly whose caterpillar is raised below ground,
entirely by ants. Dropping from a gentian, the butterfly's caterpillar 'sings'
to passing ants, convincing them that it is, in fact, their queen. Sonically
and chemically in love, they immediately pick it up, take it underground,
and raise it as their own.[2]

Such sinister fairy tales, Rob said, do not play out in the 'wild'. They happen, instead, in carefully hand-cut meadows, or those grazed by village cattle. The Alcon blue's unique lifestyle, and the life of most Hungarian butterflies, Rob said, is down to human activity. Talking to Rob, there dawned on me a new and unsettling idea. Europe's wildlife may have entirely forgotten the wild.

The Myth of a Pristine Nature

Rob explained how, in central Europe, the disturbance of human beings had become, over time, far more important to the survival of birds than the growth of forests. He pointed me to a fascinating book and told me to read it – so I did. *Species Conservation in Managed Habitats: the Myth of a Pristine Nature* is written by German scientist Werner Kunz.[3] It's interesting reading for anyone who gets carried away with ideas of 'wildness', without attention to what that means for the wildlife caught up with our landscapes today. Kunz points out that across central Europe, from Germany to Slovakia and Hungary, populations of many of the birds vanishing in western Europe are in fact stable. He argues that human disturbance and scruffy open habitats, from fallow airfields to pear orchards and military bases, have all played a role in keeping these species alive. Kunz compares the 'cleansing' model of maximum-yield farming, in western Europe, to small-scale variety in the east. But he goes further. He argues that the return of 'wild' dense forests could spell the end for many birds. Strangely, I immediately saw what Kunz meant.

For years, writing conservation pieces for *Birdwatching* magazine, as I tried to work through why birds were vanishing, the answers I had turned to were often far from wild. I'd looked to Polish villages to see wrynecks doing well, old hay meadows as paradise for shrikes, overgrown Herefordshire gardens as nirvana for spotted flycatchers, and gravel pits as safety for nightingales. Often, as I'd done my research, it had been the loss of *human* habitats, not a loss of 'wilderness', which had driven the decline of our birds. Wrynecks did not vanish from the 'wild' but from earthy wooded farmland. Shrikes vanished not from pristine grasslands but grazing meadows beside our villages.

Kunz had found a way to encapsulate what I had been trying to express in my writing. In central Europe, human action still promotes a huge diversity of birds. Little of this diversity is dependent on 'pristine' habitats such as forest, but on *broken* landscapes. Kunz's description of one such place where he made this realisation, an airfield, calls to mind some off-piste explorations of my own:

> *These ruined airfield areas were home to many birds and butterflies which had to make way for afforestation elsewhere ... On the dry areas of these airfields, I found skylarks breeding in large numbers; common snipe, redshank and crakes were nesting in the rushes of wetter areas and little ringed plovers had found suitable places to rear young on the destroyed runways.*

In his travels, Kunz detects that in eastern Europe a lack of nitrogen fertilisers means that fallow lands stay open, low in vegetation cover and rich in flowers. Such low earthy landscapes sow wheatears, corn buntings and black grouse. Kunz finally arrives at a conclusion that many conservationists would agree with today:

> *In central Europe, many species do not benefit from too many large forests ... This is because central Europe was deforested by mankind thousands of years ago, and as a result is inhabited by species that have adapted to open habitats ... The open-land species were not threatened for centuries, because they could colonise agricultural land with no problems ... However this situation has not existed for half a century.*

Kunz goes on to explain that many species, once common across agricultural landscapes, have been forced into the retreats of 'wastelands in cities, industrial areas, motorway embankments, gravel-carrying sites, brown coal open-cast mining and military areas'.

Kunz is right that many birds, and many of our most endangered species in Europe, have, over time, committed to the human world. He is also correct that as this landscape has changed, birds have been forced to specialise, as the way humans have disturbed the landscape has changed. Indeed, Britain's derelict refuges are often more diverse, more important for vanishing birds, than its carefully managed reserves. British nightingales once deafened coppiced woodlands each spring – now, the scrubby confines of a fenced gravel pit form a better refuge. Whinchats once scratched their calls across our southern farms. Now, the military-owned Salisbury Plain is the only large-scale insect-rich grassland they have left.

As Kunz rightly points out, whilst agricultural disturbance in Hungary sits in harmony with birds from tree sparrows to great bustards, agricultural 'cleanliness' in western Europe results in emptiness and extinction. Such comparisons are useful in getting to the core reasons for bird decline. They do not, however, move us to a future.

Countries like Belarus and Romania may be fascinating reminders of our past, but they do not provide a viable model for our future. House martins once thronged London's streets, feeding on flies from the dung dropped by cart-horses on cobbles. This was also a time of cholera epidemics.

We must instead look to new ways to reinstate the 'disturbance mosaic' in which our birds evolved to thrive. What is challenging is that many British birds are wedded to strange, unnatural places. Before seeking ecological restoration, we need to consider the fact that, on first inspection, many of our birds appear to have forgotten the wild.

Unwild

Deep in the coal heart of Manchester, Yorkshire and Durham, 'wastelands' of rotting elder, willow and birch hold the last viable populations of willow tits in Britain. Living under constant threat of having a super-market or housing estate built over their habitat, willow tits find their favoured floodplain woodlands growing on abandoned coalfields, not in pristine river valleys.

If you're wandering the bubbling banks of the River Tywi, in Carmarthenshire, you may spot the sprightly shape of a little ringed plover. Black, white, with a gold-ringed eye, these birds blend remarkably well with the gravel spits in the river's meandering flow. But most little ringed plovers in Britain blend even better with active quarries and landfill sites, when they arrive to breed each summer. These places recreate their native gravel river banks – on an enormous scale. Swifts, which once nested in ancient trees, have committed to our buildings. Almost every British corncrake is confined to a crofted field. Spotted flycatchers do better in flower-rich gardens than the flower-starved countryside around.[4] Blackbirds, once foraging in lawns created by wild grazers, now find worms on grass cut by lawn-mowers.

A whole range of birds have committed to thousands of years of living beside *us*. This not only makes them fragile – but also means that romantic rewilding of the 'resurgent forest' school of thought is not a clever starting point in their conservation. Battles to save the willow tit must first be fought on old coalfields. Barn owls must be saved along the very ditches that drained the fens. Birds like swallows and house martins have 'stepped out' of wild habitats so completely that they will most probably never return to them again.

There's no doubt that today, in our degraded landscape, many birds are increasingly tied to a life beside us. Most often, however, they are simply making do. And many will continue to do so – until something better comes along.

The Vanished Architects

My grandfather Fred once told me the story of a short-sighted girl who'd had a lot of stick at school for wearing big square glasses. Getting a boyfriend at last, she decided to woo him by proving that she didn't need glasses at all. She carefully placed a pin at one end of a field, noting where she put it. That evening, she and her boyfriend were sitting watching the sunset from a hay-bale, at the opposite end of the field.

'Look,' said the girl, glasses hidden in her pocket, 'isn't that a pin glinting in the sun over there?'

'Goodness, what amazing eyesight you have,' said the boy, most impressed. Carefully, the girl followed her remembered path and found her pin. Alas, her romance was not to be. As she walked back across the field, she tripped over a cow.

Conservationists like Kunz and many others have, like the heroine in our story, overlooked the enormous influence of cattle, and other herbivores, as the original agents of disturbance for European wildlife. Instead, the strange thought-leadership in ecology until recent decades has been that, until we came along and cut down the 'primeval' forests, the birds of open land were sitting around, being rare, and waiting for the trees to fall – that most of the birds in our ecosystem were *not there*. In future decades, conservationists may look back on such a position with bafflement.

The idea that early humans increased biodiversity in Europe by cutting down trees is, if given any thought, human-centric nonsense. As we explored in our first chapter, most of our birds show highly evolved feeding and nesting strategies, developed over hundreds of thousands of years, which would have evolved in the varied landscapes shaped by giant herbivores and then their smaller cousins. A turtle dove grubbing in the bare earth left by a disruptive herd of aurochs or boars, however, notices little difference if that agent of disruption, over time, becomes a farmer tilling a field edge. A marsh warbler singing from a beaver-coppiced willow transitions well into a world of human-grown withy-beds. Arguments for time-freezing agriculture to save birds become irrelevant, and boring, however, as soon as you remember how the skylark, corn bunting and cuckoo evolved in the first place.

In recent years, the emphasis has finally moved away from dark canopy forests and back to the basic rules that govern ecosystems. In seeking to restore wild landscapes, then, it's wise to avoid some examples promoted as 'primeval'. In Poland, Białowieża, often called a 'primeval forest' because of its great age, had such a depleted population of bison for centuries that the last one vanished from the wild in 1921.[5] Aurochs had died out here in natural herd sizes centuries before. A dark, closed canopy developed for centuries – its stewards, gone. Dense forests in Romania, appearing pristine and untouched, with their high densities of wolves and bears, are not entirely natural either.

Bison, cattle and horses would all have shaped and thinned out these forests – but their herds are long forgotten. Large herbivores vanished so early in our history, before bears, before *books*, that we have all forgotten the critical role they played in shaping our landscape and birds.

Fortunately, other continents have escaped such ecosystem losses. Looking at the wooded African savannah, the key player shaping the landscape is not a lion. It's a lot bigger and so is its effect. Elephants pull down trees and break up thorny bushes, and their droppings fertilise the land, sowing industrious communities of beetles. Their grazing actions create enormous grasslands, filled with Africa's equivalent species of larks, pipits, shrikes and bustards. But if elephants vanish from the Serengeti, will future ecologists call its forests primeval, and natural, in a couple of centuries' time?

India's larger national parks, such as Tadoba, Kanha and Kaziranga, contain some of the richest, most varied wooded grasslands on earth. All the big players are still in the game. Each is contributing to structural diversity – indirectly, by planting trees (predators keeping small herbivores on the move), or more directly, by opening habitat (grazing, browsing, digging), planting new trees (defecating seeds) and breaking others to create scrub mosaics (coppicing). In India, you can still observe the role of native cattle – gaur, or Indian bison, our largest living bovine. These occur in herds in size from 11 to 50 individuals. Herds range up to five kilometres a day, and average just 0.6 animals per kilometre at any one time.[6] This is what truly natural woodland grazing looks like. The result is sun-filled woodlands, broken by glades and meadows.

Alongside grazing gaur, the Indian woodland landscape is also 'opened' by large herds of deer and wild boar, far outnumbering their predators. Predators, however, create the fear dynamics by which herbivores move around, and so new trees have plenty of areas in which to grow unmolested.

The 'apex' nature of the role played by large herbivores in pristine ecosystems is also very important. Wild boar have few predators, although large carnivores will take their piglets. Looking at wild horses in Mongolia, their foals are killed by wolves, but 70% of wolf attacks take place during only the first week of a wild horse's life. Stallions can also kill wolves with a kick to the head: adult horses are no easy prey. Remaining wild cattle, like gaur, are every bit as feisty as Julius Caesar described Europe's aurochs to be in his *Gallic Wars*.[7] Even adult tigers will rarely tackle them if given any choice of prey.

This renders cattle, horses and boar dominant architects. These grazers and diggers, not tied to dens like predators, sit at the top of the 'landscape design' ladder. In India's woodlands and Africa's savannahs, the frenzied competition between trees and the animals that munch them creates the very diversity of habitats on which most birds depend. These disturbed animal–tree contests, catering for grassland, scrubland and woodland species in an extremely diverse mosaic, are revealed in some of the illustrations in this book. Take a good long look at Plates 5 to 9, showing Letea Forest, Kanha, the Serengeti, the Chernobyl wilderness and the beaver-crafted Biebrza Marshes, before reading on. These images show how our landscapes should look – and they may be quite different to what you'd think. Where native grazing animals such as cattle and horses are present, a mosaic of grassland and trees is the landscape default. Where beavers shape our wetlands, scrub, grasslands, wet meadows and ponds become the diverse landscape default. As a result, the many species of grassland, scrub and scattered trees thrive wonderfully well in ecosystems where such herbivores have either been preserved or restored. The images reveal some of the wonderful variety of our own lost, native landscapes.

In Britain today, by contrast, we worry about wetland birds like avocets requiring man-made gravel islands, because the river shingle islands formed by old beaver dams are a habitat most of us have never seen. We worry about the number of sheep needed to create the right habitat for ring ouzels, because we've long forgotten that a hillside should be stewarded by horses and cattle, or its juniper shrublands pruned back by elk. Wild processes, whether rivers flooding valleys as they should, beavers felling trees as they should, or woodlands vying with herbivores to create wood-pastures – have long been lost. In Britain, our original large-scale habitats have changed so often, and so much, that we've forgotten what we're supposed to be bringing back. Whilst African ornithologists know that elephants are the baseline architects that shape the Serengeti, most British conservationists have forgotten our wild cattle, our wild horse harems and the importance of beavers and boars.

But could we really do away with planting weed strips to save the turtle dove? Could we really have all our birds sharing the same landscape, not each one managed in isolated pockets? Can woodland and grassland birds sing side by side without a hundred management plans? It all sounds wonderful in theory. But how does it work in practice? Are Britain's birds ready for a return to the wild?

The Dutch Experiment

Species may be surviving at the very limits of their range, clinging on in conditions that don't really suit them. Open up the box, allow natural processes to develop, give species a wider scope to express themselves, and you get a very different picture.
—Frans Vera[8]

The term 'rewilding' scares many British farmers and quite a few conservationists, perhaps in part because a vision of our open lands and people's way of life being swallowed by forests has become one, incorrect interpretation of what rewilding means. In the Netherlands, however, rewilding – or ecological restoration using natural processes – has been a quietly effective conservation tool for decades – in a country smaller, and more crowded, than our own (Plate 10).

In the 1960s, a large area of coastal land along the Oostvaardersdijk was reclaimed from the sea. Originally intended for development, it fell into disuse and its potential for nature became apparent. By 1986, large areas of the Oostvaardersplassen, under the Dutch Nature Conservation Act, were set aside for wildlife: a coastal strip ten by six kilometres in extent. It was in this newest of lands, literally, that the Dutch conservationists of Staatsbosbeheer, the organisation responsible for managing the country's nature reserves, decided to try something – something very old. They decided on minimal intervention, and to restore the best available proxy for a grazing mosaic last seen in the early Holocene.[9]

The Oostvaardersplassen, whilst far from a complete or perfect experiment, was one of the first projects to demonstrate how the coastal wetlands of Europe might once have functioned, and could function again in the future. The leader in this project has been Frans Vera, the Dutch ecologist whose work *Grazing Ecology and Forest History* was the seminal text in challenging the closed-canopy forest notion in the first place.[10] Confronted with the extinction of tarpan and aurochs, the Oostvaardersplassen team populated the coastal plain with closely related proxy animals: herds of Heck cattle and konik horses. Red deer, now numbering in their thousands, were also left to roam freely and shape the landscape as they saw fit. Starvation became the main agent limiting expansion. As predicted, over time, a carrying capacity was reached, as in the Serengeti. Unlike in the Serengeti, however, no predators acted to 'plant' the landscape with trees by culling or moving the grazers. Likewise, there are no boar in the Oostvaardersplassen, so rotavation of the soil is missing. Another problem, which has reared its head in recent times, is that starving animals would, in the wild, be swiftly taken down by wolves. In the Oostvaardersplassen, however, they often linger and have been noted on occasions to suffer. So, in recent years, a humane cull has been brought into effect.

In spite of these obvious and clear limitations in recalling wildness, the Oostvaardersplassen has seen a remarkable resurgence in its wildlife and in the richness of its landscape. First and foremost, the prediction that over time this coastal marsh would 'inevitably revert to dense woodland' has been deftly disproven. What has formed instead is a state of equilibrium, with the animals determining whether succession happens or not. Indeed, long before the grazing animals were released into the Oostvaardersplassen, another keystone grazing animal had arrived by itself. We never appreciate it enough in Britain. It's big – and it's mouthy.

Huge flocks of greylag geese, arriving to moult in late summer, soon got to work 'weeding' the marshes. The worry that the marshes would give way, through succession, to reedbeds and then woodlands, was disproven not by the largest herbivores – but by some of the smallest. The greylag herds grubbed out huge quantities of marsh plants, preventing succession and saving conservationists a huge amount of time and money. The expectation at the time was that vegetation would swallow animals – not that geese, of all animals, would be responsible for determining whether that happened or not.[11]

In the Oostvaardersplassen, winged herbivores, in goose form, and four-legged herbivores, in horse, cattle and deer form, now function in tandem to create a predominantly open mosaic. Rather than people investing money in planting reedbeds or clearing woodlands, the grazers deliver a dynamic landscape free of charge. The Oostvaardersplassen remains richer for breeding birds than reserves elsewhere in the Netherlands – or almost any nature reserve of the East Anglian coastline.[12] Reedbed birds, such as bearded tits and marsh harriers, the result of management schemes in Britain, have moved in and thrived, alongside penduline tits and bluethroats, which prefer mosaics of reeds and bushes. The blend of reedbeds and pastures has also allowed colonisation by great white egrets. The reedbeds, unplanted, grow in areas of low-lying water. These are increased by large 'wallows' left by animals. The herbivores, however, often unceremoniously trash areas of reedbed and create large areas of open pasture. This is unfortunate for marsh harriers, but excellent news for lapwings.

There are more tree-dwelling bird species in the Oostvaardersplassen than in most British woodlands. You have goshawks, hawfinches and lesser spotted woodpeckers in copses and willow tits in the scrublands – nesting metres from bitterns and lapwings[13] Because the landscape is not managed, any one of these habitats, and its birds, is free to expand, or contract, over time.

In 2006, white-tailed eagles, which wisdom said would only nest in huge trees, moved in to nest in low willows: the first pair in living memory in this region. They were not reintroduced. Feeding on herbivore carrion and fish, the eagles have since repopulated the Netherlands, with over 40 breeding pairs in the lowlands. Colonising spoonbills, feeding between the pastures and reeds, grew to hundreds of pairs.

As with any initiative that takes credit away from people, and dispels a whole set of previous beliefs at once, the Oostvaardersplassen has often been feared and criticised. Whilst complaints from animal rights activists about the starvation of animals were to be expected, the thing many conservationists have most struggled with is something terrifying indeed. It's alien to nature reserves – and endemic to nature. *Surprise.*

The Art of Letting Go

If you run a reserve with targets, what happens if the targets go wrong? If you set a target and fail to meet it – you have failed. The world's stock of wildlife, however, managed without targets until around the 1960s. So it seems possible that nature has a few tricks up her sleeve for managing without targets at all.

In 1996, however, shockwaves reverberated through Dutch birding circles. The 300 pairs of breeding spoonbills in the Oostvaardersplassen crashed, in just one year – to *zero*. Then, the Vera conservationists, content with exploring 'non-linear' conservation models, whereby natural processes are cyclical, did something pretty remarkable. They did nothing.

The team were not managing for spoonbills. They were not putting dynamic nature second to reserve paperwork, designed to freeze wildlife in its current form. It seemed there had been an increase in foxes on the reserve owing to the carrion levels. Informed speculation was that the foxes, as they are prone to do, had robbed and caused abandonment of the spoonbill colony.[14] Calls for changes in stocking regimes and man-made changes in the water table – aimed at reducing the carrion, to reduce the foxes, to increase the spoonbills – came to the fore. They were ignored. In the coming years, spoonbills recolonised. But now, instead of a concentrated colony in one place, they dispersed such that birds bred across a wider area than they had done before.

One of the key tenets of Vera's rewilding philosophy is that nature does not run in straight lines. His observations are born out in any of the pristine temperate or subtropical ecosystems of the world. With a full assemblage of native herbivores in play, trees must grow through scrubland to avoid being eaten. Over time, they grow into denser groves of successful trees, which are grazed around, creating the range of birds that nest in trees and feed in open soils. But over time, even those tree glades will rot and fall. Grasslands take root, scrublands follow – and round, and round. Intact ecosystems containing large animals are not linear. They are cyclical.[15]

When researching the Oostvaardersplassen, however, even with an open mind, I was troubled by one paper in particular. It showed that since 1997, when animal stocking levels were increased in the reserve, the overall number of breeding pairs, in *all* birds, declined by one-third. As scrublands halved, bluethroats crashed from 280 pairs in 1997 to just 39 in 2012. Reedbeds also halved in size, with a predictable drop in reedbed birds. Dry grasslands have increased, as have dry-grassland birds. There had been just 246 red deer in 1997; by 2012 there were 1,898. There are now four times more horses than twenty years ago.[16]

Looking at the enormous herds of deer, horses and cattle in the Oostvaardersplassen, it's important to remember that humans, the land gods still steering the reserve, decided on a certain amount in the first place. They also decided not to introduce competitive agents such as wolves. Even if the large herds in the Oostvaardersplassen roam free, they do so within 60 square kilometres. So the effect of grazing is more concentrated than on a natural plain. To me, looking at the enormous herds of horses and cattle in the Oostvaardersplassen, they did appear larger than the natural harem groups of wild horses in Mongolia, or wild cattle in India's woodlands.

Worried by how 'natural' all this was, I put the question to Charlie Burrell, the owner of Sussex's famous Knepp Estate. His answer to my prescriptive question was a non-prescriptive answer: 'Who is to say how many are too many? The animals determine the aspect of a landscape.' For someone who had grown up with prescriptive conservation certainty, this was an interesting reply.

A lush coastal environment, for example, contains more abundant and nutritious forage than an acidic soil inland. Whilst Holocene Norfolk may have seen hundreds of horses and aurochs on its coastal marshes, nutrient-poor soils inland may only have been able to support lower densities, creating more densely wooded lands. If you cross the plains of southern Africa or any largely intact ecosystem, such as the wilds of Alaska, concentrations of herbivores, or omnivores like bears, are no constant matter. The answer to 'did nature look like the Oostvaardersplassen?' may not, in fact, be empirical at all. Herbivory depends on *desirability*. In areas of grazing plenitude, you would expect larger herds – and more grasslands. In areas around rivers, you would expect more open canopies and scrub, as trees are felled or coppiced by beavers. Nature is shaped not only by the range of animals that are present but also by the choices exercised by those animals.

Two further factors, however, would make the Oostvaardersplassen more natural. One would be if the herds were free to roam along the entire European coast. The other, if predators harvested them too. But does the Oostvaardersplassen have to be perfect, for Britain to follow its example?

This Dutch rewilding experiment has revealed, for the first time in any of our lives, the rich dynamics of a temperate coastline – one where white-tailed eagles scavenge the fallen and habitats jostle and intermingle side by side. It kicks into touch the need for dozens of action plans for individual species, with each landscape type kept rigorously apart, like separate zoo exhibits, at enormous expense. Each year, new wild animals give the Oostvaardersplassen their approval. Beavers, like eagles, are rewilding themselves. A landscape has emerged that nobody guessed would exist – colonised by other species nobody thought would return. And as Frans Vera expresses it:

> We've become trapped by our own observations.
> We forget, in a world completely transformed by
> man, that what we're looking at is not necessarily
> the environment wildlife prefers, but the depleted
> remnant that wildlife is having to cope with. What it
> has, is not necessarily what it wants.[17]

While the Oostvaardersplassen is often attacked for not being truly natural, such an attack rather misses the point. It's *more* natural, more robust, more diverse and more *real* than many conservation models, including any of those currently in place in coastal Britain – and cost-effective to boot. And that, all in all, is not a bad start.

Farmland Gone Feral

Competition between disturbance and vegetation
succession is hugely productive for wildlife, resulting
in the 'margins' where most of life lives.
—Charlie Burrell[18]

Since 2002, as documented in the bestselling *Wilding*, Charlie Burrell and Isabella Tree have overseen the largest lowland rewilding project in Britain. It's happening in southeast England, just miles from Gatwick, on Burrell's estate in West Sussex (Plate 11). The premise is alarmingly simple. Give the land over to a suite of free-roaming large herbivores in very low densities; mimicking the herds of herbivores that would have roamed Britain in the past – and let them get on with it. This project's scale and vision is unique in Britain, and it's worth taking a moment to explain how it came about, especially as other farmers, too, could follow in its wake.

Charlie Burrell inherited a mixed arable and dairy farm, but on heavy Sussex clay terrible for growing cereals. The farm tried everything they could think of to make it productive, but after 17 years they were still making a loss.[19]

Inspired by the example of Oostvaardersplassen, and the grazing ecology theories of Dutch ecologist Frans Vera, the owners turned the estate over to a naturalistic grazing system and surrendered the management of the land to free-roaming herbivores. As part of a parkland restoration around the house and the northern part of the estate, the owners sowed and repeatedly cropped native grasses – to soak up phosphates and nitrates from the farming days. In some areas, they planted a Low Weald wildflower mix to increase native plant diversity. In the 'Southern Block', however, they allowed a vegetation pulse to happen before releasing free-roaming herbivores to do battle with the vegetation.

Flowering plants started to return. With the soil purged of toxins, the health of the estate's open-grown oaks began to improve too. Below the soil's surface, fungal networks, or *mycorrhizae*, symbiotic with the root systems of the oaks, made a gradual return. Eventually, orchids, an indicator of healthy mycorrhizae networks, began to appear in the middle of former arable fields.

Into the Knepp's rewilding experiment were released fallow and red deer, Tamworth pigs, as surrogates for boar, Old English longhorn cattle, as surrogates for aurochs, and Exmoor ponies, as surrogates for tarpan. But the Burrells segregated their wildlife experiment. Into the northern block on the estate, they let loose only free-roaming cattle – into the southern and middle blocks, all the four-legged chaos they could muster. Knepp's greatest resurgence in overall avian diversity has been in the diverse scrublands where the combined force of these animals is once again in play. This has, in many ways, turned out to be the most exciting area of all – a diverse mosaic of habitats – looking much like African scrub – where the tensions between vegetation succession and animal disturbance generate truly dynamic processes, and where biodiversity has rocketed.

If you try to 'manage' an ecosystem back to life, you will almost certainly fail. You might restore a target species here or there, but an entire network of plants and insects, all co-evolved, all finding one another – that is an impossible thing for human hands to recreate. Far better, Knepp has found, to hand over such a daunting task to Britain's more experienced stewards.

Knepp's insect repopulation has, in just 15 years, set an exciting blueprint for turning things around. Since the rewilding scheme began, 600 invertebrate species have been recorded. Dung from the roaming organic herds recruits armies of dung beetles of a number of species; the favoured food of little owls. Dung beetles like the violet dor, not seen in Sussex for half a century, have recolonised the land. These increases sit entirely at odds with the remorseless decline of beetles in the rest of Britain.

Over half of Britain's butterfly species have colonised Knepp. The lost 'clouds', not seen in most counties since the pre-war days, are being seen a little more often. In 2015, in one section of the estate, 790 small skipper butterflies were counted. As you walk through Knepp in summer, you continually startle meadow browns, common blues and marbled whites from the grass. Knepp's butterfly recovery has kicked into touch much accepted wisdom about the habitats of the spectacular purple emperor – a butterfly of ancient oak woodland, it was often said. Purple emperors now thrive here in the sallow groves; their caterpillars feeding on sallow leaves.

The invertebrate ecosystem at Knepp now grows richer by the season. Each year, it continues to reconstruct itself. An extraordinary 441 moth species have now been recorded at Knepp. The ragwort-feeding caterpillar of the cinnabar moth, a species declining in many areas, is on the rise. Ghost moths once again flutter in the night. Cuckoos, feeding on a range of large invertebrates, have increased in number, against national odds. Resurgent bats, of almost every British species, including the very rare Bechstein's and barbastelle, also attest to Knepp's aerial invertebrate abundance, as well as its diversity of open-grown trees, allowed to decay in peace.

For every family of invertebrates examined at Knepp, scientists are discovering the *reverse* of what is generally happening elsewhere across the British countryside. Knepp has been rewilded for less than two decades, yet little like this has ever been seen in our country before. Management, objectives, targets or nature charities have rarely, if ever, come close to achieving such wide-ranging success, in terms of abundance and biodiversity – even if, against all odds, they have preserved invaluable 'arcs' of particular species for future generations to enjoy.

The reason this is all so vital to saving our vanishing birds is that before this, there was there were few precedents for a mass resurgence in invertebrate abundance, and diversity, in British conservation. Yet all the Burrells appear to have done, as one critic put it, is 'release a bunch of farmyard animals'. So what, ecologically, is happening at Knepp?

The Land that Heals Itself

Commentators from all sides, *Guardian* and *Telegraph* alike, remark on the 'Serengeti' aspect of Knepp, especially in the southern block. It is the first thing that strikes you on arrival. Knepp's rewilding relies on what it calls 'self-willed' ecological processes. Each animal architect exercises a different effect on the land.

Old English longhorn cattle, an old breed, have reprised many of the core functions of our lost nomadic cattle. In areas resistant to their browsing, scrublands have formed. Hawthorns, in response to being nibbled, have increased their production of tannins – and, at the same time, their density of spikes. These scrublands are now the home of nightingales and turtle doves.

When cattle are browsing an area, spiky bushes like hawthorn, blackthorn, dog rose and bramble deter them. Plants and young trees grow underneath and within these bushes – using the thorns as protection. Fledgling oaks – the seeds of which are planted by field mice and that aerial forester, the jay – grow using the cover of these thorns as protection. The cattle carry up to 230 species of seeds in their gut, hooves and fur, acting as vectors for floral diversity.

At the same time, however, the cattle also create open glades and 'control' scrub. On the move across large areas of land, the herds never mow down an area to the ground, nor do they compact the soil, but instead they open up and maintain short grasslands. The extremely low density of cattle here is key to these results, with stocking units of 0.3 livestock units (or animals) per hectare, compared to 4–7 units on conventional pasture. Knepp's stocking ratio mimics, as far as it is possible for anyone to guess, something close to our lost densities of wild, roaming herbivores.

The ecological role of free-roaming horses on the estate appears to dovetail in and around the grazing actions of the cattle. The Exmoor ponies take out the toughest and deadest grasses, and they also seize individual thistle-tops; ripping them off. Removal of these 'rough' areas creates the growing conditions for other plant species to thrive, which are favoured, in turn, by the cattle.

Knepp may one day have wild boar, but Britain's so-called Dangerous Wild Animals Act of 1976 prevented the reintroduction of these animals to the estate. So Tamworth pigs, a very old breed, fulfil a similar role. The pigs transform even ground into a rotavated rubble of chaos: a fresh start for plant and insect life.[20] The exposure of bare soils by the pigs at Knepp, in the two weeks in May when sallow seed is viable, has allowed the sallows – upon which purple emperors depend – to colonise. Huge anthills have formed from pig-rooted turf. Solitary bees have colonised the open soils. But the most striking thing Knepp's piggies may have achieved is a wilder vision for how best to conserve our vanishing turtle doves.

When Charlie Burrell took me around the estate, he was keen to emphasise the difference between the 'southern block' and the rest. It has horses in play alongside cattle – but it's also the area where pigs, as well as red and fallow deer, are shaping the landscape. Of the eighteen singing male turtle doves present at Knepp as of 2018, where there were none before, most appear to be nesting in areas where pigs disturb the soil. Turtle doves are archetypal birds of open, weedy soils. They feed on the tiny weed seeds in disrupted earth; a habitat promoted by arable farming only in the most recent of times.

At Knepp, against a backdrop of impending national extinction, turtle doves are thriving and increasing year on year. Not only do they have dense thorny scrub in which to hide their nests, but their diet is catered for by the abundance of wild weeds such as chickweed, grass vetchling, scarlet pimpernel, sharp and round-leaved fluellen, bird's-foot trefoil and fescue, which produce the tiny seeds on which turtle doves thrive. Such successes have not been sown by arable farming, but the chaos of free-roaming pigs.

Knepp is also shedding a new light on how deer, in more natural ecosystems, can play a far less destructive role when it comes to Britain's birdlife. Whereas rampant deer browsing, by unnaturally high and confined numbers of deer in our woodlands, can destroy woodland diversity, the same is not being seen at Knepp at all. Indeed, deer, famous for removing nightingale habitat by browsing young trees out of woodlands, exist side by side with nightingales. In the southern block, the Burrells allowed a large vegetation pulse before the deer were put back in. By this stage, robust plants like hawthorn, blackthorn, dog rose and bramble, capable of repelling even cattle, were more than able to repel the nibbling of deer.

Now, red deer are springing another surprise on Knepp's ecologists. We associate them mostly with our uplands, but at Knepp, given the choice, red deer show preferences for the wettest areas and swim in open water. Here, the richness of forage grows their antlers far larger than those seen on stags roaming the Scottish hills. Red deer debark trees (as bison, not currently present at Knepp, do as well, to great effect) – thereby contributing to the 'pruning' and opening of wetlands.[21]

Across the estate, as these intertwining four-legged stories run their course, the landscape grows ever more complex and diverse. Rather than being swallowed in 'forest', Knepp, the southern block in particular, remains a dynamic, shifting balance of habitats, as its grazers maintain open scrublands and meadows and prevent a shaded, species-poor forest from forming at all.

One Landscape for All

At Knepp, history is reversing in the way birds are using the land. Lesser whitethroats, it seems, did not come into existence with the hedgerow after all. At Knepp you can hear them blasting away from stands of cattle-prodded blackthorn.

Knepp's thornlands are home to healthy populations of nightingales, which appear to fluctuate, naturally, in numbers – with seventeen singing males in 2018. BTO studies in 2012 suggested 79% of these were paired; considerably higher than the national average. Any idea that nightingales are woodland birds vanishes when you visit Knepp. These scrubland maestros are thriving here – not in shade, but in dense scrub whose growth is fuelled by sunlight.

Disrupting the careful labelling of habitats, birds as disparate on the conservation red list as woodlark, woodcock, yellowhammer and peregrine falcon (the latter nesting in trees) have all moved into Knepp. Nature is rampant – defying definitions of 'woodland' and 'grassland'.

Knepp shows that Britain's birds are ready for, best suited to – and most urgently need – a return to wilder stewardship. Most of all, the Burrell ethos of 'let's see what happens' has, ironically, restored more insects, and more birds, than many areas driven by conservation targets and outcomes.

The Knepp estate also shows how even the simple act of returning livestock densities to levels last seen in the nineteenth century, can recreate, in the ecological blink of an eye, a diversity and abundance of life not seen in many areas of Britain without our lifetimes. But it also recalls how our lowlands might have looked not just centuries but millennia before, in a time before farmland. At the same time, Knepp remains both radical yet profitable. It grows wilder every year, yet, in some respects, is still a tended landscape with livestock lovers at its root. Most of all, Knepp shows how rewilding need not drive people, or farmers, off the land. Far from it, Knepp shows, to farmers from the Welsh Hills to the dairy pastures of Somerset, how farming might plough its furrow long into the future. More profitable, more diverse, more humane, more robust – and better for both people and wildlife alike.

CHAPTER 5

A Question of Scale

Only Landscapes Live

Let's start by imagining a fine Persian carpet and a hunting knife. We set about cutting the carpet into thirty-six equal pieces, each one a rectangle, two feet by three ... But what does it amount to? Have we got thirty-six nice Persian throw rugs? No. All we're left with is three dozen ragged fragments, each one worthless and commencing to come apart.
—David Quammen, *The Song of the Dodo*[1]

It's early May. Your pair of house martins has returned – the miracle of summer. Weighing barely more than 15 grams, they have flown over 8,000 kilometres from southern Africa – even now, we don't know from exactly where – to spend summer below your eaves. In navy and white overalls, they've returned to last year's nest: a gravity-defying mud castle cemented upside down, one beakful at a time, below your roof. It's a glorious summer, and you idly watch your martins bringing flies to hungry nestlings. By late June, the clumsy first brood are twittering away on wires beside your house. Then, it's time for round two. The first lot of kids are sent packing, and soon your industrious martins get to work on their second family. But in a wet spell, the second set of chicks do not survive. Still, you're pleased to see the first family have done well – and quietly wish them all the best on their journey to Africa.

As summer fades, they're gone. But the next year – the castle is abandoned. Your martins haven't come back. You're outraged. They're house martins. Your house is most certainly still there. But then, a few years before, your spotted flycatchers didn't come back. They might have had to fly from Africa, but your local starlings didn't have to. And they didn't come back either last spring.

You can't understand it. You're not spraying your garden. You're doing nothing wrong. Birds are flying the nest. You know enough to know that nothing has changed. Each summer, these little dramas of loss and puzzlement play out for many nature lovers in Britain, whether in our gardens, our nature reserves, or the countryside at large. And the reason is chilling. Your single house martin pair were, in ecological terms, already dead. Your spotted flycatchers were islanders. Your starlings were the last survivors of their tribe. By this stage in their decline, it didn't *matter* what you did. These birds were living history. Birds, like other animals, have evolved to exist in boundless populations, connected across large tracts of preferred habitat. There is no precedent for any species, anywhere, to evolve as a single pair, standing the test of time in isolation from the rest of their kind.

Populations of birds, by contrast, provide security. Landscape populations of birds allow for starvation, predation, disease, mate selection, and mitigate a whole spectrum of problems that naturally befall a species, such as poor summers or problems on migration.

Meta-populations, or meta-communities, can provide even better mechanisms for survival. In a meta-community, distinct populations are strengthened by dispersal between them, increasing survival, promoting gene flow and allowing for local extinctions and recolonisations across the geographical range.[2] Amazingly, a robust population of birds, in a natural environment, is adapted to withstand losses of up to 80% in one season – yet still survive.[3]

Isolation, by contrast, breeds extinction. It is fundamentally unnatural for that single pair of house martins in your village to be defying the community-based rules by which their species evolved. Only birds in robust populations stand the test of time. In this chapter, we'll find out how big landscapes alone can help prevent a mass extinction of our nation's wildlife, and why scale is the only language that nature understands.

Joining the Dots

The headlines tell us birds are falling into decline – both our resident birds and our enterprising migrants, arriving from Africa each summer. But in some areas of our country they're not declining at all:

The whinchat is an example of a species that was once common across England but has suffered major declines in recent years: 55% since 1995. A large, stable population of whinchats persists on Salisbury Plain.[4]

The lesser spotted woodpecker was once widespread in woodlands across the UK, but its drumming is an increasingly rare sound. In Hampshire it has also become a scarce resident. However, the New Forest remains an important stronghold.[5]

The strongest capercaillie populations are in the Cairngorms National Park. The last national survey in 2009–2010 showed 80% of the UK capercaillie population was estimated to be in the Park, with the vast majority in Strathspey. Numbers there currently appear stable.[6]

Despite national concern, willow tits seem to be doing very well in County Durham. Excellent populations were noted in the coalfields area including 13 at Rainton Meadows, 10 at Hetton Bogs, eight at Elemore GC.[7]

Contrary to the national trend, numbers of nightingales in the Cotswold Water Park have been maintained over the previous 10 years. A superb total of 20 singing males were recorded.[8]

These quotes refer to migratory and resident birds, Scottish and English, specialists and generalists. These birds have only two things in common. Firstly, they have suffered, or are suffering, massive declines and now, in most cases, face the real prospect of extinction in Britain.[9] Secondly, each has found a refuge that is not only hanging on to these birds – but defying the national trend.

When declining birds do well, you're onto a winner. That conservation gold dust, *time*, is on your side. You can begin to see what is missing in other areas, and work out why these birds are declining everywhere else. All of our good-news stories have one thing in common. And it's expressed rather well by site manager Gareth Harris, at the Cotswold Water Park in Gloucestershire:

> *Although the nightingale population fluctuates widely,*
> *it appears self-sustaining and stable in the long term*
> *… As nesting sites become too old or unsuitable, new*
> *nesting sites become available and are colonised …*
> *the size and extent of the landscape is key.*[10]

Size and extent, alone, determine whether your conservation efforts will be successful or in vain. Scale is what makes populations possible and gives them the chance to weather bad times. Narratives of scale are stories of success. But how big does a population of birds need to be to survive?

Why Critical Mass is Critical

> *The size of an island has a huge impact on the life of*
> *the creatures cast away there … Of all the species that*
> *have become extinct in recent years, around 80% of*
> *them have been islanders.*
>
> —Sir David Attenborough, *Planet Earth 2*[11]

In 1939, the RSPB bought North Warren, Suffolk, for its breeding birds, which, at that time, included red-backed shrikes. Even before the war, the shrike was on the cusp of a really severe plummet in fortunes, after already a century of decline. Buying a carefully selected piece of land was a smart move by the RSPB, and one that's proven the salvation of many birds, such as roseate terns, with the entire future of the species in Britain now safeguarded on Coquet Island in Northumberland.[12] Yet, by 1960, there were just 27 pairs of shrikes left along the coast around North Warren.[13] By 1989, shrikes no longer bred regularly in England.

Shrikes, it seems, were vanishing even in good areas of habitat, with good protection by conservationists. Their decline was operating on a scale that no nature reserve could contend with. But why were heathland reserves like North Warren unable to hold onto their 5, 10 or 20 pairs of shrike? The answer, as for all our vanishing birds, lies in something known as critical mass.

Viktoria Takács is an ornithologist who has done a lot of work on red-backed shrikes in eastern Poland, in the realisation that the species is vanishing elsewhere in Europe. In her study, Viktoria looks at a stable population of red-backed shrikes in farmland.[14] She asks a simple question: how many pairs are needed to keep things as they are? What if the food and habitat remain prime? What if you factor in a bad summer, which, in Poland, happens around every four years? What if you have 30 pairs? What if you have 200? How likely is it, in 50 years' time, that such a population will still exist?

Viktoria puts these questions to VORTEX. I pictured a small black hole, but in fact it's a piece of software for population viability analysis (PVA), a modelling tool for estimating extinction probability in a population. The more data you put in, the better and more accurate the results you get.

In Poland, the team knew things like the percentage of shrikes laying one, two, three, four and five eggs. They knew the breeding age of females, who are ready at a tender one year old but past it by nine. They knew the mortality of young and adult birds. As with many smaller migrants, just a *third* of red-backed shrikes make it back from Africa the following spring, to sing pretty badly on European soil. Armed with a lot of facts about her shrikes, Viktoria and her team calculated roughly how many shrikes you need in a single population to stand the test of time.

The study revealed that for a shrike population to have a 95% chance of surviving for 50 years, in favourable conditions, you'd need 80–90 pairs in that population. Not one, not twenty. Not even fifty. Such a population, alone, provides viability – where insurance is provided against predation, bad summers, inbreeding and other mechanisms that drive decline. Even if things are going well, then, for a bird like a red-backed shrike to survive, any one landscape must be rich enough to support at least 170 adult birds. In a British context, such a figure is alarming. It reveals how food-rich landscapes alone, not tiny pockets of amenable land, are often the *only* solution for preventing extinction. Tiny pockets of birds can persist for quite a long time, even decades. Yet, as history shows, they have no viable future at all.

Putting All Your Eggs in One Large Basket

Until the late 1800s, the corncrake bred commonly in almost every county across Britain. A century later, a national survey in 1993 recorded just 480 birds, with almost all of these on the Hebrides.[15] The corncrake would bounce back thanks to conservation work by the RSPB. But how was this work possible after a collapse comparable to that of its meadow compatriot, the red-backed shrike? On islands in the Hebrides lay the last places where tall, late-cut, insect-rich grasslands existed on a scale large enough for corncrakes to survive. But even though corncrakes were on the brink, they were staying in touch. In 1993, there were 106 calling male corncrakes on Lewis, 116 on the Uists and 111 on the island of Tiree.[16] These populations were close to one another, forming meta-populations. Numbers from the key strongholds in 1993 were, in most cases, close to Viktoria's 50-year threshold for shrikes. Like shrikes, corncrakes are long-range summer migrants, of whom only one-third make it back the following summer, to sing pretty terribly on European soil.[17] Corncrakes, however, had the decency to persist in one large population. This bought the RSPB time to consolidate their range.

A similar 'refugee' story has involved nightjars, which in the early twentieth century bred in every English county and on many village commons. By 1981, the low point for nightjars in Britain,[18] over a tenth of the remaining population was concentrated in the New Forest.[19] The nightjar was on the brink – but, as with the corncrake, its refuge was large and rich. Since this time, the nightjar has consolidated its populations, yet, like the corncrake, it has rarely moved out of landscapes where birds can network and connect. And whilst many insectivorous summer migrants are on the way down, nightjars, in these areas, are increasing.[20] With a strong enough refuge in one place, conservation efforts to improve nightjar habitat then pay off, because this work is acting in tandem with population dynamics. With nightjars, new 'islands' of heathland are now being grafted onto existing landscapes. The RSPB call this a 'stepping stones' approach. But to create stepping stones at all, your island of birds has to be large enough in the first place.[21]

If those 'stones' are placed just a short way away from your main island, RSPB research has shown that nightjar recruitment is far lower. Preserving continuous landscapes, therefore, is not an optional nicety in conservation, but a basic necessity for preventing extinction.

Facebook for Cuckoos

Plate 12 is a 'live sightings' map compiled by the website Devon Birds.[22] Each dot on the map represents someone's sighting of a cuckoo. The vast majority of Devon's cuckoos are now found on Dartmoor – most of them in its lightly wooded margins.

The first thing this map shows is hope. Dartmoor is buying conservationists the time to act. Notice the overlap between those dots on the map. As well as having caterpillars, wooded grassland, and 20,000 meadow pipit hosts, Dartmoor's cuckoos are keeping in touch with one another. There are enough birds to make up for bad summers, predation, and a turbulent migration to the Congo and back.

Whilst Devon's cuckoos have declined by 75% since the 1980s, the BTO calculates that Dartmoor's have declined by only 24%. Such is the power of landscape. With around 100 male cuckoos and an unknown number of females, Dartmoor's cuckoos still have critical mass. The other thing the map shows is that cuckoos have long stopped being the summer sound of our wider countryside. Once calling across Devon's farms and gardens, cuckoos have been forced into a moorland retreat. Only here, in the moth-rich landscape of Dartmoor's wooded margins can the cuckoos of Devon now survive.

One reason for this large-scale retreat is that cuckoos place far more demands on a landscape than many other birds. Females must prospect many host nests to lay their eggs in, roaming widely across large areas of land, whilst eating huge quantities of caterpillars. In Devon, Dartmoor alone maintains the facilities that this landscape nomad requires.

Indeed, the importance of Dartmoor stretches far beyond Devon. In the whole of southern England, the cuckoo may only truly be saveable in the foodscapes of Dartmoor or the New Forest. Who would have thought that a bird a mere 35 centimetres long might require 950 square kilometres of amenable habitat to survive? Even our smaller birds are far more demanding of space than we might at first imagine.

Joined-up Thinking

Lesser spotted woodpeckers, having crashed catastrophically in Britain, remain common in countries where the amount of deciduous woodland cover reaches above the 40% mark. Nobody in Poland, Germany or Hungary is worrying about the impending extinction of this dappled delight. There are estimated to be around 1,000 pairs of lesser-spots left in Britain,[23] but with many pairs isolated in tiny, starving woods, few places show any prospects of hope. Yet in one place they have found a haven: the New Forest, still, remarkably, home to well over 100 breeding pairs.[24]

In April 2016, on a beautiful morning, local expert Rob Clements kindly walked me around the pasture-woodlands south of Lyndhurst. Rob had cautioned, beforehand, that we might encounter an unreasonable number of lesser spotted woodpeckers. When he used the word 'common', I had a sinking feeling. I'd never met Rob and wasn't sure this could be right.

Dedicated birders invest several visits in March to see one lesser-spot in a year. Most counties now have no more than a handful of pairs. Yet in two hours, an extraordinary session passed. We heard one woodpecker, on average, every ten minutes. At one stage, four birds answered one another in the same clearing. Having studied the species for some time, I saw more lesser-spots in two hours than during the previous two years. The density here is, in fact, comparable to that of Poland's Białowieża Forest. As for poor Rob, I shouldn't have doubted him. He just suffers from having the finest local patch in Britain. So what is going on in the New Forest?

Whilst any one of the New Forest's ancient pasture woods would die, like a severed branch, if cut out and placed in the modern chemical landscape, the combined effect of having all these caterpillar-rich ancient woods in one place achieves the opposite result. In other words, the New Forest is a single, living, organic organism the size of a small county: far greater than the sum of its parts. Whilst most of Britain's woodlands are what David Quammen would call 'ragged fragments' – isolated and starved of caterpillar food – the New Forest remains an ecosystem. You cannot, viably, conserve a tiny woodland island for wide-ranging birds like lesser spotted woodpeckers. But you can, as the New Forest shows, preserve a refuge in the hundred-pair range. That requires habitat richness of a really large scale. A scale far larger than a sparrow-sized, caterpillar-eating bird might reasonably be expected to require.

Lessons from Goldfish

Until the 1950s, the classical view of the way populations worked was based around the idea of competition. Individuals in healthy ecosystems were better off if fewer others were competing with them for resources. Populations with a lot of competition were thought to experience a slower growth rate, and eventually level off.[25]

In the 1950s, however, American ecologist Warder Allee, studying goldfish, noticed something very different. He found his goldfish grew faster, not slower, when there were more individuals in the tank. He worked out that in colonial species, survival *increased* with numbers.

Cooperation aided the survival of some species.[26] A 'strong Allee effect' is now the term given to populations that require a critical population size. Below this size, population growth actually begins to *fall*. And goldfish, it turns out, are vital for understanding bird decline.

Hawfinches, for example, are birds of a hive-mind. They operate more as a 'collective' than as individuals. In areas like the New Forest, there are woods of plenty, and woods of poverty, so hawfinch roosts and colonies are usually clustered around favoured areas. This means that if there is a change in the New Forest, whether a troublesome jay, or a change in habitat, your hawfinch 'hive' can up and fly for fifteen kilometres into the next area of prime woodland. Such a strategy, however, places serious demands on the size of your landscape. Studies in the New Forest have shown that in winter, most roosts of hawfinches are placed close to active goshawk nests.[27] This smart strategy puts hawfinches a long way away from their crow predators, either fearful of the goshawk's ferocity or already transformed into supper. Even where they are very common, as in Poland's Białowieża, hawfinch populations still have nest survival rates as low as 27% after predation.[28] But by operating as a collective, hawfinches have evolved to absorb such losses. In the breeding season, hawfinches exercise communal distraction against jays. Tagging of birds in the Forest of Dean shows they also make local migrations, moving several kilometres in response to changing food supplies, or predator pressure, across a season.

Such communally minded birds cannot function in the way they have evolved, however, in most of Britain's postage-stamp woodlands. In other words, you either have a large, sympathetic landscape, like the New Forest or the Wye Valley, with a lot of hawfinches – or you don't have any hawfinches at all.

A host of other birds that we might see as individuals actually operate best in loose colonies. One we might not think of in this light is the osprey. The conservationist Roy Dennis regards ospreys as semi-colonial: you either have a cluster of birds prospering, or that cluster dies out.[29] This is because ospreys do better by fishing at communal feeding sites. White-tailed eagles, hardly a bird we'd consider colonial now, most certainly were at one time. In Russia, these communal hunter–scavengers can nest just hundreds of metres apart.[30]

As you start to look around, you'll notice all kinds of birds benefiting from close connection. The blue tits in your garden may be territorial in summer, but they'll join up in winter to form flocks. We may see blackbirds and woodpigeons as individuals, but they in fact form meta-populations across our suburban worlds.

As a rule, we always underestimate the scale and connection that birds need, perhaps because these connections have been lacking from our landscapes for so long. We also forget that aggregation is important to a whole range of species. Not just martins, swallows and swifts but wagtails, waders and many other families benefit from the wisdom of crowds. Wading birds exercise better defence against predators in large populations, alerting one another to danger.[31] This is one reason a Polish river valley can be filled with foxes, badgers, lynx, martens and birds of prey, yet still teem with lapwings. House sparrows, too, are highly colonial: you cannot have a single pair for long. Connecting landscapes, rural or urban, allows for aggregation. The collective is what wins.

The Avalon Effect

The result of inspired habitat recreation by the RSPB and other conservation NGOs, eight kilometres of wind-waving reeds now run in an east–west line across the Somerset Levels; many of these are dug from old peat workings. As the bittern flies, this forms one continuous refuge, and there are now more bitterns here than anywhere else in Britain. But the benefits of aggregation are now being seen all across the returning wetlands of the Avalon marshes.

Almost all of Avalon's birds are forming populations able to exploit this significant area – and *requiring* such an area to flourish. You can watch groups of garganey, brimming with duckish frustration, flying from wet meadows to reedbeds a kilometre away. You can find hundreds of pairs of reed warbler chattering away in the reeds. Sand martin colonies thrive in the peat. Whilst a contracting landscape enforces isolation, Avalon reveals the living opposite. The growing scale of the reedbeds now means that 'overshoot' birds from the continent have settled here to breed. Great white egrets, for example, have summered here in their first year then come back with mates to nest. Each overshoot bird arriving in good habitat has the best possible chance to establish a population.

In Norfolk, too, any young birdwatcher grows up learning the hallowed names of Cley and Titchwell, Holkham and Holme. But these are just the names we give reserves, in order to identify who owns them.

As the wetland bird flies, this coast is a connected wonderland. Between the eastern edge of the Wash and Sheringham, the web of coastal marshes extends for 65 kilometres in one unbroken length.[32]

Marsh harriers, once down to a single pair at Minsmere in Suffolk, have saturated the reedbeds of the East Anglian coast, being able to expand for mile after mile; all from a precarious single pair. When Norfolk's first spoonbills nested at Holkham, they were seen commuting miles along the coast to feed. Our East Anglian coast isn't just the result of outstanding habitat, but the combined effect of aggregation in its birds. By 'lining' our coasts or wetlands with nature reserves that join up, we combine the benefits – and multiply the birds.

Arable Islanders

Conservation support of farmland has happened as part of the fight to make the Common Agricultural Policy less terrible for wildlife. It's happened most of all our cereal farms. Other countries have not adopted this approach to such a degree, focusing instead on a clear demarcation – between areas of intense production and wild reserves for nature.

One reason for the approach here at home, however, is that a *quarter* of Britain is covered in arable land. Over half of this is given over to growing cereals. Arable has been around for a very long time, and most of us have only ever seen grey partridges and yellowhammers in such places, so this is where we seek to put them back. And as arable supplies our food and isn't going anywhere, damage limitation seems a reasonable option.

A farmer with a pair of yellow wagtails on his land, and the conservationists who've helped him regain them, can rightly feel pleased. The wagtails, however, have evolved in grazed wetlands, connected to hundreds of other pairs of their kind. They are adapted to nest in the company of fellow grazed-grassland species, like lapwings[33] – because this improves their vigilance against predators. So that pair or two only becomes significant if you have yellow wagtails on the next farm, and the next, and the next. So when investing conservation monies in arable farmland, this *only* pays dividends if farmland landscapes act as one.

The Marlborough Downs, in Wiltshire, is one landscape working for seed-eating birds. Hundreds of tree sparrows and corn buntings are doing well. Farms join up and guarantee a population of these birds. There are enough farms with strips for seeds and insects that a few 'bad' ones fail to spoil the picture. In north Norfolk, drive inland and you'll still swerve to avoid wobbly crèches of grey partridges on the roads around the villages near Ringstead or Snettisham. A lot of wealthy arable farmers in Norfolk have also, collectively, invested in barn owls. Miles of fallow strips have joined up – and barn owls are, again, a common sight.

The Lancashire Mosslands are a large area, stretching for 25 kilometres from Crosby to the Ribble. Unlike most of our farmlands, the Mosslands are still alive with birdsong each spring. You can see yellowhammers, corn buntings, grey partridges, tree sparrows, hear quail, find hundreds of lapwings, and the air is still filled with skylarks. The Mosslands hold one of the highest densities of barn owls in Britain.[34] The Mosslands are all about *joinery*. Across the crops in this area run lines of wide marginal land, scruffy willow stands and plenty of insects and cover. The original 'mosses' that remain (wide areas of long grassland and damp woods) support other birds, like whinchats and woodcocks. The net result is a landscape of mixed quality, but an overall abundance of food and places to raise a family.

Whilst 'full-scale' rewilding will in the future be able to determine outcomes for wildlife in large, depopulated areas such as northern Scotland, our arable landscapes are here to stay – and to feed us. Funding them is essential, provided we understand the how and why. 'Cooperative' landscapes, like the mosses, can speak at least a few key phrases in the language that nature understands. But support of agricultural islands most often fights the fundamental laws of nature.

What has often been pursued is a series of isolated farmland triumphs, in the hope they will add up to one big success. But birds do not acknowledge human targets. You cannot have one farmland cuckoo stand the test of time. You can only have closer to one hundred. Investing in the one is dead money – and money that could be used elsewhere.

The Thirty-three Percent Club

Able to travel widely over our country as they arrive from Africa each summer, you might think our summer migrants might be able to survive in far smaller landscapes than our resident birds. Surely they can just drop into a patch of habitat and get on with it? In fact, the reverse is most often true. Summer visitors need larger landscapes than residents. That is because of the 33% rule. On average, around two-thirds of nightingales, corncrakes, turtle doves and many other long-range migrants do not make it back to their breeding grounds the following year.[35] And what's more – this is how things should be.

Migration has always been a dirty business. With the age-old threats of predators, extreme heat and exhaustion, migratory survival rates have always been low for fragile migrant birds. Human depredation in Egypt, Malta and Cyprus, to name but three, has, for centuries, had further unnatural impacts on species like the turtle dove.[36]

If you're a small migrant, with the odds rigged against you, pumping out a lot of healthy chicks, ready for their journey to Africa, is imperative – if an equal number of birds are to return the following year. That strategy requires good numbers of birds arriving, each year, in the same place. If you are a migrant coming all the way from Africa, it is vital that you arrive to find yourself surrounded by abundant peers, in the likely event that last year's wife or husband has snuffed it somewhere between the Congo and England. Abundance means opportunity, whether online dating, interviewing multiple candidates for a tough job (see online dating) or assessing which house to buy. In contrast, the isolation effect, silently killing thousands of our birds in Britain each summer, is sometimes best appreciated from the perspective of just a few families of birds.

In 2012, I studied three nesting pairs of nightingales in Avon, as part of the BTO's Nest Records Scheme. One pair hatched young but abandoned them, as driving wind and rain chilled the nestlings to the bone. A second pair's eggs were robbed, twice, by a predator, most likely a weasel. Only the third pair fledged four young.

Had each of these nightingale pairs hatched and fledged five chicks, from five eggs, then 21 birds would have started the journey back to Africa. Even if all 21 nightingales had left my local wood, BTO data suggest that only around seven birds would have returned. This would have restored the population of three pairs from the year before. In reality, a maximum of 10 birds began the return journey that summer. With such tiny numbers involved, only three nightingales returned to my wood in 2013. In 2014, nightingales failed to nest in Lower Woods for the first time in living memory. But my local jazz maestros weren't wiped out by weasels or weather. They perished, like most of Britain's isolated wildlife, by living on an island of habitat too small to meet their needs.

Bigger is Wilder

Nightingales are birds of a successional, dynamic landscape. The spiky scrub blanket they require is ripe only between about five and eight years old.[37] Before that, it's not dense enough to form a shade-shell, under which nightingales forage like ninja robins. After that, the scrub grows towards tree height, admitting light and predators at the bottom.

That twenty pairs of nightingale still nest in Gloucestershire's Cotswold Water Park might seem insignificant. But that such a number has remained *unchanged* for over two decades is very important. This outpost is way outside the national range, holding on to these birds when almost all of western England's nightingales have vanished completely. The Cotswold Water Park, a brownfield mosaic of flooded gravel pits, is no wilderness reserve. Its secret lies in its *dynamism*.

Arriving from Africa each spring, the nightingale population here is able to disperse locally, across an area of scrubland six kilometres across. Nightingales can, as a collective, move around over the course of the passing years, as the landscape continues to evolve.

Dynamism is the most crucial ingredient that a larger landscape can provide – but that a tiny well-kept nature reserve cannot. In many cases, straitjacketing our nature reserves' pure habitats proves extremely damaging for our nation's birdlife, evolved in mosaic habitats that shift naturally across a landscape over time. The BTO, for example, observes that 'scrub is a dynamic habitat, constantly changing as it evolves into woodland. Nightingales seem to be particularly sensitive to this gradual change and will only use scrub for the few years when it is most vigorous and dense.'[38]

Scrubland birds thrive in a world *designed* for constant change. As a result, scrublands have more declining birds associated with them than any other habitat. Willow and garden warblers, nightingales and bullfinches, willow and marsh tits, scrub residents and scrub migrants alike, are all vanishing from Britain.

If, in managing a landscape, you forget that habitats are supposed to shift and evolve, then successional habitats like scrub are the ones you're most likely to ignore or fail to understand. When did you last read a press release that read: 'the processes of glorious decay, shifting habitats and the spread of scrublands is the core policy of this reserve'? Dynamism lies at odds with management. The smaller the reserve, the more dynamism is seen as a threat. By contrast, larger landscapes can absorb change, and cater for the freestyling habitats on which birds such as nightingales depend. Dynamic landscapes are not linear, but cyclical. Scale allows the space for this to happen and, most of all, the action, within those landscapes, of animals like beavers – evolved better than any other mammal to create and maintain different layers of scrub, in different places, at different times.

The loss of dynamic landscapes lies at the heart of many crushing bird declines. Every time you read about foxes preying on Britain's curlews, that entire conversation is only happening because dynamism has been lost. In large wetlands, waders simply *move* – avoiding the extremes of predator pressure, whereas many predators remain tied to their dens. Like the hawfinches that roost in close proximity to a goshawk, curlews naturally reduce predation, too, by moving closer to the territories of kestrels, which act as a powerful deterrent against members of the crow family.[39] Again, this requires large landscapes that allow curlews the chance to jump ship.

Many of Britain's curlew populations are getting foxed to extinction or their nests raided by crows, one sorry isolated pair at a time.[40] It is only the possibilities of shifting habitats, adapting populations and dynamic landscapes that will buy curlews, nightingales and many other birds a future in the British Isles.

Better One Paradise than a Thousand Scraps

Stories such as those of Avon's last nightingales or perishing populations of curlews, subject to intense levels of predation, illustrate a universal phenomenon, known to scientists as 'stochastic extinction'. This is what happens once a landscape and its birds have fragmented. At this stage, even the *normal* fluctuations acting on a population conspire to wipe them out. The smaller the number of species on an island, the greater the chance that a fluctuation, or 'bad luck' event, will reduce that population to zero.

There are tales of the last birds of populations on New Zealand islets being wiped out by just a handful of cats.[41] The smaller your island and the lower your numbers, the more luck you need to survive. Big populations can absorb a lot more bad luck. Birds like nightingales are designed to take huge losses, but only large populations can *survive* them. Once you are left with a tiny population, everything has to work *perfectly* in order for the species to survive – yet by default, nature does not cater to the perfect survival of a species. Ideal populations of British birds in one place, therefore, should not constitute ten or twenty but at least the low hundreds of pairs. Not only do we underestimate the size of landscapes needed to prevent extinction. We often lose sight of how many birds make a future, and waste critical time, and money, in propping up populations already sentenced to extinction.

With its last sympathetic landscapes now reduced to the size of postage stamps, Britain has become an island of islands. Only the largest of these are proving large enough to sustain populations of birds. If food is why individual pairs of birds are failing to survive, and bad stewardship is preventing the return of our wildlife, isolation is why birds are vanishing from the map – as populations in habitat-islands collapse all at once.

Isolation is why that spotted flycatcher hawking your bee-rich buddleia in a sterile village is a pretty bit of history. Isolation is why the situation for British birds is far worse than most of us might expect, and cannot be 'read' from habitat or distribution alone. It is the piecemeal loss of ecosystem scraps that will ensure the continuing extinction of British birds for decades to come, if conservation does not change. Birds will continue to vanish until we join the scraps back up – into the rich tapestry that nature understands.

CHAPTER 6

Memory

Rebuilding Lost Pathways

> *One couple were so irked by the swallows that insisted*
> *on nesting in their garage each year, they installed a*
> *large replica eagle owl to try to scare them off. It didn't*
> *fool the canny swallows: they just built their nest and*
> *raised their fledglings on the owl's head.*
> —Malcolm Welshman[1]

Each spring, by night, strange gangly birds, from forest clearings used by elephants in the Congo, begin their journey north to meadows in the Hebrides.[2] These birds hate flying – and seem to be pretty bad at it. Yet, somehow, corncrakes make this amazing journey each summer. They trundle and flop through the night sky – too high for owls to reach, and when other birds of prey are sound asleep.

During migration, corncrakes turn up in odd places. They are found in cabbage patches and gardens. They're occasionally seen running around bemusedly in car parks. They can even hold territory for a day or two in suitable meadows. But why on earth, having touched down in a nice little hay meadow, don't corncrakes settle down, call it a day – and breed? Instead, they pass Birmingham. They pass Glasgow. Most corncrakes, in fact, stick two claws up to mainland Britain entirely – and end up, a week or more later, on the remotest outposts of our country: the islands of the Hebrides. But why don't corncrakes just stop at a place that they like? What's the matter with them?

Birds, it seems, often remember only the landscapes in which they were never lost to begin with: the places where they were born. Once range loss has occurred, avian amnesia kicks in. As a result, birds rarely recolonise their former ranges at all, unless they can spread one territory at a time. To this day, the corncrake almost always sings in the meadows that the corncrake remembers.[3] This, alone, makes memory one of the most important factors to understand when is comes to restoring Britain's birds. And to understand how and why memory is so vital for birds, we'll need to delve down into the fascinating world of migration. And to get there, we first need to talk about turtles.

Natal Homing: the Turtle's Tale

We are what we remember. If we lose our memory, we lose our identity, and our identity is the accumulation of our experiences.
—Erik Pevernagie[4]

Night, in Gabon. Cicadas fizz under the stars. With below-average levels of sneakiness and speed, up to 40,000 Olive Ridley turtles slink out of the water.[5] Sea turtles are some of our most ancient reptiles, having visited such beaches almost continuously since the Jurassic, 150 million years ago.[6] Each year, each turtle will lay its eggs on the one beach that it knows.

Sloping away into the purple froth, the slow-moving army soon leaves, each having deposited around 80 billiard balls below the sand. Sixty-five days later, these eggs will hatch. Baby turtles, with surprised, ugly dragon faces, will burst from below the sand. Born on one stretch of sand, turtle hatchlings are imprinted with this location from birth, as they make a short but traumatic journey down the beach into the sea, avoiding predators or getting munched. Sea turtles will return here in the year they reach maturity, the year after this, and the year after that. As a result, these turtle beaches develop longevity that can last for centuries, even millennia.[7]

If you were to 'move' a turtle beach, for argument's sake, your returning turtles would not know how to find it. But how has this come about? How on earth do turtles have the ability to return to the single beach where they were born, having navigated vast areas of ocean in the meantime?

In 2008, scientist Kenneth Lohmann tackled the old mystery of the extraordinary homing instincts of turtles and salmon to the areas where they were born.[8] He picked out salmon, in particular, because chemical cues, used to identify their home rivers, could not account for their ability to cross vast areas of open ocean to get near their place of birth to begin with. Likewise, sea turtles, cast straight from their natal beach into the open sea, had baffled scientists with their ability to find their way back in future years without a map. Lohmann and his team, however, found that both salmon and sea turtles can detect the magnetic field of the earth – and use it as a directional cue.

Sea turtles were found to gain additional information about their position from two further cues. The first was magnetic *inclination angle* – the angle at which the magnetic field lines intersect the surface of the earth. This angle ranges from 0 degrees at the equator to 90 degrees at the poles. The second cue that turtles used was magnetic *intensity*. The earth's magnetic field varies in strength across its surface, being strongest at the poles and weakest at the equator. Both inclination angle and intensity may vary, but they do so predictably. This creates unique magnetic footprints for different areas – areas as precise as a single river system or a turtle's stretch of beach.

Lohmann proposed that geomagnetic cues were behind the natal homing of marine animals, and that, at a critical stage in their early lives, species like turtles are imprinted with unique magnetic signatures that allow them to return to their nesting sites in future years. As Lohmann and his team were careful to point out, however, such strategies gamble on our planet's magnetism being consistent. Big-scale reversals or drifts have, and could again, cause 'widespread colonisation events'. It's fascinating stuff. But where does this leave us with the swallows that find our barns each summer?

'Second Sight': How Birds Find Home

Birds can detect motion extremely well. As they migrate, they map their surroundings visually. But they do not see in the same limited way that we do. The American naturalist Scott Weidensaul likens the task of a returning migrant to facing the dual challenge of orientation (using your compass) and navigation (using your map).[9]

The night sky acts as a map, provided, of course, that the sky isn't covered in cloud. And a day-flying migrant can orientate by the sun or its approximate position, in the event of its being obscured by clouds. Birds thus use the sun and stars wherever they can, creating a broad-scale map that helps point them in the right direction. Hearing, however, can play an important role as well. It has been speculated that birds migrating across America's Great Plains achieve their aim of heading north, across the heart of North America, by detecting in one ear the roar of the Atlantic, and in the other the mountain winds of the Rockies. Still going the wrong way? American migrants are thought to use the seismic trundle of Mexico's tectonic activity as a third orientation point behind them.[10] Sight then returns to the fore as birds near their destination. The land below begins to read as a fine-scale map. Small features, like the welcome spire of an Oxford church, signal home for a swift. The rusty roof of last's year barn – home for a returning swallow.

The staggering revelation in this story, however, is 'second sight'. It has been known for some time that birds can detect the earth's magnetic field. This acts as a vital part of the inbuilt compass – essential to orientation, and very useful if visual navigation fails, as it often does. But birds do not just detect the magnetic field. They actually *see* it. Migratory birds use *two* senses of sight – visual and magnetic – at the very same time.

Back in the 1960s, the pioneering German scientist Wolfgang Wiltschko put robins in a steel chamber, which protected them from the influence of the earth's magnetic field.[11] For a few days, his robins were unable to orientate themselves. However, Wiltschko's ultimate findings were as follows:

> *Robins are not oriented if you keep them in a very weak magnetic field. But if you keep them for longer than three days in such a field, then they can re-orient. And if you then change the horizontal component of this weak magnetic field, so you change magnetic north, then the birds follow this change.*[12]

Wiltschko found similar mechanisms in a long-range migrant, the white-throat.[13] In the 1970s, studies on migratory indigo buntings in the USA furthered his research. Since then, the science of 'magnetoreception' in birds has become accepted as mainstream. And the question has shifted from 'does it happen' to 'how on earth does it happen?'

Four decades on, the world's foremost chemists, biologists and physicists are moving slowly forwards in working out the answer to one of the world's oldest natural mysteries. Magnetite, a mineral that can be magnetised, and therefore attracted to a magnetic point on the planet, was for a long time thought to be a key factor in second sight. This was because many migratory species have deposits of magnetite in their bills.

As it stands, however, it is believed that a bird's ability to detect geomagnetic fields, and navigate using them, is based on a chemical reaction, which creates *subatomic changes in the eye*. This brings about changes that a bird 'sees' and reacts to, in a way that humans cannot. Light striking the retina of a migrant bird's eye stimulates the chemical production of molecules. These molecules share electrons with similar properties: this is known as the 'radical pair reaction'. One key property of the radical pair is 'spin'. This property is affected by the earth's magnetic field. In the radical pair process, magnetism produces a chemical change that allows birds to see magnetic north. The magnetic receptor for this change is in the eye. This makes *light* a key ingredient needed to make the compass work.[14]

By 1992, it was discovered that certain frequencies of light, such as blue and green, aid orientation. In the last two decades, scientists such as Klaus Schulten have begun to show exactly *how* birds can turn a magnetic field into a visual impression. In 2000, Schulten proposed that molecules called cryptochromes were in fact the atomic entity behind a bird's magnetic compass.[15]

This journey of human understanding, however, is just beginning. As Schulten puts it, 'the longer you work on a problem, the harder it actually gets'. That swallows find the same barn each summer is magical in itself. That they do so aided by a quantum compass in their eye is truly remarkable.

In terms of saving birds, what we *do* know is that migrant birds are wedded to tiny pinpricks on the surface of the planet. Their breeding sites cannot simply be shifted around any more than a turtle's can – with no regard for their deep sense of home.

Home is Where the Heart Is

The result of natal homing is that birds 'imprint', at an early age, on the place where they are raised. The processes of imprinting are deep and profound: visual and geomagnetic. Indeed, 'home' is most often stamped on birds' memories from the moment they fly the nest.

In a natural ecosystem, with continuous populations of birds, natal homing has evolved to be an enormous advantage.[16] Imagine if you're a swallow that has returned to the wrong nest in your communal nesting cave. Chaos would ensue. New territories would need to be fought over and new food sources mapped from scratch. Far better to return to your own patch, avoid undue competition, and settle back into your familiar niche.

In a broken landscape such as Britain's, however, natal homing often becomes a recipe for extinction. This is because birds return to ever smaller isolated pockets of land, unable to adapt to the changing environment yet bound to the memory of the one place where they were raised.

The upshot of all this is that you cannot reasonably expect a corncrake to drop into your local hay meadow any time soon and breed – and it probably never will. This is because the world's 'stock' of corncrakes is bound up with an equal number of memories. Your Russian corncrake remembers the little forest clearing where it was born. Your Hebridean corncrake remembers its windswept coastal meadow. Indeed, Hebridean corncrakes are imprinted to make for this area of the globe even as they scuttle past elephants in the warmth of the African winter.

Once large areas of habitat lose populations of migrant birds, the geomagnetic imprint is lost forever, unless a new population of birds is reintroduced. With remaining birds assigned to areas where they have always bred and know to return to, where do your 'spare' corncrakes come from, to repopulate areas from which they have been lost? They don't.

The implications of natal homing extend far beyond what we might imagine. A study of pied flycatchers by scientist Carlos Camacho found that natal homing can even override other kinds of evolution in Europe's birds.[17] Camacho found that pied flycatchers breeding in oakwoods and pinewoods have subtly different sizes. These differences have evolved as a result of adaptation to subtle changes in the way a particular bird uses its woodland. The 'logical' decision, therefore, would be for 'pine-shaped' and 'oak-shaped' flycatchers to seek out their optimal pinewoods and oakwoods on their return from Africa each spring. *Wrong.*

Camacho's team found that birds monitored in a large study returned mostly to breed in the forest patch where they were born, regardless of whether that habitat had changed and become less than ideal for their needs. Only one-third of studied flycatchers later moved to an adjacent patch, the one to which their body type was most suited. For pied flycatchers, the urge to return to their natal site actually trumped the decision to find the place best suited to their needs.

This, in turn, brings us back to swallows. In an endearing story recounted in the *Daily Mail*, Malcolm Welshman describes his conflicted battle with a pair of swallows determined to nest in his shed, as they'd done the year before.[18] Having sealed it shut and given them another site to nest, he found one had snuck through five centimetres of open space, then chased him away as he tried to 'shoo them' into their new home. In the end, Malcolm capitulated: the swallows won. For birds, home is truly where the heart is.

Cultural Memories

Bird culture, as it might be termed, is an often overlooked yet amazing phenomenon, and the local, tactile memories of birds evolve and change over time. It would be a waste of time today to go looking for swallow colonies in British caves. If your swallow was not raised in a cave, then a cave is an alien environment in which to make its home. Swallows gravitate instead towards last year's barn.

If you're fortunate enough to study any one set of birds, over the course of some years, cultural memories begin to jump out from all sides. For the past five years, naturalist Nicholas Gates and I have studied an ancient orchard in Herefordshire, uncovering the secrets of its nesting birds. In this orchard, a lovely bee-filled place in the Malvern Hills, we're fortunate to have at least ten pairs of spotted flycatchers. One female has returned to nest, for the past five years, not only to the same house beside the orchard – but to the same missing brick within its walls. But cultural memories extend far beyond the bird world, and become far more involved and complex amongst the planet's longer-lived mammal families.

Orangutan infants, raised for up to 12 years by their mothers, come to know the nuances of their jungle as they grow up. Each fruit tree, each bee nest, the skills needed to tackle each culinary challenge, all of this is learned over many years. Territories are established, friendships formed, mistakes learned from, and the landscape mapped tree by tree. Many people giving to wildlife charities still consider rehabilitation of animals like orangutans to the wild a noble thing to do, but on spending significant time with researchers in Sumatra last year, I was surprised to learn that such acts can be a death sentence. Money is far better invested in habitat preservation.[19] 'Rehabilitate' an orangutan into virgin forest, and you demand it makes a 12-year learning curve in an hour. Such animals are often displaced, unable to find territories, or make mental sense of their terrain. This is why, as with all conservation, putting things back is far harder than ensuring they stay there to begin with.

In Namibia each year, a tiny number of desert elephants make an epic migration across a dry river valley, the Hoanib, to the coast. They are travelling to eat a bounty of *ana* berries that come out at this time of year, as thick fog rolls in off Namibia's seas and greens the desert plants. But that knowledge is entirely cultural.[20] Today, just a handful of old matriarchs know the way. It has already been shown that other elephant tribes have abandoned this route entirely. Memories have faded with the loss of key females in the group. Memory loss, as well as poaching, is now killing desert elephants. Without knowledge passed down from key individuals, the mental map of where to find food will soon founder forever.

Cultural memory, of course, is more developed again in humans. Almost any aspect of our lives can develop into a culture, one that decides who we are. But we can also change that culture, modify our diet or hairstyle, or even move away from the cultural or religious traditions passed down to us. Birds cannot make such complex decisions – and the extraordinary records of British bird ringing give us the hard proof.

Histories of Home

British bird ringing was initiated by the magazine *British Birds*, and its editor, Harry Witherby, in 1909. Its importance took just three years to realise, for in 1912 a British-ringed swallow was recovered in South Africa. An astounded Witherby wrote, 'that this swallow, breeding in the far west of Europe, should have reached so far to the southeast of Africa as Natal, seems to me extraordinary.' Only a few decades earlier, some had still believed that swallows hibernated at the bottom of ponds.

In the 1930s, the British Trust for Ornithology (BTO) was formed, and the ringing scheme migrated, too. Since this time, each year, the BTO has overseen lots of amazing citizen science. Tens of thousands of wild birds have been ringed across Britain, many as nestlings, by thousands of the BTO's licensed volunteers. The value of ringing is sometimes questioned, because most ringed birds are never recaptured for a second time. This doesn't matter. Not only does ringing reveal to us the survival rate of many endangered birds and the changing dates of their arrival on our shores – information gleaned only through recapturing the same birds – but it also reveals extraordinary stories of avian memory.[21]

Ringing records reveal the extraordinary site fidelity of many beloved summer migrants.[22] A spotted flycatcher ringed at Shimpling village, Norfolk, in 2004, was recovered there eight years later, in 2012. A nestling ring ouzel, raised in the windswept valley of Glen Clunie, Aberdeenshire, returned every spring to the same slope – for up to seven years later. A nightingale, ringed at Gadbury, Worcestershire, was found singing at the adjacent hill-fort, eight years on.

The more reports you read, the more extraordinary that story of home becomes. A stone-curlew raised on a farm near Elveden, Suffolk, was found 22 years later – on a farm just two kilometres away. A common sandpiper chick ringed at Ladybower, in Derbyshire, was found bobbing away there 15 years later – just three kilometres downriver.

Our long-lived seabirds pull top trumps for memory. A puffin, ringed on the remote islet of Garbh Eilean in the Hebrides in 1975, was looking adorable yet concerned there 36 years later. Our pocket albatross, a Manx shearwater, caught as an adult on Bardsey Island in 1957, was going strong on the same island in 2008 – over 50 years later. That's 50 return trips to South America and a global coverage of 1.5 million kilometres in the meantime.

The length of such fidelity often exceeds the length of time most of us live in any one town or village. Today, the average time between household moves in a British home is 14 years. For birds, home is, most often, an investment from the start of life to the very end. Such tiny homing details, easy to overlook, should have a profound impact on the way we think about conservation. Birds have prescribed homes and deep-seated memories for these. This may sound like a romantic concept but it is, in fact, hard science.

In Britain, home has changed, many times, for many birds, especially in the last two centuries. And mostly, once landscape populations of migratory birds have been lost, whether wrynecks 200 years ago or night-ingales today, they have seldom recovered at all.

Stately Homes

The lifespan and size of a bird, its territory size and ability to move through a landscape, are factors in how large and flexible 'home' can be. Our migrant birds of prey, arriving from Africa, offer some fascinating stories of their own. The oldest recorded honey-buzzard in Britain was ringed as a nestling in Shropshire, in 2000. Eleven years later, it was sighted on territory in Wales, 41 kilometres to the southwest. A nestling male ringed in West Sussex in 2001, was likewise seen on territory 11 years later, 22 kilometres away. These may be monster distances for a tiny passerine, but equate to just an hour's gentle flap for a honey-buzzard. In most cases, birds also disperse far greater distances after birth than they will afterwards, as adults, once their home ranges are established.

Adult honey-buzzards, for example, often return to the same wood, sometimes the same nest, year after year. Stephen Roberts, the British expert on the species, has discovered honey-buzzards nesting in woods just a few kilometres away from sites occupied not tens, but over 100 years ago – in areas documented by Victorian naturalists.[23]

In the course of its youth, a Scottish osprey will be displaced considerable distances from its birthplace, as it searches for a territory of its own. One the territory is established, however, ospreys are among the most famous of birds for the longevity of their reign. Birds return from the Gambia each year to their individual ancestral nests – such as the famous eyrie at Loch Garten on Speyside.

Brave New World

It is rare, though not impossible, for migrant birds to recolonise sites 'from scratch'. But when they do, they face formidable odds.

Red-backed shrikes have, occasionally, broken the memory rule as a vanished species. From 2010 onwards, a singing male, first stopping in a beetle-rich tract of Dartmoor in 2008, eventually found a mate. In spite of successful breeding, the chance of a foundling pair, of a species with a 30% annual survival,[24] creating a viable population, is remote.

Our position as an island also makes recolonisation far harder. In the Netherlands, 100 pairs of red-backed shrikes have recolonised a large reserve from a nearby population in Germany, moving slowly over across tracts of amenable land. Surrounded by water, Britain cannot hope for such a stroke of luck.

With migrant birds that haven't abandoned Britain entirely, the 'stepping stones' approach becomes critical to future conservation. If, for example, you expand suitable habitat in a Kentish scrubland, you could have a dramatic impact on nightingale recovery, because birds could slowly colonise across the coming years from adjacent sites. But create suitable habitat in somewhere like Herefordshire, where nightingales have been gone for just a couple of years, and your chances are close to zero. Nightingales are no longer heading to this magnetic signature on Earth. Unless they are reintroduced here from scratch, or can colonise from nearby, then the magnetic beacon of 'home' has forever been extinguished.

Resident Prisoners

No impulse exists in our resident birds to cross hostile habitats and find a better home. A capercaillie, for all its pomp, will not go blustering through the towns of Scotland in search of a new forest. Once confined and isolated, the forest where it was born becomes the forest where it dies. Such birds can only be restored through reconnecting habitat, or reintroducing birds from scratch. This is why golden eagles will most likely never dance over Wales or Dartmoor in any of our lifetimes – until conservationists put them back in.

On a smaller scale, willow tits are a species whose range has dramatically contracted in recent decades. Right now, the scrublands of Somerset's new wetlands would perfectly suit willow tits. But with no birds left in the entire county, or for many miles around, there is no impulse in willow tits to take the train from their nearest breeding stronghold: Wigan.

The BTO's ringing data reveal quite how stationary many of our resident birds can be. The vast majority of willow tits ringed travel less than a few kilometres from their hatch sites in their whole lives. Most live and die in the stretch of rotting trees where they were raised. In 2015, three birds, recaptured by the BTO, were found to have moved 'zero kilometres'.

Small woodland birds are among the most constrained in terms of their knowledge of home. Whether you're born and bred in Tyne and Wear or fly from the Congo each year, both our migrant and resident birds are confined, and endangered, by their finite knowledge of the world. Farmland birds, too, often stay put till the bitter end, with the oldest corn buntings and yellowhammers on record having moved zero kilometres in their 8 and 11 years. Ringing reveals that the oldest house sparrow on record spent 12 long years in Pontypool. Our resident birds may move around, having larger territories in winter – but most generally, they too stay put in the same area for their whole lives.

1. Constable's England? This is, in fact, Brzostowo village in eastern Poland. Small herds steward a landscape crawling with invertebrates (Chapter 3). Andrzej Gorzkowski Commercial / Alamy Stock Photo

2. Dairy desert. A typical cattle lawn in Sussex, with not a wryneck, red-backed shrike or cuckoo in sight (Chapter 3). LatitudeStock / Alamy Stock Photo

3. It's all in the soil. A black-tailed godwit on a fence post in the farmland of Texel, in the Netherlands (Chapter 3). AGAMI Photo Agency / Alamy Stock Photo

4. Food flow. The rich woodland glades of Hungary's Bukk Hills (Chapter 3). Rob de Jong

5. Europe's Serengeti. Letea Forest, in Romania's Danube Delta (Chapter 4). Nature Picture Library / Alamy Stock Photo

6. Wild parkland. Dawn in the Kanha National Park, India (Chapter 4). GM Photo Images / Alamy Stock Photo

7. Scrub management. The rich scrub-grasslands of the Serengeti, Tanzania, with zebras leading the management regime (Chapter 4). Anca Enache / Alamy Stock Photo

8. Forgotten stewards. Wild horses graze a wooded meadow in the Chernobyl wilderness, straddling Ukraine and Belarus (Chapter 4). kpzfoto / Alamy Stock Photo

9. Rivers in charge. In Poland's Biebrza Marshes, winter flooding provides an enormous stimulus for regrowth in the following spring (Chapter 4). Benedict Macdonald

10. Wild coast. A coastal plain in the Netherlands – a landscape driven by water, succession and grazing animals (Chapter 4). Europe-Holland / Alamy Stock Photo

11. English wilderness. A fledgling 'Serengeti' of our own, the Knepp Wildland Estate in Sussex (Chapter 4). Charles Burrell

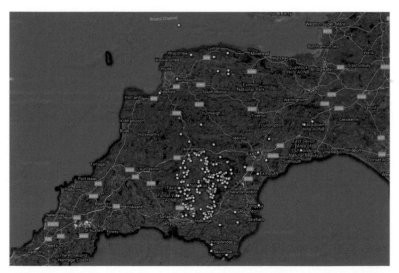

12. Last stand. Cuckoos were once birds of the wider countryside, but they have retreated into the wooded grasslands of Dartmoor (Chapter 5). Each dot on the map represents a sighting of a cuckoo. Mike Daniels / Devon Birds / Google Maps

13. Wild Scotland. The regenerating woodlands of Alladale (Chapter 8). Nature Picture Library / Alamy Stock Photo

14. Singing in the shade. The kingdom of the nightingale, Bradfield Woods, Suffolk (Chapter 9). Steve Aylward & John Ferguson / Suffolk Wildlife Trust

15. Let it rot. A maze of dank, rotting floodplain trees – one of the last UK strongholds of willow tits, Dearne valley, South Yorkshire (Chapter 9). Geoffrey Carr

16. Hazel maze. The dense joinery of low branches used by marsh tits in a hazel coppice in the Cotswolds (Chapter 9). Bob Gibbons / Alamy Stock Photo

17. Deer desert. A conservation woodland in Gloucestershire's Forest of Dean (Chapter 9). Colin Underhill / Alamy Stock Photo

18. Lost forester. A primitive-breed cattle bull wanders through the Letea Forest, in the Danube Delta (Chapter 9). Nature Picture Library / Alamy Stock Photo

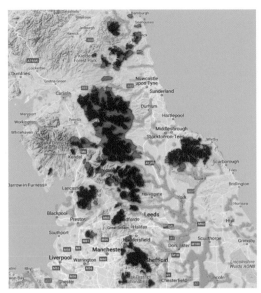

19. Bleeding out. The extent of grouse moors in England (Chapter 11). The orange areas show Moorland Association data; the red is a Friends of the Earth best estimate of the area specifically covered by grouse moors; the blue shows historic flooding. Source: https://friendsoftheearth.uk/page/map-grouse-moors-england. Adam Bradbury / Friends of the Earth

20. Grouse farming. This heather moorland in Aberdeenshire is burned in rotational patches (Chapter 11). Nicholas Gates

21. Mosaic moorland. Juniper and birch mingle with heather on the Rothiemurchus Estate in the Highlands (Chapter 11). Nicholas Gates

22. Moors without fire. Store Mosse, in Sweden. Here, with no heather burning, 'moorland' birds survive in greater diversity than on any grouse moor (Chapter 11). Martha Wägeus / Store Mosse National Park

23. Sustainable grouse moors. A man and his dog hunt willow grouse in Sweden (Chapter 11). Arterra Picture Library / Alamy Stock Photo

24. The not so famous grouse. A willow grouse, the same species as Britain's red grouse, creeps through a natural heather glade in Scandinavia (Chapter 11). Nature Picture Library / Alamy Stock Photo

25. Pelican possibility. A remote colony of Dalmatian pelicans nests within the safety of Romania's Danube Delta (Chapter 12). Stelian Porojnicu / Alamy Stock Photo

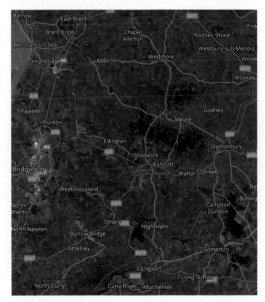

26. A question of scale. The red area sketches out an area of low-lying, unprofitable dairy farmland in Somerset that could be transformed into a wetland ecosystem large enough for pelicans (Chapter 12). Benedict Macdonald / Google Maps

27. Living with nature. A house in a Hampshire village, 1914 (Chapter 13). The Francis Frith Collection

28. Sterile Britain. The same Hampshire house as in Figure 27, photographed in 2017. Now cleansed as a result of ecological tidiness disorder (Chapter 13). Benedict Macdonald

29. Wild village. Zywkowo, a 'stork's village' in Poland (Chapter 13). Hans Winke / Alamy Stock Photo

30. Beavers know best. Without any of the usual tools in the UK conservation toolbox, beavers create a diverse wetland free of charge on Tayside (Chapter 14): a) 1 year after beaver release, and b) 12 years after beaver release. CC BY 4.0 Adapted from: Law, A., Gaywood, M.J., Jones, K.C., Ramsay, P. & Willby, N.J. 2017. Using ecosystem engineers as tools in habitat restoration and rewilding: beaver and wetlands. *Science of the Total Environment* 605–606: 1021–1030. https://doi.org/10.1016/j.scitotenv.2017.06.173

Impressive annual migrations take place within Britain each year, with the movement of birds from our coasts to inland breeding grounds each summer. But these also follow the limits of memory. A curlew nestling, ringed in 1994 on Embsay Moor, Yorkshire, was found wintering on a stretch of the Severn in Gloucestershire. In 2010, it was re-found on its breeding grounds at Low Bishopside – 20 kilometres from its birthplace but on the same continuity of moor. Like many endangered species, curlews have a dual memory, limited by their experience. Their breeding and wintering grounds are often prescribed from birth. Satellite-tagging suggests that our African migrants have a 'double map' as well. At one end, our Loch Garten osprey sits on the same nest year after year: a magnetic atom in the heart of Scotland. At the other, it returns to the same few beaches in the Gambia. Memory rules both ends of a migrant's journey through the world.

As a result, the ability of birds to fly, and to migrate, does not equate to their ability to recolonise our island. Each year, black terns dip their bills into lovely East Anglian fenlands, suitable for them to nest in, yet each is headed to a remembered home in eastern Europe. Right now, across Britain, prime hay meadows lie waiting for corncrakes, wood-pastures for wrynecks, fallows for whinchats, scrublands for red-backed shrikes and wet meadows for black-tailed godwits. And the sad but proven 200-year reality is that these birds very rarely make a return by themselves, in sufficient numbers to stand the test of time.

Even if our habitats are restored on a massive scale, memories, too, must be rebuilt. Home must be relearned. And conservation must change. What is the final piece of the puzzle in the battle to save our wildlife? Reintroduction. Reintroduction on a massive scale.

Back from the Dead

How many of us pause to think about the role played
by direct human intervention when watching a
Capercaillie lek in the Scottish Highlands.
—Ian Carter[25]

Imagine, for a moment, you were told of a National Health Service, founded in early Victorian times. Gradually, as the decades rolled by, it came to cater for fewer and fewer people. More and more people died, death rates went up, until, at last, by the time you were alive, it looked after just a few people, with a few illnesses, in tiny proportions of the country. You would be outraged. History should not run backwards. Countries should not degrade over time.

This is, however, exactly how we accept the diminished status of our wildlife. We applaud one pair of red-backed shrikes on Dartmoor; the Polish enjoy them as a common bird. Birdwatchers recently raced to see a lost Dalmatian pelican drifting around in Cornwall. With sufficient vision, they could fly over Somerset one day.

If there is a 'moral' argument in reintroductions, it is perhaps a rather simple one. We owe it to future generations to put back the glorious birdlife that farm workers of the eighteenth century could enjoy – but our educated and improving society cannot. Some of the birds we are missing are simply embarrassing. What does it say for the state of our countryside that we have no wrynecks, a bird whose key requirement is old trees and anthills? It's time to ask for and expect more.

Mass reintroduction must follow the restoration of the British landscape. It is the only way that many birds will ever make it back. But in some areas, habitats have *already* recovered – and now simply await their birds to make them more than pretty, empty landscapes.

We live in a country with Europe's largest conservation organisation, an inspiring amount of goodwill to wildlife, and amazing scientific minds that produce 16% of our planet's high-quality research from just 1% of its population.[26] If we want to effect mass reintroductions, we can. Fortunately, British conservation, when committed, has excelled in this field. We have become, in some cases, the most pioneering country in the world at carrying out reintroductions. So let's take a look at what's happened so far. The story of rebuilding memories goes back a surprisingly long time. And it starts with the last bird any sensible person would ever want to transport. Anywhere. Ever.

1837: Capercaillies Uncorked

In 1785, the last native capercaillie fell dead, shot, on the bilberry carpet of Ballochbuie Forest in the Highlands. At a time of massive deforestation and intensive hunting, our last capercaillies were hunted out in little islands of remaining pine. In 1837, however, a Scottish nobleman would successfully return the world's largest and most cantankerous grouse to Scotland. This remains the most successful grouse reintroduction in world history, and it's thanks to this man that we can still enjoy the cork-popping display of this improbable British bird.[27]

This first attempt to reintroduce capercaillies to Britain was fraught with difficulty.[28] In 1807, capercaillies released on the Isle of Arran choked to death after the stress of being handled. Other introductions failed – when all the birds were eaten. After more failures in 1822 and 1824, on two different estates, the odds were stacked against the Marquess of Breadalbane when he attempted again in 1837. His eventual success came down to two simple tactics.

Breadalbane released young capercaillies at separate but connected woods in Perthshire, mitigating bad things happening to any one population, and predicting that they would, if successful, join up over time. He also culled ground predators fiercely,[29] whilst the birds oriented themselves to their surroundings. In Victorian times, this kind of control was fairly common. In this case, there was an irate, turkey-shaped silver lining to show for it.

The fact that someone with no formal conservation training carried out Britain's most long-lasting reintroduction to date is impressive. It is also a reminder of the positive influence large landowners can sometimes wield, when they wish, to the benefit of all.

1975: Big Birds Come Back

The white-tailed eagle became nationally extinct in 1918, and, following a failed reintroduction attempt on Fair Isle in the 1960s, a second reintroduction to the Isle of Rum was initiated in 1975. With the nearest population in Scandinavia, history had proved that birds were unlikely to arrive of their own accord – let alone in happy couples. The story of the eagle's return is well known and inspiring.[30] As of 2010, white-tailed eagles occupied 94 ten-kilometre squares in Scotland.[31] In 2015, a magical milestone of 100 breeding pairs was passed. With an uncanny sense of drama, the centenary pair nested on Orkney – for the first time in 142 years. In 2019, an even more historical milestone was passed, when white-tailed eagles reintroduced to the Isle of Wight got itchy talons and began to tour lowland Britain, darkening English skies for the first time in over two centuries.

English ospreys have been restored to Rutland, in Leicestershire. At the time of writing, a move to rewild golden eagles in southern Scotland is under way. Ten reintroductions of red kites later,[32] and Oxford skies are filled with red ribbons. Raptor reintroductions have proved a consistent success and a huge boost for ecotourism. Most of all, many lessons have been learned.

With any reintroduction, there is the political side. This comes down, largely, to can we and should we – and who will benefit? There is the technical side – how do we keep chicks alive, prevent early mortality, ensure birds don't become imprinted on people or snaffled by predators? There is also the ecological side – how will birds spread? What threats will pairs face and what limitations – such as food or nest sites, landscape size, persecution – could constrain a population? Is this an attempt to regain an original home range, or to adapt the species to new areas where it may thrive instead? Britain's ornithologists have now become adept at predicting these issues.

It was learned from white-tailed eagles that if you have masses of suitable habitat, you can release territorial birds into small areas, such as Rum or Wester Ross,[33] and they will naturally disperse. It was learned from red kites that multiple releases improve your chances. Over time, populations can connect, providing insurance for the future.

In recent years, the release of great bustards in Wiltshire has shown how many years such large and long-lived birds can take to establish. But it's also shown how tendencies to disperse in winter, and return to nest in the same places next summer, become hardwired into reintroduced populations of birds over time.[34]

In May 2014, Britain passed another big bird milestone, with the hatching of the first wild crane chicks in western England for 400 years, as the result of a reintroduction scheme. Already well established in the Norfolk Broads thanks to the work of the late John Buxton, there was no sign that cranes would regain their memory for the West Country in the near future without human assistance. The Great Crane Project aimed to change this, and the project got under way from the home of modern conservation, the Slimbridge reserve of the Wildfowl and Wetlands Trust (WWT). The aim was simple. Cranes raised in 'crane school', free from human imprinting through the use of crane puppets, would be taken to the Somerset Levels and released into the habitat awaiting. Three pairs of crane, however, had slightly different plans – and headed straight back to Slimbridge to nest.

If there is any lesson learned here, it's that the hardwiring of natal homing springs into play very early. But come the end of 2017, nobody minded the cranes' errant ways too much. By now there were over 20 pairs of crane holding territory in south Wales, Oxfordshire, Gloucestershire and Somerset.[35]

On 15 May 2020, at the height of the Covid-19 pandemic, an event that would not have been contemplated even a few years before took place on the Knepp Estate in Sussex. A pair of tree-nesting white storks, one of three established here through the Project Stork reintroduction scheme, hatched chicks – for the first time in over six hundred years. Whilst derided by some experts, who had apparently forgotten that storks once graced our wetlands and villages, the scheme moved the science of reintroduction further again. This time, by releasing first-year or older birds onto the site, the project has rekindled the stork's instincts to migrate. Knepp's storks began migrating to Spain and Morocco, as other populations do, before returning to build a nest – not on the site's carefully constructed stork platforms, but in the sturdier canopy of veteran oaks.

Across the UK, bird reintroductions are slowly, like bustards, taking off. These schemes are proven, glamorous and exciting; they attract huge publicity, and put good emphasis on the wider landscape. Most importantly, when it comes to large birds that generally remain resident in Britain throughout the year, we know what works. And that's an important start.

2003: To Africa and Back

Large resident birds are robust and have a long period of time to learn, map their surroundings and take possession of their habitat. But what if you are a small, dumpy, short-lived bird, and must fly to southern Africa – and back – each year?

Assessing the future of the corncrake in England, the RSPB was faced with the memory problem in spades. Starting in the early years of the twentieth century, corncrake chicks killed by mowing machines in their natal field never reached Africa to begin with, so it's something of a truism to say they were never coming back. The corncrake's memory map for most of its haunts in England had been lost.

The RSPB set its finest minds the task of achieving a global first: the reintroduction of a migrant bird into the Nene Washes in Cambridgeshire – a wholly new start, a wholly new set of memories. For the project to succeed, birds would be required to fly to Africa – but then return the following summer. Corncrakes gave one edge to the RSPB. Like waders and ducks, baby corncrakes are nidifugous: they 'flee the nest', hatching and running soon after birth. This meant that rather than have to work out how on earth to set up a 'nest', the RSPB, working with Pensthorpe Natural Park in Norfolk, Natural England and the Zoological Society of London, was able to slip baby corncrakes, raised in captivity, into suitable fields, at a suitable age.

What was crucial to get right was the memory part of the process. Chicks were moved to pens on the Nene Washes, the release site, aged 12–14 days. Any later, and the corncrakes would have imprinted on Pensthorpe, navigating back to the wrong postage stamp of the planet next summer. Once imprinted, it was sensible to keep the birds in pens until they could fend for themselves. The birds released were at 35–40 days old.[36]

The RSPB's calculations were critical – and correct. The other end of the 'map' – Africa! – was so deeply imprinted that the released corncrakes did their thing, heading south for winter and then back the following year – right into the fields where they had been released.

At first, things were slow-going. In the first few years, only corncrakes from the release scheme, and their chicks, came back to the site. Then, in 2008, one of the corncrakes brought a mate from Poland, and more wild-bred birds have joined them on migration since.[37] Corncrakes are not only migrating back from Africa from scratch – but recruiting. Now, the RSPB has rekindled the beginnings of a dynamic migration system.

Outside the 40-kilometre length of the Nene Washes, the desert of modern fields locks these corncrakes in, preventing their expansion. But that doesn't matter. What can be done once can be repeated – in ever larger and wilder areas of land.

2004: Chick Transplants

The next issue for the RSPB was to tackle the relocation of birds that do not flee the nest, but are raised in one till fledging by their parents. This would take reintroductory science to the next level again.

Cirl buntings, their memories confined to Devon hedgerows after a huge range retraction, were unaware of habitat work being done separately in Cornwall. Whilst cirl buntings within their Devon range had been able to spread, one stubbly farm at a time, a leap into Cornwall wasn't going to happen by itself. This reintroduction therefore posed a wholly new challenge to the RSPB. It didn't involve sending a gangly bird on holiday to Africa with a return ticket – but it did require moving tiny fragile birds into a whole new county.

The target area for the reintroduction was the Roseland Peninsula in Cornwall. The tactic of choice was 'rear and release'. In 2004, the RSPB and its partners removed chicks from nests in Devon, hand-reared them, and released the fledged young back into the same site. This was partly a test – to see if the birds were able not only to survive the winter, after being fed by surrogate 'mums', but to rejoin the population the following spring. They did. In 2005, the RSPB adopted the same approach, taking small chicks and feeding them, but released the fledged young into Cornwall. Between 2006 and 2011, a minimum of 60 birds were released annually.

The RSPB anticipated the critical mass needed to sustain a population, to prevent it petering out.[38] The first breeding took place a year later, in 2007. By 2011, the population rose by 75% as 28 pairs bred; the young they fledged meant that the wild population in Cornwall now outnumbered the released parent stock. By 2015 there were 50 pairs – a self-sustaining population and one that continues to grow.

The potential of such science for national bird restoration is enormous. Over 10 years ago, the RSPB cracked two key memory milestones for rewilding. Now, it's time for such exceptional events to become a successful conservation routine.

An Outstanding National Debt

The RSPB only advocates reintroduction in situations where natural re-colonisation is not possible through other measures.
—RSPB, 2017[39]

Since cirl buntings, the cutting-edge momentum of reintroduction science has stalled, at a time when advancing such science is most desperately required. Conservationists have reverted to more kites and eagles, but not tackled the next hurdles, prepared for probable extinctions to come, or developed the technology needed to reintroduce migrant passerines from scratch.

There's no doubt that tackling the mass reintroduction of resident and migrant songbirds, and lost species like red-backed shrikes, wrynecks or black terns, is a daunting prospect indeed. There are technological challenges and cost implications in moving towards a normalised 'rewilding' process for birds.

But there are mental barriers too.

One objection I have often heard to reintroductions is the baffling idea that reintroductions 'play God'. But if you managed to get through my first two chapters without dozing off, you'll have noticed one consistent narrative. Our country, more than any other on earth, has been fundamentally depleted, tamed, and re-ordered, often dozens of times, by a Land God – ourselves. 'Playing God' is not a choice: this is the Anthropocene. We are in charge, bound to play God whether we like it or not. Deciding to restore ecosystems and hand them back to animals – playing God. Deciding what land to manage, what species get the most funding – playing God. Digging a scrape for avocets – playing God. Choosing pure heathland, for Dartford warblers, over scrub, for nightingales – playing God.

It is only *how* conservationists play God that matters – on what scale, and with what vision. As humans, we have taken away the homes of thousands of other species that once bred on our island. Conservation has accepted, by and large, that the birding 'baseline' has changed, grown to accept ever less, and aspires often towards the birdlife of just a generation ago. But for centuries, each generation has seen less and less. We are still, all of us, being robbed.

Current thinking is often that reintroduction acts as a 'last resort'. The more pertinent question would be, a last resort to what? Most birds cannot return to a home range, or landscape, unless they were raised there to begin with. Many can never effectively reconquer Britain on their own. Colossal range retractions cannot be reversed, once the magnetic glow of home has been snuffed out.

Right now, no one is preventing wrynecks being returned to raid birch woods in northern Scotland, black terns restored to the Cambridgeshire washlands or marsh warblers set loose to improvise jazz in our southern river valleys. The obstacle is an amnesia for the rich world that we have lost – one that our children should have every right to enjoy. Having played God for millennia to remove our wildlife, it's time to honour our role as the species in charge of the planet. So let's start normalising the exceptional – and plan the pathway to rewilding birds one memory at a time. This way, for the first time since our settlement in Britain, we may leave the next generation of our children wilder – and richer – than the last.

There is, however, one final factor, operating at a scale far larger than our beloved island home. It's beyond all of our control. It's beyond any British conservationist to stop. Yet if we begin to restore our ecosystems, it can be resisted – and worked with. We are talking, of course, about climate change.

The Mountain Sirens

Each summer, a single father shepherds his family across a stony hilltop. If we attributed to this father human emotions, he might be a bit cheesed off. His wife is sunning herself on the warm mountain slopes of Morocco, enjoying sunny days and varied cuisine, having left the kids at home weeks before. Dad's been left with the kids on a Scottish hillside – to face the most hostile summer weather in Britain.

Fortunately, male dotterels don't see things this way. After females lay eggs on the highest plateau in Scotland, the male incubates the eggs, and hatches and cares for his young, before leaving with them for the Atlas Mountains in August. Dotterel in Gaelic means 'fool of the moor'. In long-gone hunting days, dotterels showed remarkably few wiles – and proved very easy to catch. Theirs is a rare world, where humans are so scarce that dotterels are still tame.

Dotterels, along with ptarmigans and snow buntings, are our Arctic breeding birds. Now, man-made climate change is driving these hardiest of birds further north again. Soon, they may be driven out of Britain entirely. The dotterel's retreat conforms to climate change theory. They are moving uphill, and northward, as the climate warms. Their prey, a mountain crane fly, *Tipula montana*, is, like the dotterel, adapted to the cool.[40] Food shortages are believed to be playing a role in dotterel decline – a decline as little seen by most people as the mountains where they nest. From almost 1,000 breeding males in 1988, there are now fewer than 430. That number is still falling. And Dotterels are declining not just in the Cairngorms but across northern Europe, with a 40% decline in the mountains of Norway.

A suite of upland birds is moving north as Britain's climate warms. Ring ouzels, our mountain blackbirds, face a similar predicament. Like dotterels, they arrive from the Atlas Mountains each summer. Like dotterels, they are declining, moving up the hills and pushing north in Britain, toward the very rooftops of our mountains. On Ben Macdui summit, in the Cairngorms, you can still share sandwiches with breeding snow buntings in the summer months. Enjoy them while you can. As Britain warms, these tame little snowflakes will melt. As you breach the summit of Cairngorm, look for a surprising carpet of green moss. This is dotterel land. Be careful not to tread on a single father.

Climate change is not a 'threat'. It is real, fully under way and beyond the control of any one country. Wise mitigation of climate change in Britain must begin with the restoration of our landscapes. Climate change is such a large-scale threat to biodiversity that it can only be countered by rebuilding the ecosystems we have lost.

Protecting a tiny hillside reserve for a bird like a ring ouzel is fighting time and climate itself. Such a northerly species will simply move uphill, meaning you have designed and maintained your reserve with no regard for dynamic change. Restoring wilder uplands, where trees, and birds, move and recolonise in response to a changing climate, is far more likely to leave our wildlife with a future. The same is true in our lowlands.

But how do we get our birds and landscapes back? Where do we put the wildlands that we all want to see, and that we want our children to enjoy? Do we, on such a 'crowded' island, have the space to even try?

Now, it's time to talk about a future. More wildlife, more wild landscapes – and more jobs. And for such a discussion it seems logical to start with the places supposed to provide just that – our national parks.

CHAPTER 7

A Wild Economy

Making the Most of National Parks

*'Let nature be nature!' is the philosophy of the
Bavarian Forest National Park. Nowhere else between
the Atlantic Ocean and the Ural Mountains is such a
large area of forests and moors allowed to grow and
develop without the interference of man.*[1]

Now here's an interesting thought. We have all the space we need in
Britain for nature. We are *not* a crowded country. With over 82% of
the population of Britain living in urban areas,[2] there are vast swathes
of depopulated empty space. When did you last walk on Dartmoor,
Snowdonia or the Cairngorms and find yourself surrounded by road
traffic and the looming shapes of buildings? These are large empty areas
in which we could rekindle ecosystems, restore our birds and massively
enhance our economy. At present, however, they are the largest ecological
and financial waste of space in Europe – Britain's national parks.

In fairness, our national parks were never established as areas for the
preservation of nature, nor for their biological or economic riches. They
were, instead, established as cultural parks, with some nod to nature, and
a far larger recognition of landscapes immortalised in poetry (however
denuded those landscapes might be). In this, our national parks have
partially succeeded. The drystone walls are still standing. Architectural
heritage and iconic farmland landscapes have been protected from
insensitive development. But no other country, with such enormous
areas of depopulated space and a national passion for the natural world,
considered this model a good idea in the first place, nor the most
economic. And right now, our national parks are not only nature deserts,
compared to their potential, but jobs deserts as well.

Whilst Poland's national parks, with floodplains and bison, profit-
making birds and lynx-stalked woodlands, cover just 1% of its land mass,

England's national parks cover 9% of its surface. In Wales, that figure rises to 20%.[3] This should mean that rather than having the most degraded wildlife in Europe, we should actually have among the best. Turning to the largest depopulated areas in Britain, therefore, would seem a useful place to start, in the quest to restore our nation's lost wildlife.

The website of the Northumberland National Park states, twice, as it runs out of wildlife highlights, that it forms an area of 'dark sky'. The Brecon Beacons National Park website reminds us of the same.[4] Dark sky means a lack of streetlights. A lack of streetlights means a lack of settlements. England's national parks contain just 0.6% of its population, Wales, 2.7%.[5] It is truly exciting that our national parks harbour dark skies, because dark skies point to an absence of people and the possibility of an enormous resurgence in nature.

Quite how large our national parks are, quite how sterile and robbed of treasures, only becomes apparent when we compare them to those in other countries. Yellowstone National Park is famous for its mosaic ecosystem. At the time of writing, 95 wolves, in 10 packs, hunt 4,000 bison and 30,000 elk, whilst 500 brown bears and 280 black bears plant trees by defecating seeds and fertilise the land by dropping fish. Canada lynx, bobcats, otters and beavers all shape the ecosystem in smaller ways. Of Yellowstone's 300 regular resident, wintering and breeding birds, the 'alpha' line-up includes eagles, pelicans, cranes and ospreys. Yellowstone is large, for sure – it covers an area of around 8,990 square kilometres.[6] Scotland's Cairngorms National Park is smaller. At over 4,500 square kilometres, however, it's still *half* the size of Yellowstone: a huge, depopulated area with national park status. It doesn't, however, have even a quarter of Yellowstone's wildlife, natural landscapes or associated job opportunities.

The Lake District National Park, recently awarded UNESCO World Heritage status, predominantly for its upland farming landscape,[7] comes in at 2,362 square kilometres.[8] Snowdonia, as devoid of native ecosystems as of eagles, covers 2,130 square kilometres.[9] By contrast, Kenya's Maasai Mara covers a mere 1,800 square kilometres.[10] Now, that's quite a thought. The Mara, contrary to television depictions, is in fact highly populated. People and wild animals share the landscape, sometimes uncomfortably. But nonetheless, this tiny area of the world's surface harbours giant grazers, apex predators and over 400 species of bird in its grasslands – all within an area smaller than Snowdonia.

The North York Moors National Park, its burned hills managed as grouse farms, its forests as crops of spruce, covers over 1,400 square kilometres. Poland's Biebrza Marshes National Park, with its freestyling rivers, wolves, elk, white-tailed eagles and almost every vanished or vanishing British lowland bird still thriving, covers just 590 square kilometres.[11] The North York Moors has *three times* the space of Biebrza for wildlife, nature safaris and thriving ecotourism. Yet in its spruce and heather deserts, only the relics of life survive.

Worldwide, the International Union for the Conservation of Nature (IUCN) ranks protected areas on a scale from one to six: one being wilderness, of the kind seen in the world's most cherished national parks from the Serengeti to Yellowstone, and six being areas where the best that can be said is that resources are responsibly harvested.[12] Britain's national parks make it as far as a *five* – culture parks of scenic value. We do not even have one park that achieves a four.

One comment you will often come across is that these rural spaces are where Britain grows its food and feeds its people. This is entirely untrue. Only the South Downs National Park is dominated by arable crops. The rest of our national parks are sparse and unprofitable farms of sheep, more intensive crops of timber, and entirely optional hunting factories for grouse and deer.

Britain is *blessed* with space. But why is space in our country so wasted? Why can highly developed countries with thriving economies, such as Germany, enjoy profitable lynx, wolves and national parks left to nature and the bounties that ecotourism provides, yet we cannot?

Deserts of Opportunity

In many ways, our national parks are not national – and they are not parks. We can, most certainly, all visit them for free. Yet, on any close inspection, most are designed as timber-growing areas, scenic farms or hunting grounds, with ample space for people to inhale fresh air. Britain's Big Six crops – cereals, dairy, forestry, sheep, grouse and deer – are as dominant inside our national parks as they are outside them; a situation which to any American conservationist would seem absurd. Most of our national parks are factories. To ring the changes, the output of these factories changes as you move around the country. Several of the most promising areas for nature reconstruction in England and Wales are meeting places between intensive farms of spruce and massive areas of sheep-grazing. The Brecon Beacons, the Lake District and Snowdonia fall into this category.

On the purple heather spines of northern England, the factories change their output. Intensive grouse crops vie with sheep and forestry for space. In this category, you have much of the northern Peak District, the North York Moors, the Yorkshire Dales and Northumberland National Parks. Sadly, we have no 'spare' English uplands where natural beauty and its economic potential has space to develop. This is all we have.

Look at these national parks on an aerial map, and you can, for yourself, soak in the square scarring of bald purple hills, seen nowhere else on earth. That is the scorched heather of the grouse moor. You can trace the deep greenery of same-age spruce, or the pale green of sheep pasture. Welcome to your northern English national parks. Don't rush to visit them. Better to feel alive.

Among the moors, the postage stamps of Yorkshire's industrial valleys teem with far more life. Here you can hear willow tits buzzing away in thickets, watch long-eared owls hunting old grasslands at dusk, and enjoy courting black-necked grebes, flashing their red and gold ear-tufts on ponds that swallowed industry decades before.[13] Or you could visit the Farne Islands, with terns diving ferociously onto your head as eiders coo with quilted disapproval. Here you will not need to be sold the idea that these places are rich in nature. You will, instead, be immersed in the fullness of our deafening seabird cities. When you feel the wild, it becomes far easier to realise how you are being robbed of it elsewhere.

Further north in Britain, Loch Lomond and the Trossachs National Park pushes life to the very margins. In old pines beside the water's edge, you can find ospreys. The evocative calls of black-throated divers haunt the larger lochs. Little birch woods trill with willow warblers and redstarts. But the wider landscape is desolate. In 2017, I drove for miles past what resembled the result of a bombing raid on a timber factory. Huge areas of the park were being harvested for non-native trees, and these crops are visible, too, on the satellite photos.[14] Other areas were extensive sheep farms, taking up vast areas relative to their economic output or remaining social importance.

In southern England, some parks are quieter yet. Exmoor's ancient woodlands are full of charm and mystique, but so small and isolated that birds vanish from them each summer.[15] Its uplands are drained of grouse, birds of prey and breeding waders. A survey of Exmoor undertaken as long ago as 2012 found that 'half a pair' of curlews might remain.[16] Until recently, Exmoor's ponies were proven to create diverse habitats for wading birds. Now, large areas simply lie neglected. Sodden bracken and silence are in charge. Without natural stewardship, there is only absence – absence of curlew song, absence of resurgent trees, absence of everything but wildlife decline. Extraordinarily, each year, many of these hillsides are burned in rotation, destroying insect life and the basis of a food chain. Few uplands tantalise more than Exmoor. It appears wild. It is simply empty.

The South Downs National Park apparently celebrates the 'rolling' nature of the countryside. The 'rolling' effect is produced by the removal of nature. Farming crops is vital for our survival – but it is neither rare nor special. Surely a national park might be expected to conserve something more: something endangered, wonderful; something in need of actual protection.

Glimpses of the Wild

Some of Britain's national parks offer more sense of how a future could look. The New Forest is the only national park where a significant reminder of a rich ecosystem can still be found. Forestry blights a sizeable proportion of the woodland. Unchecked deer herds prevent the nectar layer and bramble from feeding the trees with caterpillar food. Heathlands are burned for no reason; horses are there to keep the heathlands open. But the 'core' of the New Forest has survived: a mosaic of rich ancient woodlands – the very best we have – where free-roaming herbivores still shape and maintain the land.

The Cairngorms, holding almost all of our capercaillies and a rich array of other precious wildlife, are a cautious beacon of hope for the future. Most of Scotland's native pine pastures grow here and could expand over time. In the Spey valley, more dominated by woodlands than grouse moors, the prospect of returning predators such as lynx, and woodland-edge species like elk, is far closer to being a reality than elsewhere. Inspirational long-term initiatives, such as the Cairngorms Connect project, and the work of pioneering private landowners too, is working towards the rejuvenation of huge areas of the Caledonian woodlands. In spite of their impressive work, however, much of the park remains intensively overgrazed by deer or farmed for driven grouse shooting.

The Norfolk Broads juggle interests slightly better. There are cranes, bitterns and many other birds. Otters move along the waterways and there's scope for beavers here as well. There's some tolerance of low-lying marshy land, paving the way for possible reintroductions of our long-lost fenland birds: colonies of ruff and hordes of noisy black terns. Wilder landscapes, such as the crane-rich Horsey Mere, exist alongside profitable farms.

The Pembrokeshire Coast is an interesting park because it's linear. Linear habitats in Britain, like north Norfolk's coast, tend to fare better against 'big' land uses because there isn't enough space to tame the land. Already choughs, cliff-nesting peregrines and seabirds move unhindered along the coastline, and it's not impossible to imagine a future in which wild horses have the run of the coastal strip.

If we were to achieve widespread ecological restoration within our national parks, Britain would see an infinitely richer wildlife, the kind that the citizens of an impoverished dictatorship like Belarus enjoy but we, as a nation of wildlife lovers, currently have no chance to appreciate. Perhaps more surprisingly, our country would also see a far richer jobs market and revival in its dying rural communities – and that is every bit as important. If this seems hard to believe, then read on to the end of the book.

The first obstacle to overcome, however, is partly one posed by many of Britain's conservation NGOs. Right now, many of us have little idea what we, the millions of wildlife lovers who fuel an entire sector of the economy – ecotourism – would actually *like* our national parks to look like.

The Great Wildlife Con

The fact that our national parks are factories for spruce production, loss-making sheep lawns, unprofitable dairy ranches and farms for grouse and deer is a national embarrassment and a massive waste of jobs. But this has come about in part because our nature charities are prepared to settle for the 1%, and they, in turn, take their lead from you and me. Are we really happy with these outcomes? Or are we too easily conned as to what constitutes success?

Before accepting the state of our national parks as inevitable, travel to see as many others as you reasonably can, or read about them all. Nobody in Spain, where national parks are refuges of the Spanish eagle and Iberian lynx, of wild-grazed cork oak forests or bustard-nibbled steppes, would take you seriously if you walked them around the Brecon Beacons and proclaimed it a national park. The more you and I travel, the more you and I look around at what other countries have, the less easily we will be deceived. We'll find it increasingly embarrassing that our country has sunk to a place where most of our wildlife is preserved in one or another kind of crop. We'll wake up to the degradation of it all. Then, we'll want to settle for more.

In Snowdonia, one day, you could watch the sun turn orange below skies filled with displaying snipe, cattle-grazed meadows with turtle doves, new oakwoods bubbling with wood warblers and broken birch woods with black grouse. Lynx, deer, horses, cattle, elk and eagles could one day, again, be living in those hills.

On what are now the burned hills of northern England, elk safaris could happen within your lifetime – on thriving hunting estates that benefit the wider public too. There would still be curlews, stewarded by wild grazing animals, but also many an eagle in the skies. This is an entirely possible future, and an economic one – if we work out how to ask for it, in a way that government can understand.

Farming and forestry, which National Parks England acknowledges is 'responsible for managing the vast majority of the National Park area', contribute 13,500 full-time jobs within England's national parks, just a tenth of all jobs within the park.[17] This is under a *third* of our parks' 48,000 jobs in tourism.[18] We must think about the jobs potential if these places thronged with wildlife. It is an exciting truth that, with our population for the most part living in urban areas, most of our gross domestic product (GDP) is created in just a small percentage of our land. This opens the possibility that in the years to come, money and investment, in ecotourism and rural job creation, will stay within our country, but be exported out, in part, to ever-growing rural economies.

In the following chapters, we'll envisage future models of extensive hunting, where the owners of grouse moors and deer estates make infinitely more profit, and leave an infinitely better legacy to everyone, by restoring and diversifying their estates. We'll discover how we could keep rural traditions alive, without writing off the entirety of Wales to the country's 1% of taxpayer-funded farmers. We'll envisage Somerset's marshes in 2060, expanded to ten times their present size – with beavers, elk and pelicans. We'll explore a future where timber and 'wild' forests split – so that we grow our timber in some areas, and in others we let wild animals manage truly wild woods. We'll explore how we can all help bring nature back into our towns and villages. And lastly, how our conservation charities can move from being the most cautious and paralysed to the boldest and most visionary in Europe.

As human beings, we are wonderfully positive creatures. It is easy to ignore anger about the dying natural world around us. Negativity breeds only resentment and complaining has not, for decades, won back our nation's wildlife. What is harder to ignore, however, is a vision of how our national heritage could one day look. And if our national parks cannot provide all of our spacious wildlife areas, then here's an even more promising thought. Even without them, we *still* have all the space we need. Really?

CHAPTER 8

The Wild Highlands

How Deer Estates can Save Scotland

*You see things; and you say 'Why?' But I dream things
that never were; and I say 'Why not?'*
—George Bernard Shaw[1]

Imagine if in Britain, an island of 65 million people, there were depopulated areas, stretching from coast to coast, covering 18,300 square kilometres.[2] Areas twice the size of Yellowstone, fourteen times the size of Greater Manchester and nineteen times the size of Dartmoor.[3] Imagine if virtually nobody lived here at all. Imagine if the population density was lower than in the state of Utah, with as few as two people per square kilometre in some large areas – as depopulated as Montana.[4] Imagine, on top of that, that these areas were not cropped, afforested or farmed for livestock. Imagine if virtually no roads traversed the land. Imagine if a huge amount of this area made virtually no national profit – and held just the relics of a once thriving ecosystem.

But you don't have to imagine it, because it already exists. And surely, this land should be one of the ultimate wilderness areas in northern Europe. Welcome to Scotland's deer estates – covering over 20% of the country.[5]

If we can effect real changes in deer estates, the wild future of Scotland can be as exciting and profitable as that of any depopulated upland area in the world. Income and jobs in the depopulated Highlands could soar, giving a true future to local people. The future of wildcats, capercaillies and the return of lynx and elk could be solid as rock in a generation's time. Restoring deer estates would not only rejuvenate Scotland's wildlife, but it would make infinitely more profits for its landowners, present and future, whoever they may be. But let's look first at the current economics of deer. Sadly for everyone, space and jobs are both being wasted on a massive scale.

Wasted Space

Cambridge-based economic consultants PACEC (Public and Corporate Economic Consultants) provide commissioned surveys for landowners and industries. They specialise in calculating the number of jobs and amount of revenue in particular sectors of the market. In a 2016 report, PACEC found that sport-shooting of deer in Scotland directly accounted for just 840 jobs. Indirect employment added a further 1,440 jobs. This made around 2,280 jobs in deer shooting, and an additional 240 jobs in deer management.[6] Those 2,532 deer jobs represent 0.008% of the British employment market.[7]

At a spatial scale, the economics are worse. Astonishingly, deer estates provide just one job for every 7 square kilometres of Scotland that they use. This is not even good news for gamekeepers or other rural workers. Relative to the space they use, Scotland's deer estates waste jobs and economic potential on a scale perhaps unique in Europe.

PACEC estimates that deer shooting generates £105 million per year in Scotland. No more than £70 million of this money remains in Scotland. Scottish GDP figures are available separately from England, and in 2017 clocked £150,025 million.[8] With the conversions done, that £70 million comes to 0.04% of the Scottish economy, harvested from over 20% of its land.

Each year, hunters travel to Scotland's deer estates to shoot red deer stags. A heavy cultural focus, dating back 150 years, is placed on antler size. Meat from stags that are hunted is used, more often than not, in high-quality venison sales. A Scottish Natural Heritage report indicates that 50,000 deer carcasses, worth £2 million, are sold in Scotland each season.[9]

Whilst there are many ecological problems on most kinds of farm, from chemicals to isolation, Scotland's deer farms are less intensive landscapes. The vast majority of the land is unsprayed and untidied, and benefits birds like cuckoos, doing far better on the wooded margins of Highland's deer estates than in most of England. The good news continues for most birds of prey. Prize deer stags cannot be taken down by any current British predator. This is why, whilst grouse moors remove our nation's predators, deer estates, as a rule, do not. Many are home to a good assemblage of charismatic raptors, including hen harriers and golden eagles.[10]

Britain's road network, which in England is the biggest obstacle to the restoration of large predators or herbivores, poses almost no issue in the Highlands. This alone renders the natural potential of deer estates very exciting – and far better than many areas across the whole of Europe for ecological restoration. There is, however, one enormous and well-known problem with deer estates – and that is the number of deer.

As long ago as 1872, government enquiries were made into the numbers of deer roaming Scotland's hills. In 1959, the Scottish Deer Act required landowners to take account of the impacts of deer on forestry and agriculture – the words 'ecosystem' and 'wildlife' were sadly excluded. Since these enquiries, red deer have increased in Scotland's uplands from 150,000 in the 1960s to 400,000 by 2011.[11] This has proven a death knell to its landscapes.

As you drive through the Highlands, the hills have an epic 'sweeping' quality to them. There's little doubt that deer estates are dramatic. And we've all grown used to the fact that the west-coast glens where we search for eagles are as smooth as glass. These denuded deer farms are revealed as ugly, however, the moment you visit somewhere like Glen Affric or Glen Feshie, and see the wonderful shades of pale green birch, and Scotland's deep green pines with orange trunks mingling with the open heather. It has been shown in a number of studies that where hillwalkers or the public at large are presented with a denuded, open landscape, and one containing trees, the latter is invariably considered not only more natural – but more beautiful as well.

Most Highland glens should be wooded lands. The degree of woodland would depend on a range of factors at play in natural ecosystems: the interplay of digging, browsing and grazing animals in their natural herd sizes, the actions of predators, the slope of the land, wind, water, and the relentless aspiration of young trees. These areas should not be carpets of pure heather, but lovely mosaics of willows, pines, birches, oaks and more open lands. The loss of this Caledonian mosaic woodland across truly enormous areas has rendered some of our most charismatic species little more than endangered refugees.

Each spring at their dawn lek, deep in forest clearings, cantankerous turkey-sized capercaillies, the world's largest grouse, still bubble and pop like champagne corks. But there is not much of ancient Caledonia left – nor are there many capercaillies: under 1,300 in all.[12] Most are found in places like Rothiemurchus in the Highland's Spey valley – where the last larger tracts of ancient pinewood stand.

Capercaillies, 'horses of the forest', have crashed since the 1970s. One factor has been wet summers, which inundate chicks on a wet heather floor, increasing mortality. Another has been deer-fences, which prove fatal to these low-flying birds. And then there are the pine martens. For thousands of years, these nimble predators would have munched their way through capercaillie eggs in Caledonia. But now, isolated in tiny woods, capercaillies are more vulnerable than ever before. To truly save and restore birds like the capercaillie in Scotland, there is no shortcut to regrowing the spacious pine woodlands it once called home. Capercaillies, wildcats, pine martens and many other species strongly associated with pine pastures should, in reality, be common across the valleys and hillsides of Scotland.

Ironically, the first loser from this desolate lack of tree cover, however, is not Britain's wildlife – but deer estate owners themselves. Tree roots in wooded landscapes store nutrients, creating fertile soils under a layer of fallen leaves. In this sheltered, more fertile environment, deer grow larger and healthier antlers – and so provide better trophies for hunters. It is not in the financial interests of a deer estate to have dozens of sickly, malnourished deer. It is better to have a smaller number of deer, but more prime stags for hunting. This situation is well expressed by the organisation Reforesting Scotland:

> *With a change of marketing to emphasise the stalking*
> *experience, there is no reason that hunting estates*
> *should not be as profitable as ever. In any case, stags*
> *kept at agricultural densities on the open hill are half*
> *the size of their European counterparts, making pretty*
> *poor trophies.*[13]

With even landowners losing out, some exciting initiatives have revealed how the Highlands could create infinitely more jobs and wildlife in the future. Let's first visit those places where the Highlands are getting their wilderness back.

The Lazarus Trees

Glenfeshie Estate has drastically reduced numbers of deer, but is still able to charge exactly the same for a day's shooting. This should not be surprising as sport estates are in the business of selling an experience and a lifestyle, not simply selling deer, which any deer farm could do.

—Reforesting Scotland[14]

If there is one magical place where the prospect of a wilder Highland can now be seen, it is, for me, the southern estates of Speyside. On a summer's evening, wandering the estates of Glenfeshie and Rothiemurchus, I have found curlew and snipe in wet forest clearings, snipe standing on the tops of pine trees. Woodcock chased one another like squeaking bats overhead. Tawny owls scrapped in the woods. As I sat quietly among the birches, badgers would trundle past, pine martens might sometimes race by and red squirrels were a regular delight.

In one pine forest, a revved-up male capercaillie chased me for a half a kilometre. On nearby birch moors at dawn, black grouse jumped around like black and white pompoms. The glades shivered with willow warblers, the old oaks with wood warblers, and by summer, cuckoos could always be heard. I watched one female for hours, hawking low over the heather, scattering meadow pipits as she looked for the perfect nest in which to lay her eggs.

Ospreys could often be seen travelling the length of the rivers. One male was, famously, pink: it was thought he'd eaten too much salmon. Goldeneyes swam on twinkling lochans, their brave ducklings having taken a leap of faith from nest holes many metres above the water. In big old clearings, I watched crested tits checking out a rotting stump for its potential as a nest site. One year, I found a pair of redwings. In another summer, a thriving colony of mountain ringlet butterflies. On one very special summer, in 2010, I sat spellbound as a pair of green sandpipers escorted their chicks through a pine-studded bogland that had perhaps changed little in aspect for over a thousand years. On every trip to the environs of the southern Spey valley, since 2005, I have experienced a novelty in the carefully managed world of British wildlife: the thrill of continuous surprise.

Until recent decades, estates within this region, such as Glenfeshie, operated like any other deer estate. Landowners could be 'requested' to cull deer to allow trees to regrow, but were entirely free to ignore such requests – so they did. But when nature-minded Danish billionaire Anders Holch Povlsen bought the estate in 2006, Glenfeshie began a remarkable recovery. Shocked by its total absence of new trees, Povlsen actually requested that the Deer Commission aid him in culling excess numbers of deer. Deer were targeted from helicopters and brought back to natural levels.[15] Since this time, a remarkable resurgence of trees, and wildlife, has been seen across the glen. Glenfeshie is an extensive deer farm, in purpose, but thrives on a heavy cull. It makes profits on its big stags and remains as profitable as other deer estates.

In recent years, Anders Holch Povlsen has become something of a hard-to-categorise visionary for those who associate private land ownership in the Highlands with bad practice. As his company, Wildland Limited, has bought large areas of deer estate across northern Scotland, pruning deer numbers back to natural levels, entire landscapes have been freed up to become wilder, and more wooded, for future generations to enjoy. The vision is one that will take two hundred years to accomplish: to hand back to Scotland a wooded land, filled with opportunity and life. Already in large areas here, and on other private estates that are now adopting similar approaches, there has been a huge resurgence in young birch and pine woodland. Unifying such efforts in the eastern Highlands has been the Cairngorms Connect project. Already stewarding 600 square kilometres of continuous land, this is one of the first projects in Britain showing a Yellowstone-level of aspiration and vision. The Cairngorms Connect project seeks to restore wild processes and keystone species to the Speyside woodlands and hills. If anywhere in Britain should soon fall below the hooves of elk or the padded paws of lynx, this rejuvenating land surely hosts the brightest of futures.

In other regions further west, older restoration projects, already under way decades ago, remind us what happens when visions for a wooded Scotland come to fruition. At Creag Meagaidh, where restoration began in 1986, sheep were removed from the hills, deer culled back, and fences put in place, temporarily, to allow the regeneration of trees.[16] Since this time, the feathered birchwoods have surged back. Today, you can ascend into the mossy realm of dotterels through a gradient of red grouse and greenshanks on damp heather; black grouse and cuckoos in a wavy sea of birch, into the bronze windswept realm of golden plovers, then the Arctic kingdom of the ptarmigan, croaking away on rocks that have never felt the shade of trees.

Alladale, owned by Paul Lister, has made headlines around Britain with its plans to introduce wolves, but already the reconstruction of the landscape has been remarkable (Plate 13). Some 900,000 native planted trees have been given a fighting chance against deer through fencing and a heavy cull, and juniper and rowan woodlands, forgotten in any of our lifetimes, are seen once again. Proxy animals like Highland cattle, and boar and bison, the latter for now in enclosures, have all produced a more natural effect in this slowly rekindling glen.

Glen Affric is one of the largest pristine areas of woodland remaining in Britain. Affric, each year, receives 130,000 visitors[17] – simply because it retains the majesty of an ancient pine woodland, and the wildlife that thrives in such a place. The regeneration of habitats at Affric, under the Trees for Life project founded by Alan Watson Featherstone, has been fascinating. Not only, as you would expect, have new birches and pines recolonised the valleys, but in the natural montane scrub zone, downy birches, aspens and hazels now slowly fade out into a world few of us have ever seen in Britain – a sea of montane flowers. Yet again, deer control has been at the heart of ecosystem restoration. Adjoining Affric, the RSPB's Corrimony reserve is also expanding the trees.

Birch seas are growing on the hillsides beside the open moors, and woodland edge species like black grouse thrive. But only so many people will pay to watch trees grow. The Highland ecosystem, and economy, would be made infinitely richer if exciting stewards were to complete the wilder picture.

A Much Bigger Game

In North America and Scandinavia, robust ecosystems with trees and larger charismatic animals are hugely profitable, because they combine the full possibilities of exclusive hunting with the far more profitable possibilities of inclusive ecotourism. The first step towards restoring the Highlands to a place where this could be possible, however, must first lie in dramatic reductions of its deer numbers. There are enough living demonstrations of what happens when this is done for the ecological argument to be a simple one. Yet culling deer costs estates more money than letting their herds run free.

Unlike most countries, until recently Scotland allowed a 'voluntary' approach to deer management.[18] In a new development, however, the LINK Deer Task Force recently persuaded the Scottish government to introduce a system of deer management whereby Scottish Natural Heritage can conduct deer culls if landowners refuse to carry them out. If properly enforced, this will be a promising first step for the regeneration of trees on deer estates.[19]

Another step would be the joint development of a plan for what *should* be living on the Highland's deer estates, and how this might benefit its owners, not just the wider public. Mapping out a future where animals can be harvested from ecosystems would be more likely to gain government interest and support from the estates, and lead to a change in the subsidy system. Culling deer is a negative. Exciting wild lands with woodlands – and even elk and lynx – is an incentive.

Following a 'pulse' of vegetation growth, and the return of trees to a connected web of Highland glens, the possibilities of the Highlands' empty spaces become exciting indeed. Bearing in mind that Scotland's deer estates are collectively twice the size of Yellowstone National Park, large animals, requiring very large spaces, would become possible too.

Lynx occupy enormous territories; the Scottish Highlands would provide. Lynx are the natural culling agents of the small, the weak and the young, an agent no estate owner would have to pay, and would rarely even see. Specialising in ambushing roe deer within the woodland canopy, these ultra-shy cats have never been recorded attacking a person.[20] By contrast, between 1981 and 2015, 77 people died from dog attacks in England and Wales.[21] Reintroducing lynx into somewhere like the Cairngorms' Caledonian woodlands would be the boldest and most inspired move British conservation has yet seen. With little farming, a resurgent woodland, and sympathetic management by the RSPB and others, there is wooded space enough for these quietly tufted cats. The return of lynx to just one wooded area of the Spey would also demonstrate to others the financial potential of such a venture. And that is truly worth considering.

Today, the RSPB's osprey watch-point at Loch Garten, where you can watch one pair of birds on an artificial platform, attracts 50,000 visitors per year.[22] If each spends as little as £100 in and around Speyside during the visit, one overnight plus three meals, the income surrounding that osprey pair comes to £5 million. Lynx, an animal so sexy the world's leading deodorant is named after it, would, very stealthily, kick Loch Garten's ospreys into touch. Wild lynx would take several visits to actually see, on telescope stake-outs or driving safaris. Such a 'hunt' would draw in Britain's millions of wildlife watchers over the course of just a few years. If 50,000 ecotourists pay for ospreys served up on a platform, a very conservative estimate would be that four times as many, at least 200,000, might join the hunt for lynx. If just half of these tourists stayed one night in the Spey valley, at £50 a head, and spent £20 on food and £20 in local shops, as they wandered the woodlands of the Spey, that alone would bring £9 million to the local economy. If the keenest quarter paid £100 for a lynx safari, with expert guides, this would bring a further £5 million. That's £14 million invested locally, before we think about the extra jobs.

No deer estate can compete with 'safari' economics. But they could. If just five estates set out to truly rewild cooperatively, embracing a government-incentivised 'full ecosystem package' over twenty years, and each in time offered competitive safaris alongside deer-stalking, Scotland's lynx income could be closer to £75 million. That is currently the sum of the income from *all* deer estates, in their current form, that stays in the Scottish economy each year!

Ecotourism doesn't threaten the Highland's economy, culture or landowners – it would instead be transformative. A practical three-step approach – cull deer, grow woodlands, replant charismatic animals – is entirely in the interests of most rural parties in the Highlands.

At the moment, just 1.3% of Highlanders work in agriculture, forestry and fishing, but 12% work in accommodation and food. It is the ecotourism sector that would further expand these jobs, before you even start thinking about the tour operators, conservationists and specialist ventures involved. With lynx back in the Cairngorms National Park, landowners would be acting *against* their own interests not to want in.

The ecological arguments are just as appealing. If lynx were restored into the most wooded of the Highland's glens, deer would soon be culled quietly at no expense at all. Deer fences, devastating to low-flying caper-caillie and black grouse, would be rendered obsolete. New pines would swiftly regrow, extending the habitat of the capercaillie. With robust woodlands growing back tough birches, resilient willows and healthy young pines – a gentle giant could also make a return, perhaps within as little as twenty-five years.

Elk are long overdue a nibble on the fringes of places like Abernethy Forest. Resilient birch, aspen and willow woodlands are designed for elk – and the other way around. Whilst lynx 'plant' trees by culling roe deer, elk 'open' habitats by browsing, forming mosaics for birds like black grouse and maintaining open moors for endangered species such as curlews.

For deer estate owners, the most exciting hunting experience currently available for shooters is one of shooting weedy red deer stags on a hill so open you can hardly miss. Deer estates currently have virtually no income from ecotourism, and very limited income from the hunting of red deer. So the restoration of wooded uplands, and the possibilities of elk, in particular, would work in their interests too. Britain's hunters would pay huge money to visit Scotland to bag elk, but you would not see elk being 'grown' in denuded hill farms, as deer are in the Highlands at the moment. A Highland landscape filled with willows and birches could be one filled with elk, as indeed is the upland landscape default across much of Finland, Sweden and Norway.

Elk are more exciting, profitable trophies for stalkers than red deer, as well as enormous magnets for ecotourism – an animal we last saw in Britain in the Bronze Age. The 'harvesting' of elk by hunters and tourists alike is not a pipe dream either, but standard practice in the uplands not only of Fenno-Scandinavia but Russia, Canada and the United States.

A wooded Highlands, making profits and increasing jobs, would finally allow for the widespread return of wildcats. Expanding wood-pastures would increase the chances of a successful reintroduction of the wryneck. Mountain scrubland birds we have forgotten, like the bluethroat, might also return. In all possible ways, a wooded Highlands, with some of its original stewards back in play, is a winner for Scottish wildlife. Right now, with deer estates wasting over 20% of Scottish land for less than 1% of national profit, it's in the interests of Scotland to have a wild economy too.

The future for jobs, for capercaillies, for the wild Highlands, has to begin with what we ask for – not just from landowners, but from our NGOs too. Until our nature charities show to government what wild wonders *should* be in the Highlands, and what money that would make, then we'll walk on sweeping hills with all the space in the world. They won't be wild – but they will be empty.

CHAPTER 9

New Forests

Bringing Back the Wild Woods

*Planting trees is thought of as the essence of
conservation, rather than an admission that
conservation has failed.*
—Oliver Rackham[1]

So lost and damaged are the relics of our once wild woodlands, that
perhaps the last great act of conservation for native British trees took
place in 1088, almost a thousand years ago. Since William the Conqueror
isolated and protected the New Forest as a hunting preserve, there has
been almost no attempt to safeguard, or restore, the true beauty and
diversity of the glorious wood-pastures that once covered most of the
British Isles. This is as wasteful as it is strange. If the identity of the USA's
Pacific coast is its giant redwoods, then the lost identity of our country
is surely an island of oaks – once unparalleled, perhaps, in their majesty
anywhere in the world. Yet even though this treasure trove was largely
lost to us by the Bronze Age, we are still feeling the repercussions of
woodland damage to this day.

Of all British birds arguably facing most imminent extinction,[2] five
of them – turtle dove, wood warbler, lesser spotted woodpecker, willow
tit and nightingale – are all associated with the varied world of scrub and
trees that once filled much of our island. Given that around 60% of Britain
once lay under oak-led woodland, restoring this largest native biome
of the British Isles is the biggest challenge facing conservation – and
we have to begin almost from scratch. Since the 1970s, woodland cover
has actually increased, owing to increases in commercial forestry. But
of 1,256 woodland species studied since then, 60% declined still further
over those four decades.[3]

Before even thinking about restoration, however, our woodlands
face a growing problem – and a strange one. It involves a familiar and

much-loved animal – wreaking the kind of desecration few might imagine, not just in the Highlands, but in the woodlands close to where most of us live today.

A Story to End All Storeys

There are now around 2 million deer in Britain. And those studying these numbers at the University of East Anglia believe that to return Britain to natural deer levels, around a half of British deer, in the south and east of England in particular, would need to be culled *each year*.[4] This is just to allow our woodlands to regenerate.

Deer, however, have been around a long time. So what has happened, in the last four decades, to render these animals so harmful? The Deer Initiative, a leading voice in calling for the control of deer, has explored several reasons.[5] Milder winters, it says, have played a role, allowing greater survival than in centuries past, when weak deer would be weeded out by the harshness of the seasons. Winter crops, too, have decreased mortality, but most of all, increased woodland cover, such as forestry, has allowed deer to move around. Today, deer not only have no wolves or lynx to hunt them – no new state of affairs – but their weak and sick are better able to survive.

In recent decades, a deer invasion has been waged against Britain on two fronts. Firstly, there has been a huge expansion in the range and numbers of native roe deer and red deer, and large numbers of feral fallow deer, present since the Norman Conquest. Fallow deer are grazers, feeding on ground vegetation, but where they are concentrated in a small area, they can remove all flowers from a woodland glade. Roe deer, in high densities, are more devastating. Browsers, not grazers, they nibble away buds, shoots, leaves and shrubs, stunting the growth of future trees and gardening away the scrublands on which so many birds depend. The disappearance of bramble – fuel for the woodland butterflies that once carpeted our woodland glades – is largely down to its wholesale removal by deer over time.

Secondly, non-native muntjacs have brought another level of devastation to our woodlands. Having originally escaped from areas like Woburn and Whipsnade in the 1930s, they are now increasing by 12% each year, according to bag figures from the Game and Wildlife Conservation Trust (GWCT). As long ago as 1995 there were 40,300 in

England alone,[6] and that number will have expanded considerably by now. Muntjac, adapted to the forests of southeastern China, not our own, may be less numerous than roe deer but are, per animal, even worse for our woodlands, nibbling out the smallest shoots and buds. They prevent embryonic trees, shrubs and flowers from ever forming. As you eat your breakfast, armies of these satanic, sapling-wrecking pig-hamsters are destroying our nation's next generation of trees. They will continue to do so for ever more, unless brought under control. Before we think about solutions, let's take a look at what this nibbling invasion means for some of our vanishing woodland birds.

Deer Devastation

To understand deer devastation, you first need to see for yourself what is missing in our woodlands. In Plates 14 to 16 you will see photographs of the kind of dense, shady places beloved by three of our vanishing birds – nightingales, willow tits and marsh tits.

In Plate 14, showing nightingale habitat in Suffolk's Bradfield Woods, the shrubbery on either side of the path is so dense, it's as if you are gazing down a tunnel. Rose, bramble, blackthorn and buckthorn buckle together – creating places you wouldn't want to stick your head. The area in this photo has been deer-fenced, allowing a pulse of thorns to grow. Here lies the shade where nightingales sing.

Willow tit habitat, by contrast, is usually so rotten and brittle that moss and mushrooms cover decaying branches (Plate 15). A chaos of scrub trees rots fast in wet soil. Deer devastate such a habitat by browsing out the shrubs. The soft rotting stumps used for nesting by willow tits require a sheltered forest floor. Deer disturb it. The post-industrial woodlands of northern England, however, being largely sealed off from deer damage by roads and canals (and densities of deer being lower in the northeast), still provide willow tits with the rotting scrub woodlands they need: an unlikely refuge indeed.

In Plate 16, you can see the networks of low branches that allow marsh tits, a close relative of the willow tit, to move low through the woodland, one branch at a time. All of these dense tree mazes are vanishing from British woodlands, along with the birds that specialise in them. We cannot watch it happen year by year, but visit most woodlands in southern England and you'll soon see deer devastation.

Now examine the photo of the fern-filled forest in Plate 17: a conservation woodland in the Forest of Dean. The empty spaces in this picture are the end result of decades of accumulating deer – undeterred by hunting, predators, or fences. You can see for a long way indeed. All that is left is a carpet of ferns below cathedral boughs. Long gone are the brambles and butterflies; the new bushes with the warblers; the next generation of saplings.

Perhaps the clearest single proof that deer play a huge role in such acts of woodland robbery was uncovered by Chas Holt, a doctoral student at the University of East Anglia. Studying radio-tagged nightingales in Bradfield Woods, he found that densities were *fifteen* times higher in deer-fenced areas than outside.[7] The choir was singing behind a carefully constructed fence. Keats and Coleridge would be turning in their graves.

In southern England, hunting bag results show that fortunately the muntjac has yet to make serious incursions into low-lying marshland areas of north Kent, where the largest nightingale populations remain. Scrubland nightingales now survive only in areas where the deer armies cannot conquer. Indeed, maps of nightingale density correlate closely to those areas in southeast England where the muntjac, in particular, is less abundant.

Though nightingales have become the flagship birds of deer desolation, many vanishing species have felt the impact of deer. Willow warblers thrive in deciduous bushes in those early, wavy stages: they have vanished in their hundreds of thousands from much of southern Britain. Garden warblers, which share a bushy world with nightingales, have disappeared from many of our woodlands.

One of the reasons deer have impacted scrubland birds so badly is that, for a long time, scrubland trees have been restricted to growing below the shade of our woodland canopy, as wild scrubland landscapes and extensive grazing lands have vanished from our island. Yet given any chance, a lot of our native scrubland trees, from hazel to elder, are actually best adapted to thrive in sunlight. Growing in the sun, their leaves and thorns grow faster, and are better able to repel the advances of herbivores. Because our native ecosystems have been long forgotten, however, these scrublands have become rare. So the effect of deer on nightingales (which thrive best in sunlit scrub) has become even more pronounced. Whilst nightingale colonies now often thrive in fenced, post-industrial wasteland habitats, from the Cotswolds to Kent, there are many areas where deer cannot simply be fenced out. And the most immediate solution for this problem could be practical, lucrative – and tasty.

A Wild Harvest

Enormous culls, regrettably, are now necessary if the nectar layer, and the insect food in our woodlands, let alone the next generation of trees, is ever to replenish. Such a cull, too, cannot be a one-off. If you carry out a deer cull and then walk away from it, compensatory breeding kicks in. Within a few years, your woodlands flood with young deer once again.

As a rule, Britain does not like large culls, but perhaps that's because they're framed in the wrong way. After all, each year, at great cost and great stress to the animals involved, we cull millions of sheep and cattle in abattoirs. Deer, however, if properly marketed, and responsibly targeted by well-trained expert marksmen, could prompt a sustainable shift in the British diet. Harvesting deer from our woodlands, especially muntjac, would create a bountiful wild harvest, providing both respite for our woodlands and an ethical source of red meat. Many people express strong concerns about the treatment of domestic livestock, from their confinement in trucks to watching their fellow animals being killed. Deer hunting, professionally executed, provides no such trauma. A skilled marksman kills a deer on the spot. Deer is the most ethical source of red meat that there is, and the lack of stress in the meat would only enhance the taste as well.

At this point, many will argue that the widespread return of the lynx would solve the deer problem at a stroke. If only that were true. For the foreseeable future, not only are the woodlands of southern and eastern England far too small for a predator dependent on vast wooded lands, but there is another constraint. The road network of southern England is among the densest in Europe. Even the New Forest is so bisected by roads and tracks, cyclists and dog-walkers, that the peaceful environment required by lynx is simply not a viable prospect. Whilst therefore lynx are eminently suited to the Caledonian Forest, we face little choice but to play 'predator' in the small, fragmented woods of southern Britain – whether we like it or not.

The first step in this process is to fathom a way to harvest our deer in an ethical, sustainable way, with specific developments needed in the hunting of muntjac, a challenging quarry for hunters. Any business that could lead on this with government, and make this achievable on a national scale, could employ thousands of people and prove profitable indeed. Nature charities need clear plans that courageously advocate eating deer as a way of saving the nation's woodlands. Until they do, most other woodland restoration tasks are, in essence, a total waste of time. It's time for a new campaign: 'Bring the birdsong back.' It's time for the wild harvest to begin. In doing so, a new sector of rural jobs would also be created from scratch.

Green Silence

Natural Resources Wales … now that's a name and a half. Resources … That to me is something to be used and abused. Something to be exploited.
—Iolo Williams[8]

Just 12% of Britain's land is now wooded.[9] This is still one of the lowest percentages in Europe but an increase on any time since the last known 'high' of 15% – extrapolated from the Domesday Book in 1086. Yet, as Britain's woodland cover has increased, Britain's woodland birds have continued to vanish. This is because most of our birds do *not* need more trees. They need trees that they can understand.

In most European countries, the largest wooded areas are also the richest. The oldest trees, the most diversity, and the most birds are found within national forests. In Germany, just 5% of trees are non-native.[10] The scale of forestry blight in Britain, however, is enormous – and more damaging than we might imagine. Of the large expanses of woodland in Britain, only the New Forest stands out as a single expanse of woodland, 270 square kilometres in size, where aged wood-pasture trees have survived not in remnants but in large areas of the landscape. Surrey, with its old estates, contains more deciduous trees than any other county.[11] In all, Surrey, Sussex, Kent, the other Home Counties and the New Forest harbour our last extensive deciduous landscapes.

In 2017, the Forestry Commission released a breakdown of the woodland area in England, Wales and Scotland.[12] It makes for depressing reading. In England, the private sector manages 9,030 square kilometres of deciduous woodland – the Forestry Commission, 630. By contrast, the private sector has 1,510 square kilometres of conifer, while the Forestry

Commission manages 1,890. So the vast majority of England's native woodland, intelligible for birds and wildlife, is not even in public hands. There is no issue with private estates protecting invaluable reserves of oak: it's sad that our public bodies don't.

In Wales, where the native biome is broad-leaved woodland, the situation is worse. The country has 1,500 square kilometres of conifers and 1,560 of broadleaves: half and half. The private sector provides most woodland life, with 1,360 square kilometres of deciduous woodland, three times the conifers. Natural Resources Wales, however, manages 980 square kilometres of conifers, to just 190 of deciduous. Welsh woodlands of any scale are not wild woods but dark timber factories.

In private Scottish woodlands, there are twice as many conifers as broadleaves: at least closer to a natural balance than in England and Wales. By contrast, Forestry Commission Scotland manages *ten* times more conifer plantations than broad-leaved woods. And most of Scotland's conifer forests aren't lovely spacious places filled with native pines, wildcats and capercaillies. They're silent Sitka spruce crops – planted unchallenged across Scotland's uplands and national parks.

The Forestry Commission confirms that across Britain, conifers cover 51% of the woodland area. That figure is entirely disproportionate in a nation whose native biome, except in the far north, should be oak-led deciduous woodland. Overall, a shocking 40% of Britain's trees are non-native. Just 2% of Britain is covered in ancient woodland.[13] Even as recently as 2016 and 2017, over half of the new trees being planted were conifers.[14]

Of all these conifers, Sitka spruce species accounts for a *half*. In each country, the Forestry Commission manages more Sitka spruce than any other conifer – 3,260 square kilometres of it in all, an area larger than the Lake District. Even in Scotland, where pine woodland is most desperately needed for wildlife, the Commission grows six times more Sitka spruce.[15] There is, in truth, no larger single crop for wildlife prevention in our country.

Forestry literature will often explain that fast-grown softwoods like Sitka are invaluable to the British timber market. It is noticeable, however, that no other European country has these ratios of trees, yet all produce viable stocks of timber. In France, over 60% of forestry consists of broad-leaved trees, of which a third is oak.[16] Yet even as a nation rightly proud of its heritage, somehow it has flown under the British radar that almost all our national woodlands and 'forest parks' are now little more than alien crops.

If this sounds far-fetched, have a look now at the satellite images on Google Maps. Zoom out so you can see all of Britain. It's mostly pale green – that's where the woodland is gone. But there appear to be glimmers of hope. You will notice dark green beacons spread across the map. To the east, you will see Thetford Forest, in East Anglia. That's a crop of planted pine, a tree that grows well on sandy soil. Look closely, and there's a blob of forest in the centre of Britain: the remains of Sherwood. Mythical, wild and full of oaks? On closer inspection, Sherwood is now mostly a plantation.

In western Britain lies the Forest of Dean. As you zoom in, you'll see the Dean comes in a series of squares. Forestry dominates. The Dean, a place of rich heritage and woodland history, is now mostly a crop. But deciduous woodlands hang on. Often, forestry operations will leave a buffer zone of deciduous trees alongside the road, creating the illusion of a natural woodland. The aerial view, however, does not lie. Our shared Crown Forests, once glorious and spacious, lie mostly in harsh, planted squares. How quietly our prized natural heritage has been grubbed out, whilst many of us might believe it is still standing.

Fifteen percent of Wales is covered in woodland – but its trees grow in dark shades. From Swansea's Neath valley to central and northern Wales stretch deep blankets of conifer. Only on much closer inspection can you discover the softly wooded valleys of places like Dolgellau, rich in ancient oaks.

Heading north across England, as you reach the North York Moors, the landscape reads like an ugly quilt. In between the scars of grouse moors, you find the scars of planted timber – no soft woodlands here. In the Lake District, woodland has largely been driven out, but around Windermere and Rusland you can find the freestyling remnants of some lovely ancient woods. From here onwards, life gets truly bleak. Britain's most visible band of crops, bridging England and Scotland, are the commercial forests of Galloway, Kielder and Ettrick. Across central Scotland lie the enormous crops of Queen Elizabeth Forest Park, Loch Lomond and the Trossachs, all the way down to the far reaches of western Scotland – Knapdale and Campbeltown. North of Perth, another conifer crop grows in Tay Forest Park. At last, the Spey and Dee valleys, in the Cairngorms, with their pine and birch pastures, offer our wildlife a native wooded landscape at last.

You should then look on Google Maps at the forests in France, Germany or Poland, to see that our situation is quite unique, in terms of both our woodland cover and our alien forests. Almost all of Britain's wooded landscapes, the New Forest and Spey valley excepted, are profoundly damaging to wildlife.

The Trees that Swallow Birds

In recent decades, press releases and media campaigns have urged us to see forestry as good for birds, with particular focus on clear-felling, a process far more rarely seen in Europe, where trees are usually harvested selectively from within native woodlands. Clear-fells, in their early years, resemble poor-grade heathlands, or regrowth in the wake of a fire. Woodlarks and woodcocks briefly find a home. Tree pipits and nightjars sing on isolated trees, but nest amid the woody chaos on the ground. But a clear-fell is a time-bomb. For a few years, the young crop is open enough to sustain small populations of the birds of our lost grazing woodlands. Then the conifer crop is replanted. Instead of self-sustained woodlands, you have humans with chainsaws, creating a habitat that will last, for a few birds, for a few years. Ironically, most forestry is best for birds after it has been cut down. The instability of clear-fells as bird refuges compares unfavourably with ancient grazed pastures in the New Forest. In Thetford Forest, since 2000, nightjars, once thriving in the forest's clear-fells, have declined by 43% as forestry practices have changed.[17] In the New Forest, where the species hawks over moth-rich heathlands, many grazed by horses, many populations remain stable.[18]

Dense crops of alien trees can also act as barriers to the movement patterns of deciduous woodland birds. Woodland specialists like lesser spotted woodpeckers, for example, will move happily through a riverside line of willows just one tree thick, but are ill-adapted to work their way through kilometres of spruce, or feed in its silence.

Most plantation conifers come with no useful insect package, and do not 'compute' for Britain's native wildlife. Of our dominant non-natives, only European larch, introduced to Britain in the seventeenth century, comes from a continent where our own bird species have adapted to use it. Finches and tits thrive on its winter crop; goshawks nest in its sheltered forks. Sitka spruce, however, is virtually unreadable to the majority of our wildlife.

Worse, in the twentieth century, 40% of Britain's ancient woodland was converted into conifer plantation,[19] a kind of national desecration. These areas today are known as PAWs: plantations on ancient woodland. One day, they should not be known as anything at all. Even now, our original species of tree lie dormant in the soil – waiting to emerge. So as soon as the Forestry Commission wishes to repay the national debt, fell the conifers and regrow native trees in our national forests, it can.

In the few areas where oak-led landscapes have remained, an exciting range of birds reappear, as if by magic. In areas as disparate as the New Forest's wood-pastures, the Dolgellau valley of northwest Wales, the Rusland valley of the Lake District and the long, thin oaklands of the Malverns, in Worcestershire and Herefordshire, hawfinches, redstarts, spotted flycatchers, honey-buzzards and many other woodland birds pop back up on the map. Where old wildwood trees dominate the landscape at scale, many of our wildwood birds can, even now, find a way to survive.

An enormous increase in our native woodland – and birds – will happen as soon as the conifers come down in our national forests, and a new generation of native trees begins to grow. To get our precious woodlands back, we first need to ask for something – unusual. We need to ask for the existing trees to be cut down.

The Lost Foresters

Eight thousand years ago, as the last ice age ended, there was infinite woodland variety, but only one form of management: the four-legged kind. No one was planting oaks in rows. Foresters will justify the two-legged approach and talk of ride creation and coppicing, managed and 'non-interventional' woodlands until the rewilded cows come home – but none of these methods are natural. Forests, dense blocks of trees growing without disturbance are not natural either, being predicated on the removal of large animals. Britain did not, after all, have a separate ecological history, one with the animals fenced out. Beavers, not dozens of men with axes, create the most natural and diverse woodland coppices of all.

Whilst forestry deserts and deer pose two enormous problems for wildlife in our woodlands, the third has therefore been the loss of natural processes over time. We have forgotten how woodland works. Beavers, boar, cattle, elk and horses are the five missing foresters in most British woodlands. They once acted as a complementary force, shaping our woodlands into configurations of maximum benefit to wildlife.

Take a hazel. Growing in a world devoid of animals, hazel grows straight. Only in its broken form, coppiced by animals or humans, do its branches spread to create a low protective maze beloved by marsh tits and dormice. Take an apple. Growing in a dense, planted orchard, an apple is a thwarted tree. Growing in space, and nibbled around by grazers over time, the apple thrives. Redstarts, mistle thrushes or starlings feed in the pasture around the apple, but nest within its boughs. Take a willow. Beavers rapidly coppice waterside willows into dense bushlands, whilst creating adjacent ponds and meadows. Within an otherwise homogeneous woodland, habitat is rapidly created as a triumvirate of wood, meadow and pond. The dynamism created by large herbivores is what would once have made many of our woodlands most intelligible to our native wildlife. But without the stewards returned, in their natural densities, our woodlands will never reach the diversity of times past.

At present, grazing animals like cattle and horses, in very small herds, diggers like boar and river engineers like beavers – all designed to maximise our native biodiversity by promoting the growth of varied grasslands, flowers and trees – are absent from the vast majority of our woodlands. But they *should* be there. Browsing animals like deer, in enormous armies, however, are present in unchecked megaherds that are entirely unnatural too. But if we can return our woodlands to places governed by the right stewards, in the right configurations, then our woodland wildlife has an exciting future indeed. If herbivore rewilding schemes are happening in areas as urban as the woodlands around Berlin (see Chapter 13), they could happen in our larger national forests. But big changes require big stimuli. Perhaps as big as an aurochs.

The return of woodland cattle as landscape architects may seem exciting, but aurochs are extinct – and have been for quite a while. Enter Operation Tauros.[20] The aim of the Tauros programme is to 'create a true replacement for one of the most impressive wild animals ever seen by men'. The programme, with scientific backing from Trinity College Dublin, is slowly breeding back animals similar to the aurochs. It started by discovering 'hidden' genes in older cattle forms, those that withstood further out-breeding 'thanks to the stubborn farmers that held onto their old breeds'. A number of ancient cattle breeds have been involved, and the first generation of Tauros cattle is looking good. They're currently a little short on the horns front – so the operation is using Croatian Boskarin to breed longer horns back in. The genetic core is provided by a range of old breeds, from Portuguese Maronesa to our own Highland cattle.

Sensibly, Operation Tauros will not wait for a 'perfect aurochs' – nor do they expect one to arrive. Their goal is more practical for woodland conservation. In their words, 'we want to get cattle back at the other side of the fence, as a wild species' and a flagship return of wild cattle to a few British woodland areas in years to come could be hugely significant.

The return to the wild of Britain's most impressive wild animal of recent times could legitimise the forgotten role of our most important woodland steward, prove hugely profitable for ecotourism, and tap into an existing British love of bovines. Where necessary, very large enclosures, protecting the animals from roads or settlements, would be in keeping with methods used in the Netherlands for bison, where the animals have already played an exciting role in landscape restoration, in an area far smaller than many of our Crown woodlands.[21]

With such master foresters back in the game, the true native woodlands of Britain would finally be able to begin their comeback. This, of course, would still require the oversight of the Forestry Commission. So how might this work?

Timber Forest, Wild Wood

Splitting the Forestry Commission into two, separating 'timber forests' from 'wild woodlands', would be one way to commercially realise a project of this scale and ambition. 'Timber forests' and 'wild woods' would fuel a dual financial sector in British forestry. Timber forests would remain as lucrative crops. National or 'Crown wildlife' forests, however, would protect native woodlands, wild processes and large animals for the nation to enjoy.

In large areas of Britain, the timber crop could carry on as before. Britain, of course, needs timber, and better to grow it sustainably in our own country than, as in times past, to raid the priceless forests of other countries. In national parks and national Crown-estate forests, such as the New Forest, Sherwood and the Dean, restoration would begin. Ecotourism, conservation and wild game would become a driving source of revenue and jobs.

If any single organisation has a responsibility to begin such an initiative, it has to be the Forestry Commission. Not only is it Britain's single largest landowner, but it has an outstanding national debt to pay, in turning some of our shared national woodlands from blighted factories back into thriving ecosystems that we can all enjoy. Across Britain, the Forestry Commission manages 7,720 square kilometres – an astonishing area and one that proves, again, that we have all the space we need for nature restoration. The economics, however, are also worth considering.

Even in its present form, forestry is, as you might expect, a serious player in our economy. In Scotland, there are over 12,000 jobs in the forestry and wood-processing sector.[22] In England, around 14,000 people are engaged directly in forestry,[23] but the wider market for jobs around it comes closer to 34,000 people.[24] In Britain, forestry is seen as an expanding market, and that, in itself, is good news for everyone. With a strong employment market, there's room for diversification.

Forestry contributes around £4 billion to the UK's GDP, but at present, recreation visits account for just £484 million.[25] Given the area covered by the Forestry Commission's woodlands alone, and the nation's thirst for spending on nature, that number is incredibly low. Increasing it would increase jobs. But first, there would need to be great wildlife to see.

Rather than envisage a future where lines of trees are planted, the Forestry Commission could, in national forests and national parks, lead the way in national restoration. With timber profits no longer driving operations in some woodlands, such as those owned by the Crown, enormous additional revenue, through ecotourism, would come to these forests through interest in wild cattle, horses, boar and perhaps even the prospect of forest cats. Indeed, nobody benefits from lynx more than foresters. Lynx do not damage trees but aid their recovery by culling deer. In Poland, lynx are strongly welcomed by foresters – they prevent tree damage free of charge, saving millions in fencing costs.[26]

This wilder woodland model, however, is nowhere near as radical as it sounds. Lynx aside, the idea of grazing forests is, in fact, wholly in keeping with far older traditions of royal land management in Britain. Originally, our royal forests were configurations of grazing animals in very low densities, creating spacious woods. Restoring native grazing animals to manage woodland is neither ecologically nor historically a new idea. If, by 2030, one large Crown forest, like the Dean, could follow such a route, you would see extraordinary reversals in the fortunes of its wildlife. Native diversity would begin to self-restore. Native trees would grow in the formations that nature understands.

Such a future has never been asked for. This is, perhaps, the main reason why it has never come to pass. Few visions of Britain's landscape future are so well aligned to both liberal and conservative vision. Of all 'rewilding' visions, a return to the oak-led grazing forests of our past is surely the one most compatible with the beliefs of most people, and of how our existing woodlands could one day look. There is, and should be, a deep-seated patriotism in the idea that Britain was once a kingdom of oak lands, unparalleled, perhaps, anywhere on earth. That, surely, is something we can all aspire to see return.

Without such radical change, an ever-growing desert is the only future we can reasonably expect from our woodlands, as their birds and wildlife remorselessly vanish each year. After fifteen years of wandering my local Forest of Dean, I now visit its ever-emptier woods with growing trepidation. I know, because I have watched it happen, that each year, fewer and fewer birds will sing in its boughs. Marketed as a forest famous for its wildlife, nothing is now further from the truth.

Its famous pied flycatchers, for all their nest boxes, have all but vanished. Wood warblers have almost gone. Turtle doves, once purring in clear-felled woods, have fallen silent. Willow tits, of which fourteen males sang in 2010, are now extinct. Nightingales in the main forest are virtually unknown. Lesser spotted woodpeckers cling on in just a few places. Most walks through the nectar-starved and cramped plantation trees yield silence more than song.

With great effort, birdwatchers in the Dean today visit the same few sites to see, in many cases, the same few pairs of birds. And a once proud woodland, standing for thousands of years, has become little more than a zoo. In many ways, the Dean acts as an emblem for all our dying woodlands. Deer rampage through the trees, browsing out its nectar supplies, killing its saplings. The loss of woodland caterpillars starves its birds. No clouds of butterflies dance along the woodland edge. There are no natural processes and few wild stewards except the returning boar and, in recent months, a few closely-monitored beavers. There is little space or light, and too few native trees, of too few species. Worst of all, most of the forest is a factory.

Yet, loved by tourists and rich in heritage, the Dean is profoundly suitable for restoration. By taking out the conifers, driving a lucrative trade in ethical deer meat, restoring flagship animals and allowing trees to be governed by their older stewards, the Forestry Commission would truly turn a corner. It would reveal to the British public, for the first time in any of our lives, the wonders of a wild wood. For the first time in millennia, our woodlands would grow richer. Then, the full beauty of our native oaks, and the fullness of our woodland wildlife, could once again bring jobs, delight and wonder into Britain. It only needs one demonstration. One leap of faith. In thirty years' time, we'll all wonder why it didn't happen thirty years ago.

CHAPTER 10

The Golden Hills of Wales

How Wildness Gives Wales a Future

Adfyd a ddwg wybodaeth, a gwybodaeth ddoethineb
(Adversity brings knowledge, and knowledge –
wisdom)
—Welsh proverb

In most areas of Britain, democracy holds sway. We might not like our politicians but we can vote them out. We may not have full democracy, voting for each issue, but we have broad democracy that's kept us from extreme forms of government for hundreds of years and ensured centuries of broadly continuous improvement. So what happens in a democracy when sharing breaks down? Our country gets poorer. Poorer economically, poorer socially, poorer environmentally. So let's talk about a situation where sharing isn't working.

Let's envisage a situation where one industry – a respectable, hard-working one – has, because of market forces, fallen into terminal decline over more than a century. It contributes 0.7% to its nation's economy. It employs a little under 1.9% of its country's population. Yet, in spite of all this, this industry occupies 88% of the land surface of that entire country.[1] Without taxpayer-funded support, which gives the industry's workers up to 80% of their income, the average business, according to trade unions, would now earn just £2,600 per year.[2] This enterprise works hard, but it's not a business by any recognisable standards – nor proportional to the country that it occupies. Welcome to livestock farming in Wales.

For any analyst, an agricultural sector that contributes 0.7% to a country's economy and employs 1.9% of a its population, dominating nine-tenths of its land, on 80% public subsidy, is a recipe for no rural future in that country at all. Eighty-eight per cent of Welsh land (17,530 square kilometres) is given over to agriculture, of which 75% (13,260 square kilometres) is sheep pasture, containing 9.6 million animals. Those sheep occupy an area almost twelve times the size of Greater London.[3] Whilst arguments are made that sheep farming makes Britain more self-sufficient, this has not been true for quite a while. In 2012, Britain exported 94,700 tonnes of sheep meat, and imported 86,100 tonnes. Three-quarters of this came all the way from New Zealand. In 2013, the UK was the second largest importer of sheep meat in the world.[4]

Whilst farmers have, overall, been given special status with subsidies because they produce our food, only 1.2% of the calories in the British diet now come from lamb.[5] Lamb is an enjoyable but optional food source. Red meat consumption is nationally on the decline, partly for ethical reasons, but partly because of scientific worries that its consumption increases the risk of certain cancers.[6]

A better arrangement is required not only for wildlife to return to Wales, but for its rural jobs, even a generation from now. But surely, without farming, the nation would collapse? In centuries past, that was most certainly the case. Now, through no fault of Welsh farmers, the times have changed.

An Honourable Decline

During past centuries in Wales, sheep once powered a nation. At the start of the nineteenth century, in 1801, the majority of Wales's 587,000 people were still rural-based and working in agriculture.[7] This itself was the end result of a lucrative wool trade in earlier centuries. As early as the twelfth century, Cistercian monasteries had been granted special rights to graze sheep, and the country's woollen industry sprang from the enterprise of monks. Water-powered mills turned wool into one of Britain's first mass industries. Between 1350 and 1500, around fifty fulling mills, where wool cloth was beaten with wooden hammers to form finer cloth, became serious business in Wales. In 1372, the mills of the southeast produced 18,500 fleeces in one year. By 1660, wool made up two-thirds of all Welsh exports.[8]

From 1800, towns such as Welshpool began to industrialise – but new technologies were not properly routed across to Wales from the modernising industry in England. Weaving towns in Montgomeryshire, for example, only had four power looms by 1835, at a time when Bradford and Leeds were already thriving on steam. In winter, Welsh farmers employed their labourers in spinning, but Welsh wool was never really brought into the industrial era. Most wool manufacturing declined in Wales by the 1860s, and weaving did well until the 1920s. By 1947, there were 24 active mills. By 2013, there were nine.[9]

This decline has since been mirrored in sheep farming as a whole. Before 1900, the original nomadic sheep-farming traditions had already been replaced with large, enclosed farms. In 1851, 18% of the Welsh population still worked in agriculture. That has fallen, across 150 years, to just over 1% today.[10]

As early as the 1850s, labourers were also moving to the cities, beginning a 160-year narrative of rural evacuation. Large numbers of nomadic shepherds once wandered the Welsh hills, moving with their flocks. Today's large farms employ fractions of the pre-Victorian workforce. Yet even now, almost nine-tenths of the country is dedicated to an industry that, economically, is no longer alive, yet demands enormous monies from our taxes to survive.

It is now difficult to imagine a future – a fair or viable one, at least – where 88% of a country's land area lives on subsidies indefinitely; time-frozen jobs propped up like roles in a museum. This, in truth, is disrespectful to everyone – not just to British taxpayers but to younger farmers too. Today, the culture of sheep farming is a culture endemic to the 1.9% of Welsh people who still practise it, and that number continues to fall.

For sheep farmers, however, hill farming is the profession that they know. Suddenly, in their eyes, the economy and the environmental movement have both turned against them. They do not wish to see a long tradition in the hills vanish, and understandably struggle to envisage a future beyond the life they know. So why do those in the ecological movement push so strongly for change? One answer lies in the devastating silence of Welsh wildlife – the other, in the growing silence of its dying rural communities. Let's look at the wildlife first.

Feeling Sheepish

Choughs are what would happen if you took your local crow to a stylist. How they remain glossy and sleek in the face of Welsh weather is a mystery. Their coastal grass is no longer cropped by wild horses, but choughs now probe for insects with their curvy bills and dig for victory in the wake of sheep, which, on these wind-battered coasts, function in a similar way to horses.

Conservationists have become accustomed to adapting bird protection to the presence of sheep. If you watch cheeky wheatears foraging on the Pembrokeshire coast, or choughs searching coastal lawns for food, it's clear that our coastal grasslands are one area where sheep can form a more natural part of the picture.

Remote from the gardens of lowland England, ring ouzels, our mountain blackbirds, are birds of upland lawn. On their barren rocky slopes, ouzels hide their nests in heather but forage for food in mountain pasture: grass once kept low by horses or elk – and today by sheep. Studies in Scotland show that having the right amount of sheep is now important if you're going to get your lawn–heather balance right.[11] Too few sheep, and your pasture can be swallowed by heather. Too many sheep, and your heather is munched away – so too are your berry-rich plants, essential for young ouzels to feed in.

Lapwings are waders of soggy pasture. Sheep are selective feeders that avoid coarse vegetation, leaving open lawns for lapwings to feed, but some taller cover for chicks to hide in. If you travel to places like the Outer Hebrides, you can often find tiny sheep-fields alive with starlings and lapwings. Add a few more sheep, however, and things get worse. Dead stock, afterbirth and supplementary feeding draws in buzzards, crows and foxes. These prey on lapwing eggs and chicks, finding these by sitting on sheep fences. So only on rushy, low-intensity sheep farms, like some of those seen in some parts of the Yorkshire Dales, can birds like lapwings really thrive.

Recently, hen harriers on Orkney surged back when sheep numbers were reduced. Too much grazing of heather, and ring ouzels vanish. Only extensive grazing, by tiny flocks, benefits the black grouse.[12]

In very traditional sheep-farming cultures, like the truly nomadic transhumance shepherding still practised in the hills of Romania, or even the uplands of the French Alps and northern Spain, the relentless herding of animals across the landscape can lessen their effect. After the sheep have been moved on, upland meadows take root. In addition, in many European uplands, a large proportion of sheep farming takes place within heavily wooded lands. This creates a very different outcome, with short-cropped wood-pastures playing host to a far wider range of birds than a penned, cropped field. Indeed, looking to the wildlife of the Lake District in 1800, when it was home to red-backed shrikes and wrynecks, two species adapted to such landscapes, it is clear that the ancient shepherding dynamic would have created very different results in the countryside to those we see today, even if sheep, as a non-native animal, would never have boosted biodiversity in the same manner as our native, free-roaming herbivores.

In modern times, industrial sheep densities, confined and penned, or allowed to graze in one hillside all summer, have conspired to remove many, if not all, the nuances of nature. In Wales, this problem has now occupied a country – and created a landscape unintelligible to most native wildlife.

The Silent Hills

Being, originally, agile specialists of mountain slopes in the far southeast of Europe and the Middle East, sheep don't increase our native tree and plant diversity as our native herbivores once did, but actively degrade and prevent it. They are not, truly, the best animal for upland farming either, being infinitely more susceptible to ticks and other diseases than hill cattle, particularly old breeds like the Belted Galloway. The manner in which sheep feed, in particular, is alien to British grasslands, or the formation of native woodland, because they evolved in a far sparser, higher and rockier world than our own.

Cattle, by comparison, are unselective grazers, tackling vegetation tufts with their tongues. In natural densities they do not deplete nectar sources. They also avoid areas of their own dung, and these fertilised mini-grasslands grow longer as a result, creating variety in the grass, whilst dung recruits up to 200 species of invertebrates. Horses, more selective than cattle, create a mixed grassland, nibbling down some rough grasses, whilst others are left – creating a mosaic habitat for grassland birds. Wherever old-breed upland cattle have vanished from a landscape, as has happened in the hills of Galloway, we have realised, often too late, how these animals once created the rich diversity of sward needed by our vanishing curlews, and allowed the proliferation of diverse grasslands used by the black grouse.

Sheep, by contrast, whilst often seen as rugged animals, are surprisingly fussy in their eating habits. Rather than promote insect life, they remove nectar sources for invertebrates. They create a tight, uniform lawn, leaving the least tasty elements of a hillside, bracken, last. Sheep also produce a great deal of soil compaction when confined in a field. What remains, therefore, is a nectar-free lawn, often interspersed with the least valuable of all vegetation: ferns. Worst, and perhaps least known, when short of calcium reserves, a common state of affairs for a herbivore in a denuded landscape, sheep will eat the eggs and even chicks of ground-nesting birds.[13] Whilst culturally beloved of some farming communities, the use of sheep grazing as a tool in bird conservation, therefore, might be regarded as rather unwise. With large sheep herds grazing in one place, no ecosystem can rebuild from such a basis.

As a result of this playing out both in penned farmlands and across supposedly 'wild' landscapes like Snowdonia, Wales is now largely a country of two alien extremes: deep green forestry and pale green sheepery. Almost all native Welsh ecosystems now exist as remnants, scattered between these two shades of green. If this sounds unfair, satellite mapping will again provide the evidence. Here, on very close inspection, between the two seas of green, you can discover the postage stamps of gnarled Atlantic oakwoods. There are clear smaller rivers, with goosanders and otters; rock faces, with ravens and peregrines; and just a few birch moorlands, in north Wales, that still bubble with black grouse. Where human agency falls away at the coast, we have our wonderful seabird cities offshore. And that, really, is that.

In the entirety of Wales, fewer than 600 pairs of curlew survive,[14] compared to 68,000 in Britain as a whole. Curlews have declined in Wales by up to 80% in just the last fifteen years – often, as hill cattle have vanished and herds of sheep have increased, removing the taller grasslands in which to hide their chicks from predators. Of Britain's 5,000 black grouse, just over 300 are found in the uplands of Wales.[15] Of Britain's 500 pairs of golden eagles and 100 or more pairs of white-tailed eagles, not one soars over its mountains.[16] The story of silence goes on and on.

A recent *State of Nature Report* found that of 3,128 species of plants and animals in Wales, 60% have declined over the last 50 years, with 31% of these declining strongly and one in ten facing extinction.[17] The green shades of upland Wales may now constitute one of the most bird-free areas in Europe. It is hard to imagine an emptier area for wildlife. This comes down, in turn, to near total trophic collapse.

Right now, the upland estates of rural Scotland and northern England, for all the issues outlined in this book, still contain rushes, mosses, heather and other plants from which some kind of food chain can build. Small insect communities exist to feed wading birds. From this damaged but persistent food chain, species like grouse and hares feed, in turn, the Highland's golden eagles. By contrast, compacted lawn, covering enormous areas of western Britain, creates the greatest wildlife silence of all, one we have grown accustomed to over time. In all truth, most of Wales is now, in one form or another, lawn.

If the ecological assessment of the country cries out loud and clear for change, considering the human impacts of rewilding is every bit as important. If large areas of Wales were to be restored to a more natural state, what would this mean for its rural communities?

Facing Up to the Future

Deuparth gwaith yw ei ddechrau (Starting the work is two-thirds of it)
—Welsh proverb

Rural communities are the backbone of Wales. Far more than sheep, it is cohesion and common purpose that rural Welsh communities fear losing the most if things were to change. When words like 'rewilding' become so toxic that large charities, such as the National Trust, dare not mention them in Wales, it's clear that a war, a fairly heated one, has begun.

On the one side, those who wish for wildlife, ecosystem restoration, carbon absorption by woodlands, or a diversified rural economy in Wales, find themselves paying taxes to keep unproductive forms of hill farming alive. They express frustration when farming claims to be a culture beyond reproach, stifles the return of profitable wildlife and jobs – and yet relies on the taxes of others. In a democracy, taxpayers are entitled to ask for change, especially if their hard-earned money is on the line.

On the other side, sheep farmers, just as understandably, are sceptical that ecosystem restoration could benefit their way of life. Most farmers I have spoken to believe, in good faith, that somehow, in spite of being reliant on subsidy, hill farming will find some way to survive. Many also forget that just decades before, they too enjoyed more diverse landscapes by using smaller herds and more mixed farms. Many upland farmers fear, however, for the future of the next generation, and the wider social collapse that would happen if farming vanished.

The problem is that social collapse is already under way, and has been for well over a century.

With Wales having seen the collapse of community industries like coal-mining in recent decades, however, fierce protection of its remaining ones has understandably sprung to life. But with sheep farming on taxpayer life support, looking to other rural futures would seem a good idea for everyone. In Wales, some taxpayers may, through cultural or family ties, be sympathetic to sheep farming. But few will want to pay, collectively, billions of pounds for 1.9% of the people of Wales to use 88% of its land.

In January 2018, it was intimated by the UK government that future subsidies will not only pay farmers for environmental services, but pay those farming against the odds – to *stop*. In Wales, most farming is conducted against such odds. Eighty per cent of the land is officially deemed unfavourable to farming.[18]

A targeted, specific payment to wind down in certain areas would finally break a cycle of economic and environmental collapse for everyone in Wales. Farmers could be well incentivised to change course. Stopping would allow new rural economies a chance to establish over a period of time.

Wales's most depopulated spaces, where sparse sheep-grazing runs at an epic market loss, scream out the solution. In Snowdonia and the Brecon Beacons national parks, true ecosystem restoration could take place. Other areas, virtually depopulated and making no significant agricultural contribution, such as the Cambrian hills, could also be incentivised to wind down. In other areas, such as more productive valley farmlands, extensive farming methods, widespread tree planting, and the increased use of native animals such as cattle and pigs could also play a beneficial role.

No one I have met wants to see Welsh communities collapse – nor does the half-Welsh author of this book. So here's how a wilder economy could benefit communities in Wales.

Culture on the Farms, Nature in the Parks

The Brecon Beacons, Snowdonia and the Pembrokeshire Coast, the three national parks in Wales, occupy a combined area of 4,095 square kilometres. Snowdonia, the largest, is an impressive 2,132 square kilometres. In a restored wooded state, Snowdonia is so large that it might hold up to 21 territories of golden eagle[19] and 30 home ranges of individual lynx,[20] not to mention an ecosystem large enough to halt extinction in the many other Welsh species requiring smaller areas to survive.

At present, however, whereas Germany's national parks cover just 0.6% of its land area[21] but contain lynx, wolves and thriving ecotourism economies, Welsh national parks cover a fifth of its land, yet contain few ecotourism magnets that would be recognisable to visitors from another country. They are merely the continuation of degraded upland farming.

At present, according to a 2013 report, *Valuing Wales' National Parks*, the majority of the 'environmental' jobs in the national parks are in 'primary industries ... who derive their value from use of the environment'.[22] In other words, the only semblance of an environmental economy in Snowdonia, and the Brecon Beacons too, is a small number of people who harvest the park – like any other kind of farm. Few other countries' national parks exist merely as scenic resources for extraction.

Welsh 'environmental' jobs exist, at present, without a natural environment. They exist without Wales having a single charismatic predator or herbivore, or even more standard tourist spectacles such as eagles. If Wales's national parks were replete with amazing animals, the requirement for rural jobs would be *greater*. An entire sector lies untapped.

Right now, what's more, the unemployment rate is higher in Wales than in England or Scotland: the number of people out of work is 76,000 and rising.[23] This is a higher number than all the people working in Welsh agriculture. Virtually none of these people can now move into farming. Whilst the farming lobby opposes ecological restoration at every turn, it has no solutions to unemployment. The wider interests of Wales would be served by more jobs – and in part, we must look to nature, not only farming, to provide them.

On Mull, 23% of the island's 350,000 visitors cited eagles as a reason to visit, and Mull is far more inaccessible than the coasts or valleys of Snowdonia. On Mull, white-tailed eagles, or 'flying barn doors', have created 110 jobs in and around eagle tourism, in a sheep-driven economy deprived of opportunity much like upland Wales.[24]

Across a more significant area of land in Wales, the return of white-tailed eagles, in the lower valleys, wetlands and coasts, and long-overdue golden eagles, into the higher crags of the Brecon Beacons and Snowdonia, might reasonably be expected to generate more jobs again. The allure of eagle watch-points where viewers could, for the first time in centuries, watch golden eagles dance over Glyder Fawr, would create jobs not only for those directly involved, but in the accommodation, catering and wider tourism industries that arise around such spectacles.

In 2015, 3.89 million tourists visited Snowdonia National Park.[25] If, as on Mull, 23% of these were drawn by eagles, and each of these spent £10 by visiting an eagle nest-cam, and bought a coffee and a sandwich, that alone would inject £8.9 million into the local economy. Of course, if we factored in just one overnight stay in a B&B for a third of these people, at no more than £40, we add another £11.9 million and our eagle expenditure in the local economy increases to just over £20 million. It is such simple calculations that remind us why rewilding Wales' lifeless hills to even a moderate degree would be a good idea for everyone.

In Scotland, nature-based tourism is estimated to produce £1.4 billion per year and maintain 39,000 full-time equivalent (FTE) jobs.[26] English tourists, who, of course, also visit Wales, spent an estimated £21.1 billion on their trips in the financial year 2013/14.[27]

If the Welsh government considered a wild economy in Snowdonia, the only obstacle would be that 76% of it falls under agricultural land – but that land is unproductive, with just tiny flocks of sheep. Of 25,000 or so people living within the park, just 870 – the equivalent of one person every 2 square kilometres – work in agriculture. This figure includes forestry and the fisheries, representing 7.2% of the jobs in the park.[28] The number of farm workers in Snowdonia has fallen dramatically in the past fifty years.

Fisheries jobs are not threatened by ecosystem restoration. Indeed, animals such as beavers increase fish stocks, by increasing spawning habitat. And you need more foresters, not fewer, to help in the restoration of oak forests. Only failing sheep farms prevent the thriving of a rural economy in Snowdonia. And with the economics as they are, the Welsh government might wish to think about what other jobs could be thriving here.

Our quick-fire example of the potential value of eagles in Snowdonia really takes account of just a fraction of the economic benefits of a full wildland economy in fifty years' time. Excluded, of course, are the villages secured against flooding with the restoration of native woodlands, which break and absorb the flow of water off the hills – and the incalculable cost of lives saved. Excluded are the full range of catering, accommodation and transport jobs that ecotourism provides. Excluded are the economic possibilities of other iconic species such as lynx. Excluded are the sales of prime wild meat and the jobs associated with its production. Excluded are attractions such as bugling elk ruts, or the potential of beaver-watching excursions. Excluded are the tens of thousands of visitor nights that slowly increase, each season, as more animals, worth travelling to see, make a return.

In ecotourism economies, the 'draw' of people is far wider than in national parks with farming, even if farming unions continue to peddle the myth that farming creates some kind of mystical draw for tourists. For example, whilst a handful of local people may quietly appreciate a sheep being 'hefted' uphill in Snowdonia, you could poll the British public and not find one who has travelled to see such a sight. Many Welsh people I know have no idea what 'hefting' is. Once told, they do not drive 240 kilometres to see it. But 240 kilometres is the average distance that a British ecotourist travels to see Mull's white-tailed eagles. Specific attractions like these act as catalysts for wider spending. If there is one thing I most emphatically believe in writing this book, it is that the best way for sheep-farming cultures to survive, in the future, is to exist *alongside* ecosystems and ecotourism, and benefit from their revenue – not to stand in angry opposition against them.

Rural Wales lies in desperate need of jobs. Restoring Snowdonia and the Brecon Beacons, and returning at least two charismatic mammals, most suitably elk and lynx, would certainly drive unemployment numbers down. As a rule, the more visionary and exciting the venture, the more interest, money and jobs, are drawn to it over time.

Younger people in rural areas, finding themselves without a farming future, yet loathe to abandon the land, could be paid incentives by the Welsh government to diversify. Communities would thrive. Extensively grazing native animals, increasingly seen as a key part of ecological restoration, would provide for the continued husbandry of livestock, albeit in a different way from that we see today. Wild national parks across the world benefit from the investment of their local people. Around the world, rural populations have kept their cultural traditions alive through increases in tourism.

Perhaps, one day, our grandchildren might spend a morning in Snowdonia, learning to shear sheep in enclosures in the valleys, learning the importance of shepherding and all it did for Wales – then spend the afternoon seeking eagles, lynx or elk in the wooded hills above. It would still be the community-driven Wales of today, but the difference would be summed up in a single word – *proportionality*.

Right now, there is none. Less than 2% of the population of Wales decides the odds for its entire landscape – and the rural future of everybody else – using other people's money. Communities, grown reliant on subsidies for survival, will suffer more than most when, one day, the subsidy rug is pulled from under their feet by a government strapped for cash. Accepting wilderness in the parks, however, would be a gateway to the future.

The Living Valleys

In the end, only Wales can decide between farming for tradition's sake and jobs that can support themselves. But, in my view, the future of rural Wales does not pivot on lamb meat. It hangs on community. Now that coal-mining has collapsed, nobody laments the coal. But to this day people lament the loss of shared values, cohesion and purpose in the valleys where coal was once king. In a couple of decades, the same could be true of sheep. Or, things could go a different way.

For there to be a future in rural Wales, there have to be jobs – real jobs that markets demand, not only those that taxpayer benefit provides. Even if sympathies run high for livestock farms today, in fifty years' time it might be asked how one loss-making concern was so keen to relive earlier times that it forgot about all the generations to come. It might be asked how, with ever-rising unemployment, future communities were killed off before they had the chance to flourish in the first place.

Yet the day, fifty years from now, that radio-tagged Wmffre the wolf gains followers on social media as he pads through the cwms of Betws-y-Coed, the day that songs are written about the tufted cats living like ghosts in the hills, may also be the day rural communities in places like Snowdonia are most likely to pulse once more with human life as well. And in a hundred years' time, if *that* path is followed, nobody will lament the loss of villages and people in the valleys. The reason will be simple. This time, there will be nothing to lament.

CHAPTER 11

A Grouse Moor Wild

Restoring England's Uplands

The genuine sportsman is by all odds the most
important factor in keeping the larger and more
valuable wild creatures from total extermination.
—Theodore Roosevelt[1]

If only the above lines, written by one of the great hunting patrons of early American conservation, were true of sporting estates in our own country, then we might all walk in all the hills of northern England and enjoy the glorious sight of eagles in our skies, wildcats in our woodlands and an ecosystem teeming with life.

In Britain, during the mid nineteenth century, the upland estate owners of the day took a different route to that prescribed by Roosevelt. They decided on total extermination of the many large and beautiful wild creatures on their estates, and of wilderness itself, to make way for the intensive farming of one quarry – the red grouse. Regrettably, to this day, in our ever more educated society, this model continues to dominate our uplands.

The creation of Britain's grouse farms, the heather wilderness that now covers much of northern England's uplands, and is mistaken by many for our natural landscape, has been as destructive as it has been unique. In centuries past, however, the earlier royal 'forests', or hunting areas, were nothing like this at all. On a morning's walk through the New Forest, few will curse William the Conqueror, or the royal tradition of protecting hunting land. And the reason for this is simple. In centuries past, hunting estates were areas of native wilderness, protected forever against agricultural development. Into these landscapes, rich in many forms of life, wealthy hunters would enter – kill – and leave.

Across the planet, first-world countries with large depopulated areas would recognise Britain's original royal-forests model as common sense.

From Alaska to Scandinavia, hunters, alongside ecotourists, bring considerable money to sponsor the natural world and its charismatic animals. Excited intruders in the wild, hunters take a quota of animals from within their number – and leave the rest. No one farms their uplands for mass killing sprees conducted once a year.

In Britain, in the past two centuries, it is not hunting, or hunters, that have left us with the most bird-robbed uplands in Europe, but the farming of grouse to be killed in their thousands each year. This simple change in British hunting must now be reversed – if Britain is to be left with any chance of beautiful uplands in the future.

After two hundred years of destroying our national heritage, the grouse moors have left our country immeasurably poorer. Not only has the landscape been burned and denuded of most life, and a rich heritage of trees, animals and birds of prey robbed from all, by few, but these are economic deserts on a scale that few people realise. Most arguments against driven grouse shooting start with the wildlife – so let's start instead with hard cash.

Grouse Versus Jobs

In England, according to the Countryside Alliance and the National Gamekeepers Organisation, when reporting to parliament in October 2016, the country's 175 grouse moors support 1,520 full-time equivalent (FTE) jobs.[2] Only 700 of these jobs are directly involved with grouse moor management. A further 820 are in related services and industries. It is not clear reading the report what these jobs are, or indeed if they exist.

In Scotland, a report written by the Fraser of Allander Institute in July 2010 stated that 140 'core' estates in Scotland, who responded to their enquiries, supported 493 jobs.[3] The report's authors estimated that if this held for all Scottish grouse estates, these would support a total of 1,072 jobs. Adding this total to the English one makes for no more than 2,592 people employed in grouse shooting across Britain, of which no more than 1,772 are directly involved with the management of grouse moors.

As of September 2018, the number of people in employment in the UK is 32.4 million.[4] Grouse shooting, if we include the 'related services' jobs too, therefore accounts for 0.008% of the jobs in Britain – that's one job in every 12,500. In terms of ownership, the 175 English grouse-moor landowners currently represent 0.0003% of the English population.[5]

Astonishingly, however, the grouse moors of England and Scotland cover around 8% of Britain: an area of 16,763 square kilometres.[6] To put this in context, this is an area almost twice the size of Yellowstone, seven times as big as the Lake District,[7] and ten times the size of Greater London.[8] Such an area, alone, could cater for the full restoration of Britain's wildlife, even if every other land use stayed the same.

Those grouse jobs, however, equate to only one job for every 6.5 square kilometres of land used by grouse shooting. No other European country has, or would tolerate, such extraordinary wastes of national land, in employment terms alone. Preventing most forms of ecotourism, grouse moors block jobs on a massive scale. The economics get worse, however, when you look at what grouse moors contribute for their 8% coverage of our country.

Back to the Fraser of Allander Institute's report, commissioned by the industry-friendly Game and Wildlife Conservation Trust. The report states that in Scotland just 43% of estates made a profit on their grouse-shooting activities. Only 46% of employment on studied estates even came from grouse-shooting jobs. Scotland's 1,072 grouse jobs, it estimates, cost £14.5 million in wages and contribute £23.3 million to GDP.[9] Millions always sound a lot, if taken out of context. The UK, however, the fifth largest economy in the world by GDP, was worth, in 2016, $2.647 trillion, or £1.88 trillion at the time of writing.[10] That Scottish grouse-moor contribution, with the conversions done, therefore comes to 0.001% of UK GDP.[11]

In England, the Moorland Association estimates that the economic value of the grouse-shooting industry is £67.7 million.[12] That's under 0.004% of GDP.[13] This brings the combined contribution of grouse shooting, in England and Scotland, to less than 0.005% of the United Kingdom's GDP, a contribution made from 8% of its land!

It gets a little worse again. In 2014, Friends of the Earth used Land Registry data to identify 550,000 acres (223,000 ha) of grouse moors across northern England (see Plate 19). Of these, just 30 estates covered 300,000 acres (121,000 ha) of our uplands. It was found that these estates received £4 million in taxpayer-funded subsidies, in 2014 alone, from the EU's Common Agricultural Policy.[14] Not only are jobs and wildlife being prevented from returning across England's uplands – the public are paying for this to happen. So with grouse moors using 8% of Britain to no profit – often, even their own – how does ecotourism stand up in comparison?

Grouse Versus Tourism

National Parks England provides for some 48,000 full-time jobs in the tourism sector.[15] The grouse moors in England create 1,520. Tourism-based jobs in our national parks, even before there is charismatic wildlife to see in them, trump the jobs economy of grouse moors 32 times over. Whilst England's national parks harvest at least £4 billion in tourism and recreation expenditure,[16] grouse moors in England, adding £67.7 million, contribute 44 times less to local economies.

In contrast, the grouse-moor money supposed to reach rural communities appears to form part of a closed 'loop'. Some £52.5 million is spent on land management, which deprives our nation of beautiful wild uplands, birds of prey and such basics as trees. A further £15.2 million covers travel and accommodation.[17] But this accommodation is for the tiny minority of the British public who pay to shoot grouse in the first place.

The British Association for Shooting and Conservation (BASC) estimates 40,000 shooting visitors to grouse moors annually across the UK,[18] which represents around 0.06% of the British population.[19] The Moorland Association claims that 6,500 'visitor nights' were generated by grouse shooters in 2010.[20] If 40,000 shooters visit grouse moors, this means just 16% of them spend a grand total of one night each in a bed and breakfast. By contrast, dolphin watching on the Moray Firth alone generates up to 52,200 overnight stays in a year.[21]

If you look at the revenue of grouse moors, it's irrelevant to most businesses in the countryside around. A shooting day with thirty beaters (grouse chasers), nine loaders (gun-helpers) and drivers for five vehicles may cost the industry around £1,800, and a really large shoot can cost up to £30,000.

The only jobs created in this interchange of money are for gamekeepers – at the cost of the tens of thousands of jobs that arise in ecotourism economies, which could incorporate these rural jobs, too, if an area ten times the size of London were not used for grouse shooting. By contrast, in 2009, the RSPB's reserves brought £66 million to local economies, and created 1,872 FTE jobs. This is comparable to all of England's grouse moors, but in just a fraction of their land area.[22]

Compared to 40,000 visitors to grouse moors, Scottish Natural Heritage finds that 240,000 tourists visit western Scotland for whale-watching each year: a mere snapshot of one local ecotourism economy.[23] An estimated 290,000 people visit osprey watch-points around the country. A minimum of 250,000 people fuel local economies by visiting our seabird cities. Around the RSPB's South Stack reserve on Anglesey, in 1998, 43,000 visitors spent £418,000 in the local area – money that actually feeds into local pockets.[24] Even today, the English adult population make just over 3 billion visits to the natural environment each year, spending £21 billion as they do so. In Scotland, nature-based tourism is estimated to produce £1.4 billion per year, along with 39,000 FTE jobs.[25] That not only shoots down grouse-moor economics, but all the grouse jobs in Britain come to just 6% of this regional total.

The RSPB has over a million members. The National Trust has a membership of 4.24 million. 'Nature-minded' people in the UK, it would seem, are up to 11,000% more abundant than grouse shooters. We are very much more profitable to our economy – and the proof is out there for any government to see. Ecotourism economies trump grouse economies on a staggering scale, even before we have restored most of our nation's wildlife.

A government report, *Valuing England's National Parks*, states that 'national parks will be recognised as fundamental to our prosperity and well-being'.[26] Replacing a destructive grouse-farm economy with a profitable wilderness economy would seem a sensible thing to do – not only for the innate wonder of our children, but for national economic sense.

Right now, the northern Peak District, the North York Moors and the Yorkshire Dales are dominated by grouse moors. The Forest of Bowland and North Pennines 'Areas of Outstanding Natural Beauty' are entirely taken over. The Land Registry map created by Friends of the Earth (Plate 19) shows how one strange twist on hunting, adopted nowhere else on Earth, dominates the uplands where jobs and wildlife could all thrive in the future.[27]

Eight percent of a nation's land, burned by 0.0003% of its population, for the creation of 0.008% of its jobs and under 0.005% of its economy, is a destructive national embarrassment. There is a second cost, of course. And that is the wildlife.

Fake Conservation

Over two hundred years, our extraordinarily rich heritage of wildcats, golden eagles, hen harriers and many other species has been taken from everybody, to enormous national detriment, by the spread of the grouse farms in our uplands. The scale of wildlife removal has been as breath-taking as it was entirely optional, as we saw in Chapter 2. Whereas arable farmers removed biodiversity by expanding large crops for a very good reason – growing our food – grouse-moor owners burned our uplands to the ground so they could farm grouse to kill, in enormous numbers, for no valid reason at all.

With almost all the wildlife already gone from our uplands, and just the relics left, a clever language of 'fake conservation' has been developed by the grouse-moor industry to conceal this from the public – and it's been surprisingly successful. The rhetoric has focused, largely, on the wildlife capable of surviving in the wreckage. So often repeated has this rhetoric become, it has allowed grouse moors to claim a 'conservation' role in our national parks, whilst despoiling them of most of their wildlife.

If you visit the ever-vanishing rainforests of southeast Asia, you will often come across areas of jungle felled for palm oil. These clearings are not entirely lifeless. Around twelve species of bird can thrive in an oil palm plantation, in place of around a thousand in a pristine primary rainforest. But those that thrive in these burned lands do very well indeed. Munias chatter in the bushes. Palm swifts hawk through the treetops. In Britain, of course, we have become accustomed to burned lands. These are our driven grouse moorlands. And in place of unnatural densities of sunbirds, we have unnatural densities of red grouse.

Grouse moors, covering such a large percentage of our country, inevitably create huge areas where certain birds can survive. Curlew, lapwing and, of course, red grouse, are three birds that thrive in a simplified world of heather and grass, where predators are entirely wiped out.[28] As a result, grouse-moor managers have focused heavily on the last survivors of our uplands, claiming that without their protection such birds would be unable to survive.

It must be remembered firstly that intensively managed grouse moors are not only unique to Britain but have only existed in large numbers since the mid nineteenth century. Birds such as curlews, however, have existed for millions of years.[29] Whilst most upland birds evolved alongside large herbivores, and curlews show many signs of being best adapted to a life beside cattle, no species on earth has had the time to evolve on a grouse moor. Indeed, Swedish and Finnish curlews are found largely in extensive meadows or wind-crafted moors with trees – and not on burned lands at all.

In addition to claiming strong protection of a small selection of birds, the second thing that the grouse lobby sets out as 'conservation' is its burning of the uplands. Moors are often burned in April – possibly the least natural time heather would catch fire, just after the winter. Wild fires in temperate Eurasian Russia come around August, at the end of summer. This doesn't work for grouse moor owners, because this is when they choose to shoot grouse. Furthermore, pure drained heather is not a habitat, but an absence of one. In other words, it would not have existed to catch fire in the first place. Finally, it is deeply unlikely that lightning striking a blanket bog in the early Holocene would have kindled fire. An undrained, natural, birch-crowned moorland is a virtually non-flammable habitat. In southern Sweden, for example, in a comparable biome to northern England, deer, wind, water and tree growth shape the uplands. Fire on these bogs and moors is a very rare occurrence indeed.

The natural upland ecosystem of Britain, similar to the upland ecosystem of southern Scandinavia, should in fact be a wooded mosaic, with heather and wet bogs jostling with trees like willow, oak, juniper and birch. Cleansed purple heather, rolling for mile after mile, is as natural as an oil palm plantation. Compare the two Scottish landscapes shown in Plates 20 and 21 – one, a burned grouse farm in Aberdeenshire, the other, the wonderfully protected Rothiemurchus Estate in Speyside.

The final aspect of Britain's grouse-moor 'stewardship' is to rig the odds against predators. This works in favour of the breeding success of ground-nesting birds – most of all, the red grouse. Gamekeepers shoot members of the crow family – particularly carrion crows and magpies, which predate large numbers of grouse and wader nests.[30] Keepers further improve wader breeding success, though this is not their primary goal, by killing foxes, weasels, stoats and, in some cases, illegally killing golden eagles, peregrines, buzzards, goshawks, badgers and hen harriers.[31] An unusually high incidence of illegal poisoning in birds of prey, or the untimely disappearance of tagged birds, is directly mappable to the presence of driven grouse moors.[32]

With virtually all predators removed, a treeless grouse moor becomes a prime, unnatural crèche for red grouse chicks, farmed in the densities needed to be shot in their thousands. If you destabilise an ecosystem in favour of birds that are prey species, however, that's not great news for their predators – which have evolved to keep the ecosystem in balance.

Habitat-mapping studies show that 300 or more pairs of hen harrier should float across England's uplands.[33] This calculation is supported by numbers elsewhere. Orkney alone, for example, its moors managed largely by the wind, averages around 100 breeding pairs – in a fraction of the area of the English grouse moors.[34] Often described as being on the brink of extinction in England, this is no new state of affairs for hen harriers. The peak of hen harriers in recent decades has been just twenty pairs. In recent years, it has ranged from six – to zero.[35]

In southern Scotland, one study has explored the dynamics of grouse, predators and their human hunters to a forensic degree. The Langholm Moor Demonstration Project straddles the border of Dumfries and the Borders. Here, a combination of gamekeeping and bird conservation interests engaged, for years, in an experiment. The question at the heart of the study was simple: could birds of prey share a landscape with the intensive rearing of grouse?

Throughout the experiment, Langholm was managed as a grouse moor. It was burned and its common predators controlled, but its hen harriers were fed diversionary food in the nesting season. Between 1988 and 1999, this was shown to reduce the number of red grouse chicks caught by hen harriers, the reason for their intense persecution, by 86%. It was also found, however, that 78% of satellite-tagged red grouse found dead were killed by birds of prey. This is what birds of prey do. It is also why birds of prey will never be economically possible on intensive grouse moors.[36]

Langholm's keepering experiment concluded that whilst red grouse and hen harriers could co-exist on a moor without one wiping the other out, as indeed they do across Eurasia, there would never be enough *surplus* of red grouse for driven grouse shooting to make its money. Driven grouse moors making profit, in other words, *preclude* a nation's heritage of birds of prey.

Large areas of the central and eastern Highlands of Scotland, some of the most prey-rich areas in Britain, have no golden eagles for a similar reason. Given overall increases in Scotland, these areas should be regaining their eagles too. Overall, it's estimated by the Scottish Raptor Study Group that there should be 600 pairs of golden eagles in Scotland – 100 more than at present.

Every scientific study reveals the nature of the golden eagle's problem. Firstly, eagles are shot, both as adults and as dispersing younger birds, enough times to prevent recolonisation. Secondly, they are disturbed, should they occupy home ranges where their presence is not welcomed. This occurs in areas statistically likely to be managed for grouse, rather than deer. The latest of several studies concluded that 'in the central and eastern Highlands where grouse moor management predominates, the eagle population continued to decline to levels where increasingly large areas of suitable habitat are unoccupied by breeding pairs.'[37] It is a backward country, for us all, when a bird as globally respected as the golden eagle, as beneficial to tourism revenue, must give way to a tiny minority of people farming birds to kill. Deer estates, by contrast, are by and large supportive of golden eagles, and a proof that 'hunting' is not the problem – intensive grouse farming is.

When it comes to protecting even tiny quantities of our nation's once-plentiful birds of prey, globally unique compromise stances have, for decades, been pursued to no avail by conservationists. Even such tiny compromises as feeding a bird of prey with rodents have been ignored. Yet the persecution of birds of prey is, arguably, not the problem – it is merely one terrible symptom of a far wider malaise.

Even if you wander a burned, treeless moor in future years, knowing you might see a hen harrier or golden eagle, knowing it has, graciously, been kept alive, what a damaged world you would still live in. The current grouse-moor model doesn't just preclude birds of prey. It prevents wildness itself: charismatic animals, natural landscapes and ecotourism. Because most of our birds evolved in wooded mosaic habitats, grouse moors, being treeless and burned, with just a fraction of native food plants, stifle most wildlife – most of the time.

Black grouse, for example, thrive best between trees and open, flower-rich grasslands. Intensive grouse moors denude their treeline and deprive them of their food. Whilst black grouse can make do on heather and cottongrass, they have evolved to move around and eat a whole range of plants, leaves and shoots at different times of the year, from different parts of a varied landscape. Very oddly, grouse-moor owners often claim to benefit black grouse. Yet there are now none to be found on the majority of northern England's grouse moors, in areas where they bred commonly two centuries ago.

In harsh winters, hawthorns, alders and rowans, growing on valley sides, all provide critical bounty for the black grouse to survive. Treeless moors in the Pennines rip out the natural diet of the black grouse

– increasing rates of winter mortality.[38] Like any species evolved before 1850, black grouse do not require grouse moors at all – simply, the presence of some heather buds in their diet. Wooded moors, filled with birches and rowans, not burned treeless lands, are what actually increase black grouse numbers.

Many mixed-landscape species, like cuckoos, founder on grouse moors: they need to watch their host species from trees. Our two species of 'eared' owls hunt rough grassland quivering with voles. Intensive moorland management drains and desiccates the wet, rough grassland where voles hide. Long-eared owls require stands of hawthorns or pines in which to nest. Intensive grouse moors burn those trees into submission. Where they are common, as on Orkney or the Hebrides, short-eared owls do not hunt heather moors. They hunt grasslands, with some heather in between. A monoculture of heather removes the habitat of these graceful daytime owls. Indeed, away from the lower-intensity moors of Teesdale in County Durham, short-eared owls have become rare across many northern English uplands.

Grouse-moor managers often claim that tree-free, predator-free heather benefits birds like merlins – but this too is now being brought into question. Merlins show unusually high rates of ground-nesting in Britain, compared to the continent, where they favour old crow nests in stands of pine. Where British merlins do nest in trees, their broods are larger, not smaller, than on the ground, revealing, again, how they have evolved – rather than how they are now forced to survive.[39] On grouse moors, the natural treeline, where merlins would naturally nest, has long been burned out.

Studies in the Lammermuir Hills in southeast Scotland showed that merlin numbers dropped on two estates by 50% after intensification of the burn.[40] Repeated burning of merlin habitat removes their meadow pipit prey, which nest in tussocky grass. The study, with supporting photographic evidence, also showed that one merlin, several dippers and two young ring ouzels were all killed by rail traps put down to catch predators on a grouse moor; a specific spring-mounted trap placed on a wooden beam over a stream, a key habitat of dippers. Moorland 'conservation' doesn't get much worse than killing moorland birds, especially fragile species such as merlins and ring ouzels, which grouse moors explicitly claim to protect.

If our moors grew wilder and were studded with trees, black grouse, cuckoos and many other birds would surge back. Merlins would elevate their nests off the ground. Short-eared owls would return to hunt for field voles, as they do on the moors of Orkney. Ring ouzels, a treeline species whose young are evolved to seek out juniper berries, would increase. And not one wild bird would end its life in agony, dying in a trap.

Oddly, in spite of all this, simply banning grouse hunting may not be the solution. The absence of a negative does not restore nature. And here's what happens when you stop something bad, without a decent plan of action for something good to take its place.

Lessons from Berwyn

As we have seen throughout this book, landscapes with an absence of large mammals are inherently unnatural and species-poor. *Neglect* of the uplands is not the same as restoration. If grouse moors reform, there must still be wilder stewardship – and new forms of mammalian management must come to the fore.

This is where Berwyn Moor, in Wales, can teach us some valuable lessons about how desolate an abandoned grouse moor can be. After the Second World War, large areas of Berwyn Moor were managed for driven grouse shooting. But then, by the 1990s, driven grouse moor management slowly tailed away until it stopped completely – but nothing was put back in its wake.

The declines and increases of birds were stark.[41] Red grouse declined as heather gave way to rank grass. Between surveys in 1983–1985 and 2002, pasture-loving lapwings vanished entirely from the moor, along with 90% of golden plovers and almost 80% of curlews. Driving some of these declines, nest-robbing carrion crows increased by over 500%, and ravens by 300%, in the same period. Berwyn became a silent wasteland. Only the crows, it seemed, had won the day.

Various conclusions have been drawn from Berwyn by the increasingly few organisations that now defend driven grouse shooting. The main argument has been that if you take away predator control, predators go haywire – and wading birds vanish. A lot of discussion has been devoted to whether these claims are correct. In my view, they are correct. They are also entirely irrelevant. The argument should not be about driven grouse moors at all. It should be about stewardship.

In recent decades, conservationists and grouse-moor owners have become locked into arguing about something very odd and entirely unrelated to nature: how best to manage a treeless wasteland. If you have no wild grazers, no natural processes, then you need some kind of opening agency in its place. Gamekeepers, armed with fire, guns, and a few sheep, had become stewards on Berwyn – and their loss created, simply, a vacuum. The elk and wild horses had long gone. The Victorian stewards, the keepers, had gone too – so the crows and foxes flooded back in. The crow's natural hunters, such as goshawks, were in short supply, and others, like the golden eagle, centuries departed. The fox's natural hunter, the wolf, was consigned to children's fairy tales. Nothing was there to keep the ecosystem in balance. The crows and foxes grew to huge, unnatural numbers. They raided nest after nest – until the lapwings vanished. Nobody brought native grazing animals back to maintain plant and tree diversity. Sheep, having played no role in the evolution of Britain's ecology, could not rejuvenate the moors.

Berwyn should be a salutary lesson for anyone who believes that stopping bad practice brings wildlife back, or that stopping people doing bad things to wildlife is a magical recipe for restoration. It isn't – at all. Instead, you need active agencies to rebalance and rejuvenate ecosystems. Today, if you drive across much of upland Wales, there is an eerie silence, punctuated by the rasp of crows. This is *not* a return to the wild – or something to which any of us should aspire.

If grouse-moor owners simply stop managing the land, but if conservation cannot afford to buy it and government has little interest, which usually it does not, then either the moorland will degenerate to rank grassland, as at Berwyn, or one of two equally disastrous futures might emerge – forestry, a death-blanket for birds, or new forms of agriculture. This is why a 'ban' on grouse shooting may not, in itself, lead to a wilder future. Better, perhaps, to envisage a future where we put the wild processes back in. True upland restoration, beneficial to wildlife, the public and game hunters alike, must now come to the fore. It's time to restore the moors.

Beyond Canned Hunting

In Britain, grouse moors are only profitable when 'driven' – managed as intensive farms, where grouse are driven in their hundreds over the heads of high-paying shooters. Even then, the profit margins are often small. When operated in a traditional manner, with hunters stalking skilfully on foot, these moors rarely make a profit. What is unique about grouse moors, among the world's hunting economies, is their economic reliance on just one species, killed in a strange and specific way. Diversifying the hunting harvest, coupled with ambitious ecological restoration, could replace a destructive economy with a far more profitable model.

At the moment, grouse-moor owners have to maintain an almost impossible position. To the nation's 40,000 grouse shooters, 120,000 people recently signed a petition calling for the sport to be banned completely,[42] and grouse moors are increasingly forced to defend the indefensible. At huge management cost, they feel compelled to burn our hills, wiping out tens of thousands of native animals and create heather farms: all for just a few days' hunting per year. A century ago, a few hunters used to feel compelled to shoot ten tigers in a day – and today most hunters look back on such people with pity and regret. Hunters from other countries, too, appear confused and embarrassed by Britain's canned hunting of birds. Indeed, clay pigeon shooting already caters for those who prefer to down hundreds of flying objects in a short space of time. And it is simply because grouse are farmed to be living clay pigeons that Britain has been left with the most robbed uplands anywhere in Europe.

So what, then, is the difference between an intensive grouse moor and a wild upland? Once you know what you're missing, it's enormous. Other countries maintain far greater upland diversity than Britain without a single grouse-moor gamekeeper acting in the way they are paid to do here at present. Scandinavia preserves curlews without fire and traps. 'Open' landscapes are retained without burning – an example is shown in Plate 22. But how do these landscapes survive?

The Hunted Wild

If you're British, the word *fjelljakt* ('mountain hunting') does not slip off the tongue. And if you're passionate about saving nature, what Fjelljakt does, arranging expensive hunting holidays in Sweden, might not appeal to you either. Yet Fjelljakt has a lot to teach us about how Britain's uplands could one day look. In Scandinavia, as in Alaska and elsewhere, the uplands are populated – by ecosystems. Large broken woodlands – dominated by pine, willow and birch – mingle with open mires, heather and grasslands. But this varied wilderness is hunted too (Plate 23). It's hunted for grouse, and it's hunted not only for deer but also for elk and boar.

This is where Fjelljakt comes in. It's a lucrative hunting enterprise that, together with ecotourism, places commercial value on these wild Scandinavian uplands in the same way that some huge areas of African wilderness are protected against agriculture.[43] It may do something that some people disapprove of, but it doesn't trash Sweden's uplands, or Sweden's national heritage, for everybody else. In fact, Fjelljakt operates a hunting system used across most of the developed world's wilder areas. It's known as sharing.

Rather than create grouse farms to the obliteration of all else, Fjelljakt makes serious profits by having its wealthy customers pay to harvest from a varied landscape. This modest change means that whilst the hills of Lancashire are alive with red grouse, burned heather and little else, the private uplands hunted by Fjelljakt's customers are home to a richness not seen in Britain for three millennia.

It hasn't even occurred to anyone in Sweden to burn their uplands to the ground, in order to populate the ashes, entirely, with one type of quarry. Instead, their diversified hunting enterprises appeal to adventure, wilderness, long slogs, and *skill* – the skill of hunting wild, elusive animals with a few guns, and a few dogs, as true hunters do. Into the wild go such hunters, drawn, often from cities, by the challenge of the wild. And this wilderness is varied indeed. The areas around Mattmar, in western Sweden, where Fjelljakt operates, are filled with willow and black grouse, capercaillies and woodpeckers. Its open marshy areas bugle with cranes and curlews. Its birch-dotted moors are hunted by hen harriers: Sweden has twice as many of these daytime ghosts as Britain. Across the open woodland, upland predators such as golden eagles, purged in vast numbers in the 1800s to make way for British grouse hunting, still hunt beside the hunters hunting grouse.

If you visit the uplands of southern Scandinavia, you enter a softness where habitats mingle and merge. The largest steward of these wooded uplands is the elk. Elk act against woodland formation in Scandinavia. They feed predominantly on willow, creating stands of trees where otherwise there would be dense scrub and forest, and so maintaining open areas of moor. Competing against the browsing power of elk, as well as deer, is the forest itself – pines, willows and birches, all pushing upwards. In some areas of Sweden, wolves manage the elk, scattering them so that they defecate more seeds. In others, where wolves are still absent, elk are managed by hunters. This lightly stewarded wild forms your typical Scandinavian upland, most of all within its national parks.

In such places, the behaviour of willow grouse reveals how red grouse would have fared prior to burning by gamekeepers. These grouse, of the same species as ours, shelter and feed in stands of willow, which increases their chances of survival through the winter. It is from between these willow stands that the guests of Fjelljakt hunt small numbers, at high prices, with trained dogs, great skill and much enjoyment. On the private lands owned by Fjelljakt, up to fifty elk are harvested each year, bringing important money in to sponsor the ecosystem. No wildlife is trapped or wiped out. Checks and balances work out. Democracy is at work.

A world away from the lifeless uplands we share in Britain, this hunted wild is no pipe dream. It's real, it's profitable and it works, from conservative Alaska to the centre-left countries of Finland, Sweden and Norway. In a mixed mosaic landscape, with large animals present, hunting of all kinds becomes more lucrative and diverse, and bestows greater prestige on the owners of the land. Both hunters and conservationists win, the moment that landowners become the custodians of *wilderness*.

The Two Roads Ahead

The year is 2060. You are walking through the willows of the Monadhliath mountains in the Cairngorms National Park. You hear a sharp crack – but there is little terror in that sound. This is not a golden eagle being shot out of the sky because it threatens surplus grouse stock, but a crack shot, a paying member of the public who has stalked for hours to bag a tiny quota of prized, expensive grouse, as they flush, nimbly, from between the stands of trees. It's not your cup of tea, but Britain's uplands are shared, after all. No one has it all their own way. Long gone, however, are the days when burned lands robbed your country of wildlife, jobs and billions in income.

Soon, the shot rings away. The birdsong fades back up. The light gleams on the silver birches and russet heather and the 'go back, go back' call of the red grouse grates in your ears as it flees a cluster of willows. Above the curlews yodelling beside you as you walk, you spot the perfect silhouette of a golden eagle overhead, scouring the hills for the grouse that flies too late. A pair of cranes sweep close overhead, bugling to one another. Your father would have considered these lowland birds, but you know now that they thrive just as well in wooded upland bogs. Hen harriers chatter, black grouse bubble, mountain ringlets dance and, just for a moment, you see the shadow of a shadow. Perhaps it was a lynx. Perhaps it was a wildcat. You will never know.

What's extraordinary, wandering these hills, is how many local enterprises and how much local income has come from people simply watching elk nibbling a birch tree, chasing the ghost of a lynx or wildcat, or visiting eagle feeding stations here each winter. There are the elk-view cafes, the camera shops, bike trails through new woods, and, like it or not, the carefully governed hunting operations, where wealthy clients harvest elk, boar, deer – and grouse – from the wild. A whole sector of the British public is travelling abroad less and less, spending their money instead on the wilderness at home.

Or perhaps, one day, like today, you'll be walking over burned heather in the Monadhliaths and stumble on a perfect line of mountain hares, laid out like strange little toys in the heather. Each year, thousands of these delightful animals are slaughtered in the Cairngorms, not for sport, but because it is believed a tick they carry threatens the farming of red grouse.[44] As a result, the mountain hare population has crashed.[45] Yet you know, by now, that this is how upland conservation works. Without grouse moors, without fire and traps, how would you see your curlews? Even so, you don't really feel like walking on. The red grouse scold from all around: 'go back, go back'. And go back you will.

Both these futures lie ahead. Every other country, except Britain, has decided on shared nature in their uplands. If grouse farming ends its 200-year purge of our country, then the profitable hunt for wilderness will begin. Then, to the benefit of all, the uplands' rightful landscapes, wildlife and jobs may finally return.

CHAPTER 12

Pelican Possibility

Rewilding Our Wetlands

In a country where nature has been so lavish and where we have been so spendthrift of its beauty, to set aside a few rivers in their natural state should be considered an obligation.
—Senator Frank Church of Idaho[1]

Your electric car powers you silently through the Somerset countryside. You have seen spoonbills, shrikes and a white-tailed eagle from your car – but you are restless. At any moment, an enormous pelican could fly across the road in front of you. Your father, a keen naturalist, would never have thought such an enormous fish-hoovering distraction would be possible in Britain.

You know a pelican is on the cards because in the last few decades the wetlands of Somerset have expanded ten times in size and grown larger than at any time since the Roman period. Newspapers and radio stations reliably inform you that its wetlands now rival those of the central Danube Delta in scale, as government, local people and well-paid farmers have collaborated to create the largest and most profitable wetland in western Europe.

Forty years ago, in 2020, the plan was to 'rewild', but the plan was costed, too. Farmers, who had worked the land for centuries, would not lose out. Some were paid handsomely to wind down, and even then their cattle would not vanish. Older breeds, instead, would be set free. Others were paid to farm very extensively with small herds, in the name of restoring western Europe's largest wetland: a task future-proofed against further crashes in the ever-failing dairy industry. For such a pioneering scheme to benefit the lives of local people, it was realised that the off that the local economy would have to prosper. Welfare would be put before wildlife. The area around villages would remain, as today, protected and drained. But in the lands beyond, no longer used as dairy lawns, wilder water would return.

To overcome centuries-long fears of free rivers, charities, working with local government, paid for local residents to be taken to several valleys in eastern Europe over the course of one winter. Here it was demonstrated that unconstrained rivers, meandering and slowed by gravel and wood, and dammed by beavers, were far less of a risk to low-lying communities than rivers locked in deep concrete channels – prone, like bathtubs, to sudden floods.

Over a decade, and not without many a heated discussion, the Levels began to regain their natural waterways. Without the need to protect static livestock, the spill of water onto grass was nothing to worry about, for those in villages. Their homes were protected. A handful of roads needed to be raised. Beyond the safety of village life, however, some interesting changes were under way. For the first time in centuries, large areas of the Levels, carefully separated from human habitation, were governed by the crafting power of water. Reedbeds no longer needed to be excavated by JCBs. Sedges no longer needed to be cut by hand. With rivers such as the Parrett and Huntspill set free, water lying on the land generated extraordinary pulses of new life.

Beavers got to work. As the trees fell, blocking waterways, small oxbows and lakes were formed, holding back the flow. Each year, meadows regrew after the flood, feeding starlings at the start of spring and corncrakes by the end of summer. Whole landscapes forgotten by anyone alive, from bushy willow beds with marsh warblers to wet fens alive with reintroduced large copper butterflies, came to thrive in the beavers' wake.

In place of neat green fields, feeding hundreds of Friesian cattle, all to sell milk at a loss to their owners, came profitable chaos. Each evening, fens rang to the whiplash calls of spotted crakes. Each year, a few more black-tailed godwits, ruff and snipe arrived and began to breed. Wintering white-tailed eagles, fleeing a harsh winter in Scandinavia, were soon nesting, and scavenging the carcasses of wetland-dwelling deer. The Levels had slowly outgrown the need for manicure. Lapwings no longer nested behind electric fences, designed to keep the foxes out, but simply moved across the landscape to avoid them. Footpaths still existed – but it became far more exciting, and profitable for local people, for tourists to explore the growing marshlands by boat.

Management was still happening – it was just happening in a different way. With small herds of wild cattle, horses and elk now stewarding the marshes, the Levels took on a character not seen in millennia. Far from endless open fields or unbroken stretches of reed, there came diversity. There were scrublands with willow tits, reeds with bitterns and shallow fens with ruff. Over time, the beaver-coppiced willows filled with colonies of herons.

By 2050, huge fisheries had established on the Levels. Some birds, however, had long forgotten Somerset. They would never arrive on their own. So, with huge public support and consensus, the RSPB brought, from Romania, ten clutches of very large white eggs. It was time for the lost giants of Avalon to return.

Hatched in an incubator, young Dalmatian pelican chicks were assiduously fed by eager volunteers, in pelican dress, with tasty cocktails of carp, perch and pike. Swimming around in enormous reed-filled bathtubs, they formed an incongruous, confused and angry sight as they grew ever larger – and more hungry. Three months later, to everyone's enormous relief, it was time to let them go. The pelicans were released into the very middle of the marshes. For two years, they were tracked by satellite telemetry. Three years later, a drone image revealed something not seen since Roman times: the nest of a Somerset pelican. The first year was unsuccessful, as a white-tailed eagle skimmed the single chick from its island home. But the second year, a tiny punk-haired trooper would survive.

Ten years on, with twenty breeding pairs established, your journey is rife with pelican possibility. Now, in 2060, hundreds of thousands of visitors, including a good number coming from overseas, and thousands of new jobs in the rural tourism and fisheries sectors, attest to the bounty of Somerset's wild audacity.

Pelican Practicalities

Whether the scenario described above remains a pipe dream, or can become the kind of reality that other countries, such as Romania, already can and do enjoy, comes down to whether such a dream works in the interests of most jobs and most people, and whether government will listen – and then provide the funding.

The wildlife benefits of floodplains are enormous and self-evident, and you can, any time, visit the Biebrza Marshes in Poland, or the Danube

Delta in Romania, as an example of what Somerset could be like in the future. But could such ambition, could a profitable floodplain, really be possible in modern Britain, now that such places have been lost?

The Dalmatian pelican's European stronghold, comparable in scale to Somerset's original marshlands, is the Danube Delta in Romania (Plate 25), home to 400 pairs of these prehistoric-looking birds.[2] This wetland measures around 60 kilometres from east to west, and 60 kilometres from north to south.[3] The delta may be large, but it's no empty wilderness. There are villages in the middle of it, as well as thriving artisanal fishing and ecotourism industries. In total, 25 villages lie here, and over 14,000 people call the delta home.[4]

Looking back to the humbler marshes of present-day Somerset, the main stretch of nature reserves, around the RSPB's Ham Wall, now measure 8 kilometres from east to west, and 1.5 kilometres, on average, from north to south.[5] These reserves may form a paradise for bitterns but are mere postage stamps of ecosystem size. So pelican possibility comes down, first of all, to scale.

In low-lying areas of the southwest, the dominant industry covering the land is dairy farming. Agriculture as a whole now contributes only 1.2% to the gross value added (GVA), or regional economic output, of southwest England.[6] Whilst shifts in other economic sectors lead to shifts in tenancy, the dying dairy industry of Somerset occupies just as much land as ever before. Historically, between 1861 and 1931, the agricultural workforce declined by 63%. Today, just 2% of all people in Somerset work in agriculture.[7]

An impartial analysis of the Somerset economy by its own county council reveals the economics in more depth. Agriculture is not within even the top ten sectors of productivity for Somerset's economy, and its income declined by almost 9% from 2009 to 2014.[8] Irrespective of use of the land area, the local economy is dominated by manufacturing, accommodation and catering.

Such a decline is part of a wider picture. The most recent figures from the National Farmers Union suggest that there are now 2,885 dairy producers in the whole southwest region – half as many as in 2002. At the smaller scale of individual farms, NFU case studies of two farms in the southwest show that for a dairy herd of over 200 cows, the operation can now support between three and six full-time equivalent (FTE) jobs.[9] It is clear that at the moment, Somerset, and its rural communities and ways of life, is not headed for an exciting rural future, filled with either prospects or jobs – but a gradual extinction, and a collapse in the rural way of life.

In spite of all this, as of 2013 the southwest of England was still home to 432,000 cattle.[10] Yet in the past twenty years, the nation's dairy consumption has fallen by a *third*.[11] There have been growing fears of a range of health complaints associated with milk for some users, from asthma to lactose intolerance.[12] Each year, based on market demand, ever smaller areas of floodplain land should be required for dairy farming. So, with a continuing picture of falling employment in dairy farming, what might a more diverse rural sector look like?

Somerset's Danube

You hear politicians talking about rural jobs for rural people, but all too often this is no more than an idle promise. In part, this is because government has yet to realise that ecotourism is one of best models for the future of low-lying communities, or realise that farming on a floodplain is the opposite.

The map shown in Plate 26 outlines one possible area in which, with little impact on homes or infrastructure in Somerset, a national park could be created for the restoration of wetlands and rural jobs. This could only happen if farming within such an area were incentivised to stop production, in some cases, or, in others, to collaborate in conservation. The red line on the map encompasses an area equivalent to the central Danube Delta.

The more ambition brought to such a project, the more jobs would be created. The maintenance of the landscape would never end. Even rewilded, everyone from local councils to the Environment Agency and forestry interests would find increased opportunities in a landscape with more going on. Angling, too, is one of Britain's most lucrative hobbies. Danube-scale wetlands in Somerset would generate fishing possibilities not seen in a thousand years. These would become some of the nation's best places to cast for gold.

If the aspiration was that by 2050 there would be elk on the Somerset Levels, as in the Biebrza Marshes, and, by 2060, Dalmatian pelicans, it is the ecotourism potential that would bring the most jobs. At least fifty rangers, ten guiding outfits, five boat safari ventures, a hundred conservation jobs and fifty new bed and breakfasts would be just the start. What would really kick-start Somerset's trundling economy, however, would be the income.

In Romania, access to the Danube Delta, heavily restricted, accounted for 100,000 tourists in 2015. Having a Danube of our own would be a different matter entirely. The bracken of Exmoor receives 1.4 million visitors a year, without holding so much as an eagle. The Norfolk Broads today receive 8 million visitors a year, but you could reasonably expect double this number, 16 million tourists, to visit Somerset's Danube if fully rewilded with a host of A-list animals and birds – and an iconic landscape to boot.

Each tourist spending just £50, with no overnights, would bring £800 million to the local economy over the course of a single year. That kind of income, feeding into local jobs and businesses in a relatively small area, is transformative. For hundreds of years, not one more subsidised generation, it could form the kind of rural backbone no dairy farm could ever hope to provide. Yet local people would remain at its heart.

Whilst only carefully regulated numbers of people would enter the protected areas, wildlife on this scale generates all kinds of income. There are the people who simply want to look at the view. There are the people happy with a telescope view of a pelican. There are others who pay more for an immersive experience: a journey into the very heart of Avalon. And all of these people are feeding, constantly, into the surrounding sectors – food, accommodation, transport, guiding, fishing and more.

And if you're thinking, 'This is jumping the gun, what about saving the birds we have?' – only robust landscapes give such birds a future at all. The Danube Delta, for example, does not only host Dalmatian pelicans. It purrs with thousands of turtle doves. It is thronged with red-backed shrikes and white-tailed eagles, herons and cuckoos. Virtually every other bird we have lost, or are losing, is thriving in that rich mosaic refuge. Pelican Possibility provides the vision needed for landscape-level wildlife recovery. But most of all, it could provide the *economic* stimulus to make this a reality.

Compare this vision to the present state-supported dairy silence of the Levels, running at a loss, losing jobs each year, requiring incredibly expensive drainage of the land, all for a fraction of the nation's milk. At the moment, so lost are Somerset's floodplains that its lapwings must nest behind fox-proof fences, water is feared and the wider land is desolate. It's time to replace that dreary, jobs-poor world with a vision for the future.

Listening to Rivers

'In Britain, this would be a golf course.'

My guide, Waldemar, smiled at the puzzled look on my face. I had no idea what kind of place this was. It indeed appeared for all the world to be Europe's largest golf lawn. I'd seen habitats across Britain and Europe, but nothing quite like this. We were standing in the Bug Valley, an hour's drive from Warsaw, in eastern Poland, at the very onset of spring.

The golf course, on closer inspection, was alive with birds. Starlings and fieldfares were getting down to nesting, foraging on its fertile lawns. But there were stranger inhabitants. The bushes were filled with yellow-hammers, but the lawns were filled with lapwings and godwits. It was as if someone had released lots of disparate birds, which in Britain you'd travel to many separate places to see – and set them loose on lawn.

As it turned out, we were standing on a floodplain. The 'golf course' was the recovering ground: the fresh, levelled grasslands left in the wake of the river's winter flood. Above the lawn soared white-tailed eagles. Some birds had arrived a little too early. A corncrake made a bad job of standing, motionless, in the middle of the grass, as if waiting for its meadow to grow.

The impression was that someone had spilt a lot of water, but that no one really minded. Having never truly considered that wild rivers once created habitat for 'farmland' birds, seeing dozens of yellowhammers, starlings and a red-backed shrike sitting in the same tree was exciting but odd. The combinations seemed random – yet, as Waldemar explained, everything had been beautifully worked out – by the river itself.

Each winter, the Bug bursts its banks – except 'bursts' implies some kind of accident. The annual water cover means the valley is too wet for trees to flourish in the river's wake, so the woodland in the valley is restricted to floodplain bushes. In place of hedge is natural willow scrub, coppiced by beavers. What is left, in most places, is a fertile wet grassland. Houses are built above the winter flood-line.

As the waters of the Bug withdraw, Poland's resident birds get down to breeding. Starlings are the commonest birds of these floodplain valleys – literally everywhere. They nest in trees and feed on pastures, which, during their April nesting season, are still short and low after the flood. Godwits, cranes and lapwings are early breeders too, feeding in the pasture as it grows. By early May, the grasses begin to rise, along with the temperature. In come the spotted crakes to colonise the shallow fens. The electric trills of snipe and the yodels of redshanks fill the air. These species, each requiring open areas of lawn to feed, but denser areas to hide their chicks, nest a little later than the pasture-loving birds. By June, the meadow rises like a phoenix from the lawn. Whinchats, corn buntings and corncrakes get down to raising families in the longer grass. As insect abundance peaks, red-backed shrikes hunt beetles from nesting bushes, and cuckoos sweep the meadow for caterpillars. It's all part of the seasonal plan. British action plans would dictate that starlings are 'pasture' birds, snipe require 'tussocky wet grasslands' and yellowhammers 'arable farmland', but in a dynamic river valley such distinctions are redundant. Each bird picks its moment to perfection. This is why, in one Polish valley, you can find most of Britain's vanishing 'farmland' and 'wetland' birds, common, side by side. Variety is not created by management. It's created by rivers – and *time*.

The scale of the Bug Valley is then what cements such diversity. Within the valley, there are really wet areas that never dry up, even in summer. These reedbeds are where you find your bearded tits and bitterns. There may be fewer bitterns in the Bug than in Somerset, but that's because the river isn't managed for their needs. Bitterns just occupy a portion of this rambling floodplain. Over time, a particular haunt of bitterns may become unsuitable as the river subtly shifts its course, or an oxbow dries up. But at such scale, there are always new places to move into as new wetlands spring into being each season.

River valleys like the Bug act as living reminders that we do not need to reinstate the transient habitats of our farmland past, nor micro-manage our wetlands as fishponds for bitterns or fox-free crèches for lapwings, provided rivers are allowed enough power in the landscape. Such large-scale natural systems can feed a staggering range of birds, which we have long forgotten could exist in one place, without endless policies, cutting, digging, scraping or other costly forms of management. Travel to Warsaw, drive two hours east, and you can stand in the Bug Valley for yourself.[13] It's a magical place and well worth the trip.

With the river freed, it crafts habitats at a fraction of the costs of a reserve – with skills beyond the reach of any human conservationist. And this is why it is worth holding onto what other countries have, and what we could one day have as well.

Living with Water

Whilst focused government incentives for dairy farms to wind down would break the cycle of ecological damage in Somerset, and pave the way for a future based in ecotourism, home owners, too, would need to realise the economic benefits of floodplain restoration. In its natural state, the entire area of the Levels should be inundated by a variety of rivers and hold low-lying water across the winter. Presently, this water is locked away, we are told, for reasons of national safety. In recent years, however, the story that deep, dredged channels lock water away, and keep people and livestock safer than if rivers were left to run free, has taken something of a surreal twist.

In 2014, when a large area of farms, built below sea level, flooded, as in some 'do not' fable from the Bible, the instinctive reaction was that rivers were to blame, as they had not been properly dredged. The government sprang into action. Sixty-two pumps worked for days to remove 1.5 million tonnes of water from the Levels.[14] Dredging certainly makes rivers deeper – and they accommodate more water. So the knee-jerk reaction, if your farmland is flooded, is understandably to call for dredging.

Firstly, however, the silt that clogs up rivers such as the Parrett is a direct result of farming practices, washed in from the fields. The worst field type identified by scientific studies for losing soil is maize. Three-quarters of studied maize fields in southwest England have a degraded soil structure. This means more surface water runs off them, taking silt with it into our rivers as it does so.[15] Maize is not planted during the rainy reason, and the fields in which it is are deep-ploughed. This means that at the time when soil most needs vegetation to anchor it, it is, instead, being washed into our rivers. Growing food for cattle contributes towards filling Somerset's rivers with silt. This silt causes flooding,[16] and causes dairy farmers to ask the taxpayer for more financial support – and more dredging.

It is hard to envisage a worse case of social irresponsibility. As Franklin D. Roosevelt realised eighty years ago, 'the nation that destroys its soil destroys itself.'[17] But you hardly need be an environmentalist to consider such practices a waste of your taxes, and a presumption on the many people active in the economy who do not fill our rivers with silt. Indeed, Somerset's dairy farms, at present, demand a double fee from the taxpayer. They require regular assistance to be drained, yet require the canalised rivers that flood them to be maintained. The heavy irony, of course, is that these silt-filled, canalised rivers largely exist because of dairy farming in the first place.

Rivers, in fact, are often much safer when neither canalised or dredged. A study of the Cumbrian floods of 2009 compared the devastation wrought on people by two rivers: St John's Beck, flowing from Thirlmere, and the River Liza in Ennerdale.[18] St John's Beck was a typical British river. Canalised and controlled, it filled with rainwater and catapulted it downstream without a brake, and that pulse of floodwater was devastating. The Liza, which had been released to run its natural, shallow course, was filled with accumulations of shingle and wood. All of these acted as brakes for the floodwater. By the following day, the Liza was back to its usual levels. In hydrology terms, it's pretty simple stuff.

Rivers running their natural course is what rivers have done for millions of years. The reason people fear rivers in Somerset is in part because they are prevented from doing what they do best. Dredging may drain your farmland quickly, but only unleashes a more powerful jet-stream of water further down the river's course, to the detriment of your neighbours. The Environment Agency dryly summarised the situation during the winter floods of 2014:[19]

> *The concept of dredging to prevent extreme flooding*
> *is equivalent to trying to squeeze the volume of water*
> *held by a floodplain, within the volume of water held*
> *in the river channel.*

In Britain, we have overlooked the self-regulating dynamism of nature, locked our rivers up, and now express surprise when they break their narrow confines. By contrast, free-running rivers, naturally expanding and contracting, with many breaks and meanders, are a far safer bet – not only for creating wonderful wildlife habitats but also for protecting people's homes.

Without intensive dairy farming, without soil runoff and the need to control rivers in linear form, the natural course of water in lowland Somerset would create a safer living environment. Rather than sudden bursts of intense flooding, 'spills' would become commonplace and, in the absence of dairy farming, of little threat to people. As long as Somerset's rivers are locked away, protecting the failing farms that we pay for, almost everything will suffer. Animals, unable to roam the levels freely, will drown in gated fields from which they are unable to move.

People's homes will flood. Rather than rise by centimetres at a time over a whole season, rivers will suddenly burst their banks and push floodwater outwards with ferocity.

The key 'culprit' in recent floods, the Parrett, would no longer fill with silt so quickly, washed off farmlands growing maize. Its canalised state could be slowly broken down, so that the gentler river that it once was resumes its natural winding course. Affording full flood protection to homes is achievable. It becomes even more achievable, however, if rivers are no longer set up to flood those homes in the first place.

However clear the arguments, though, there is a very long way to go in reversing cultural attitudes to wild rivers in our country. In eastern Poland, for example, houses were sensibly built above the winter flood-line. In Somerset, some houses are built below sea level, and very much below the natural flood-line. In Poland, some rivers were never canalised or their marshes drained – in Somerset, you would need to 'reverse' this process on a massive scale. In Poland, cattle roam extensively. In Somerset, they are confined intensively in fields. In Poland, where the land has not been drained, people live alongside the river, farmers too, and benefit from its bounty. In Somerset, as in Britain as a whole, such a sharing arrangement hasn't been seen for hundreds of years, but Poland's Biebrza Marshes is increasingly being seen as a viable template for how things might one day work here too.

Whilst more than 40% of the land in Poland's Biebrza National Park is privately owned, farmers agree to let the park administer their lands. This is because rather than use state money to fight the laws of nature, farmers consider the floodplain land unproductive for intensive forms of agriculture. Extensive grazing of the valley's herds has therefore become an iconic, cultural part of village tradition.[20] The river, the local people and the herbivores work in tandem to steward an extraordinary wetland.

For the nation to truly enjoy a Danube or Biebrza in Somerset a generation or two from now, floodplain restoration would require several intelligent government actions to make it a reality, not to mention a large degree of cultural change. It would require the end of intensive dairy farming in agreed areas. It would require the economic vision to realise that the 'reversal' and unwinding of Somerset's waterways is worth the effort. Strong incentives would be put in place to farm extensively in conservation areas, and rural job guarantees would be needed. But if all this happened – imagine what a future this could be.

Trickles of Change

Reclaiming dairy lands for conservation, using powerful incentives for the next generation, is an economic possibility waiting to happen. For now, though, a few leaks of wilder water are starting to trickle into Britain. And saltmarsh is one place where the water is finally getting a little bit more wild.

In Essex, between the rivers Crouch and Roach, lies Wallasea Island. By 2025, the RSPB, working with Crossrail (a new high-capacity rail service for London and the southeast), intends to restore wild saline marshes here, governed by the tide. Crossrail is providing large amounts of clean-spoil from their tunnelling operations. Ingeniously, this spoil will raise the land on Wallasea.[21] Controlled breaches in the sea wall have already seen the land inundated to various degrees. Each level of inundation will create a different kind of habitat, from the drier areas favoured by marsh harriers, to the damper ones inhabited by redshanks. That, in itself, marks a historical move in British conservation: a partial welcome to the forces of nature.

This may be more exciting than other wetland conservation projects in Britain, but there is still an innate caution at this early stage, compared to schemes like the Oostvaardersplassen, reclaimed from the sea in 1968. Whereas the Dutch experiment has recruited 100 regular breeding bird species, including hundreds of spoonbills, and white-tailed eagles, through the intercession of grazing and flood, Wallasea remains an experiment in altering the heights of water. It is, however, part of a most welcome beginning.

Alongside other 'managed retreat' schemes such as those along the Humber, and at Tollesbury Wick in Essex, the Wallasea project welcomes water as a force that can shape the landscape by itself. In Cambridgeshire, the Great Fen Project is a fifty-year plan that aims to restore 30 square kilometres of the land around Holme and Woodwalton Fens to something closer to its original state, a mosaic of reeds, fens, pastures and woodlands.[22] It is too early to assess what kind of landscape and birdlife this project will eventually restore, or how much freedom the landscape, and key mammal stewards, such as beavers, will be given in the future.

If the fenland restoration mentality can expand in future years, and rivers can be freed to shape areas of land, aided by beavers, then perhaps one day we won't have to travel to Poland to see valleys glowing with godwits, or to Romania to find pelicans gurgling with pike. In the years to come, anything that involves 'reflooding' land may face more opposition than any other kind of restoration. But that is where strong ambassadors come in.

Millions of us pay charities like the RSPB to act as Nature's Voice. Nature charities never need to apologise for putting forward an idea like Somerset's Danube. If they are to leave outstanding wild places to our descendants, they, too, must ask for the right things, at the right time, with a proper scale of ambition. In doing this, charities have an enormous ally of their own. The economies of ecotourism – the possibility of a whole new sector in a countryside starved of wildlife and jobs – are the natural allies of conservation. It's time we put such arguments to better use, in our quest for a richer world to live in.

CHAPTER 13

Our Birds

Sharing Our Homes with Nature

*What is the use of a house if you haven't got a
tolerable planet to put it on?*
—Henry David Thoreau[1]

The roar of traffic rises within you. You long for the wild. You feel the tingle of the hunt. The prospect of wild boars with their stripy summer piglets. The red-eyed fury of a goshawk flashing through the trees. The towering sound of skylarks. The grisly larder of a red-backed shrike. Ill at ease in your house, you decide to take action. You leave the house. You get on the bus. And straight away the game is on.

Facebook messages tell you that in the local park some cute wild boar piglets have been sighted near a children's playground. The police have not been called: this is a welcome and unremarkable sight for many children. On arrival, sure enough, a band of furry little JCBs are digging away beside their mother. Pedestrians stop for a moment, but the boars aren't hanging around: they've got business to attend to. It's true, not everybody likes them. A few cause traffic accidents, some are killed by hunters – but the boars, everybody knows, are here to stay. Most people accept them. Some have even created fan groups for them.

It's a crisp spring day. A female goshawk rises above you, clapping her wings in slow motion as the fearful male, half her size, circles overhead. Her harsh chatter punctuates the traffic noise below. It's another unremarkable sight. With the morning ahead, you are free as a bird. Next stop – the city's abandoned airport. It's been left to grow fallow: not by chance but by design. As you reach the airfield, the hum of traffic falters, and the thrum of crickets fades up. Red signs inform you that skylarks have made their home here, and soon you can see them as your eyes adjust to the wide open sky. Kestrels hang motionless, stiller than the kites on strings flown by the free-range children below. Given time, the sparrows will take food from your hand, but right now you're feeling a little more wild.

Your next bus journey plunges you through tree-lined streets until, at last, you're on the shores of a forested lake. It feels for a moment like you've skipped the city. In fact, you're in the middle of it. Around you, newly planted trees heave oxygen into the city's lungs. Ahead of you, the graceful shape of a black tern hovers, breaks the mirror water, and catches one more fish for its chick, sitting far out on a nesting raft, placed in the middle of the busy lake.

Down by a large river, a strange kind of 'service station' has been designed – for beavers. With their lodge just a few kilometres from the city's centre, beavers were getting tired crossing the river. Now, they have a pit stop made for them by the local council: a floating platform where they can rest in peace. There aren't any there today, but you can't have it all your own way. These are wild animals, after all. A city is not a zoo.

Next stop, an old military area to the north. Here, amid a sea of purple heather, yellow gorse and downy silver birches, the song of warblers fills the air. For over twenty years, this suburban wonderland has grown wilder and wilder. You cannot walk everywhere in this wild place. The wildest areas are fenced off. Nature is colonising. Dogs are not welcome. You know, however, that if you were standing inside the fenced heart of this urban wildland, you would come across the grumpy walking carpets of European bison. They're not truly wild, the city's confines are too small for that, but they are shaping the landscape nonetheless. Flowers and ponds lie where the bison have wallowed. All in all, it's been quite a morning.

You need to get back into the city, in time for lunch. You have not been dreaming. And the morning just described is no uncertain vision of a future. It is, in fact, a daily springtime experience available to any of the residents in one of Europe's largest and most profitable cities – Berlin.

Home to wooded heathlands, open lakes, flower-filled airfields and grazing woodlands, there is more landscape dynamism and more birdlife in Berlin than in most of southern England.[2] Berlin has eagles and ospreys, shrikes and wrynecks. Birds that have vanished from Britain are thriving here. Many animals, such as wild boars, beavers and goshawks, live right within the city. At least 170 breeding species of bird, more than in most British counties or any British nature reserve, thrive around the city.[3]

Somewhere along the line, we attributed aesthetic value to the green deer-lawns of places like Richmond Park, but Berliners decided instead that nature should come to them. Today, anyone in Berlin, should they choose, can breathe and feel alive in a city not entirely owned, managed and dominated by its human inhabitants.

Berlin's dereliction doctrine can be seen in many areas of Europe, where the land around towns and cities is not required to make profit. Because there isn't a financial outcome, the outcome becomes 'recreation'. But recreation doesn't mean local councils spraying chemicals around the base of each aspiring tree, as it does in many parts of Britain. Recreation means enjoyment of what the city lacks. The wild.

Nature on our Doorstep

Just 6% of Britain's land is covered with human structures of any kind.[4] But that 6% is now more invaluable for wildlife than ever before. A whole range of British species now find themselves tied to the Built Life. These are the most fragile of our wildlife hostages.

Thousands of tiny actions, writ large, can determine whether our villages and cities are filled with wildlife as Berlin is, or sterile deserts tidied to the last degree. Whilst the long-term future of most wildlife will be decided in rural landscapes, there is a peculiar exception in the ecosystems that you and I can shape.

British gardens, as a collective, have the potential to form a network of nature reserves unparalleled in Europe. Our gardens account for 18% of land use in urban areas. Around 22 million people – and 87% of all homes – have access to a garden.[5] Birds in gardens are often fed, whereas those in the countryside often starve. Higher temperatures in our towns and cities increase winter survival. Gardens are to twenty-first-century Britain what the hay meadow was to the eighteenth century – a massive life-support system, tipping the odds greatly in the favour of some birds.

Today's fat-balls and feeders provide blue tits with an ecological anomaly: a tied-down food supply. Blue tits are certainly helped through the winter by fat-balls.[6] Blackbirds, on a warm summer evening in April, provide you with the finest song on earth free of charge. If you're wandering through your local village or park, the blackbirds are probably singing at a higher density here than in the New Forest. Suburbia's blend of lawns and fat-balls stacks their food in a series of predictable parcels.

The British garden network is thriving in many ways, but it's odd how gardens have changed in the species they support – and could change again. Just a very short time ago, gardens were the perfect life support for starlings – two centuries earlier, for wrynecks. Hedgehogs, common in my childhood, now face the widespread prospect of extinction.

Gardens change and mutate – reflecting our changing social conditions, and the conditions of the wider countryside. But most of all, they reflect our changing states of *mind*. Chemicals, decking, concrete – we all have a whole array of tools to wipe birds out, quickly, and on a massive scale. These tools are ours to use, or withhold, as we see fit.

Our seemingly humble shared spaces – our gardens, verges and parks – are a habitat we need to maintain wherever we can. But in all of our personal quests to save our local wildlife and rewild a little of our country, at however small a scale, there is a critical enemy in the battle. It's a creeping disease, rotting the minds of many, and especially our local authorities. It's deeply un-British – and it's terrible for wildlife.

Ecological Tidiness Disorder

Our policy on grass cutting along urban roads requires Cheshire East to cut verges to a higher standard, once a fortnight, during the summer months, to keep areas tidy.

—Cheshire East Council, in 2012, explaining the removal of roadside wildflowers planted by a resident.[7]

The staggering neatness of every inch of Britain is only driven home when you return from a visit to Poland, Hungary or even many of the villages in Spain, France or Germany. Our alarming obsession with order and sterilisation has yet to arrive in most European countries. Nobody in the rural villages of Spain seems to worry about 'creeping' vegetation. No council workers are spending taxpayers' hard-earned money strimming a daring profusion of flowers on a verge. No children are waking up in a cold sweat, screaming at the prospect of unlevelled hedge.

In Britain, however, many of our local councils act, using our taxes, as the sterilising force of shared public places. Unlike farmers, who tidy their farms to increase yield, there is no economic reason given for the prevention of life in our towns. It reveals a new nastiness in Britain: a compulsive need to cleanse. Ecological tidiness disorder, or ETD, the growing compulsion to tidy Britain down to the last weed – appears to infect new sufferers each year. For a growing number of people, the need to sterilise appears to have become a hobby in itself. To beat it, we need to question not only our council's actions but our own.

Is that untidy bush, chippering with sparrows, overgrown? Or is it just a bush? If you have a rural garden, is that ivy an indicator that you're not in control – or is it the home of a spotted flycatcher? That muddy 'waste ground' down the road may be the only place a house martin can find building supplies. Is that loose roof tile a social disgrace – or an opportunity for a swift?

Given that almost all our insectivorous birds thrive in 'scruffy' margins, then expanding areas of natural profusion around our towns is now more important than ever. The future of the cuckoo will be decided by scruffy places rich in moths, the future of the house sparrow by how many untidy corners are left around our homes. And whilst the rhetoric of garden bird conservation has become about bird-feeders and nest boxes, far more useful to birds are gardens where trees, bushes and 'scruffiness' are rife. That gooseberry bush is an insect powerhouse. That 'pernicious' ivy is what supports your sparrows. But in many modern gardens, if there is one supreme victim of ETD, it's not a bird at all. It is an animal that, at current rates, our grandchildren will never see – and may even struggle to imagine.

Snuffling out of our lives, hedgehogs, of which around 30 million existed in the fifties, numbered less than 1 million by the 1990s. In the last thirteen years alone, a further 66% of them have vanished. Losing a fifth of their population every four years,[8] hedgehogs are now set for extinction by 2025. The hedgehog's beetle prey has vanished from the countryside at large, but for a long time afterwards, our garden networks, rich in native flora, messy areas and old log piles, became their refuge. Scruffy piles of wood helped hedgehogs snuggle through the winter. Lawns were once unsprayed pastures; the relics of ancient ecosystems. Allotments and gardens were food-rich corridors through which hedgehogs could move and hunt creepy-crawlies in the grass. ETD has changed much of this. Many of our gardens, devoid of nuance, disconnected by impenetrable fences, with a high chemical input and the poison of blue metaldehyde slug pellets filtering upwards in the food chain (endangering our own domestic dogs as well as native wildlife)[9] no longer harbour hedgehogs. In my home city there is just one hedgehog refuge left, and that is the Bristol allotments. Here, an organic maze of earth, bramble, beetles and wood piles recalls the wildlife-friendly gardens of fifty years ago. Hedgehogs snuffle on.

The absurd tragedy of these walking doormats being faced with national extinction is a very British problem. Countries such as Germany are not worrying about hedgehog extinction. In Hamburg, studies have shown that hedgehogs have adapted their home ranges to the city, making them smaller. During the day, they hibernate under 'brush' in people's gardens. By night, they forage the green spaces in its towns.[10] Other countries don't have plenty of hedgehogs because they're doing something remarkable – they have hedgehogs because they're *not*. It would be deranged and odd, in most countries, for city parks to be cleansed of life, lawns sprayed, bramble banks cleared and wood piles removed to the last degree. The idea in most countries is that gardens and parks provide an *antidote*, not a continuum, to the order of the indoor home.

At present, the micro-potential of our roads, villages, gardens and shared communal areas is wrecked by small agencies writ large. We do not need to return to unpaved roads, like those in Romania, to save our swallows and house martins. We just need to adjust our cultural attitudes – and open up our minds to letting nature a little closer to our homes.

The more we travel, the more we can learn from other countries' tolerance of nature, the more we will realise that our destruction of it in our urban areas is entirely optional. It's entirely against our interests, too. Insects pollinate our flowers. Hedgehogs remove a surfeit of garden pests. Sharing with wildlife benefits us financially as well as emotionally. But if one form of ETD has reached chronic plague proportions in recent years, it must surely be the desecration of our roadside wildlife havens.

The Vanishing Verge

If you are watching a barn owl ghosting along a scruffy roadside verge, you are enjoying the bounty of fallows. Where fallow land exists, it provides hunting corridors for some of our most special village birds.

Barn owls used to nest in stone and timber barns, with large, dark attics. These, in turn, resembled large tree cavities. Then, those barns were taken down. Now, with the invention of the barn owl box, barn owls are present once again. They're using human cavities, made of plywood, and hunting voles on roadsides we've left alone. The roadside verge has now, often, become the habitat that decides if your local barn owl lives or dies. A good aspiration for any village would be to have a pair of barn owls. It would show that squeaky verges were making a comeback. Barn owls should be common, living beside us all. They bring universal delight to people, and benefit farmers, as well as residential homes, by removing rodents.

Kestrels, the barn owl's daytime business rivals in the world of pest removal, are also vanishing. And whilst the scruffy edges of our motorways may appear unchanged, they contain fewer rodents than at any time before. This is one reason why your kestrels, short of food and short of cavities in buildings or trees, are vanishing from your daily drive. Kestrels, like barn owls, need Britain to be both scruffy – and mousy.

Rodenticides, applied widely, devastate the kestrel's key prey. The fact that most kestrels tested in the UK now contain rodenticides is a shocking verdict on the way our countryside is poisoned from the bottom up.[11] Alongside this assault on our verge-side hunters, verges themselves are relentlessly tidied. Those verges, if left, would harbour voles. Those voles would recruit kestrels. Those kestrels would remove thousands of rodent pests for our farmers. Yet, in most areas, the sterility of our council-mown roadsides means barn owls and kestrels struggle to survive.

In spite of such an assault, over 700 species of British wild plant, 87 of which face extinction, such as the man orchid, can still be found on our roadside verges. Bird's-foot trefoil, one of the ultimate insect generators in the British countryside, can thrive here. And these areas are not insignificant in size. In total, our road verges cover over 1,000 square kilometres of Britain, an area larger than Middlesex.[12] If all these roadside verges were left unvandalised, just think about the recoveries of nature around our towns – and the reconstruction of our pollinator populations.

Ironically, it takes no money at all to leave a verge, and for local residents, in turn, to benefit from the bounty of flying insects, and birds, that verges can provide. In recent years, the charity Plantlife has pointed out that *plants*, not chemicals, are, in fact, the best natural killers of grasses – not local councils or herbicides.[13] Species like yellow rattle effectively attack grass roots, halving their growth. Plantlife praises several councils, such as Dorset, for promoting life along its roadsides: using plants, not chemicals, to control invasive grasses, and even within a few years of this policy being adopted, a drive across Dorset is one filled with far more kestrels than you will see in comparable drives across many British counties. Sadly, other councils have yet to follow suit, reserving their money for worthwhile endeavours and not for the desecration of the natural world around us.

Few places feel the life-removing force of councils more than my home county, South Gloucestershire. Around the town where I grew up, life is not only silenced in the farmlands. It is thwarted for no economic reason around our homes. Each year, South Gloucestershire council sprays the chemical glyphosate, banned in public places from Chicago to Paris, banned in eight countries, onto any promising green surface and around the bases of trees.[14] The impact on people is still not fully known, but it is increasingly considered by some experts to be cancer-causing.[15] There is no doubt about the impact on birds: it is devastating. Glyphosate dramatically reduces plant diversity and vegetation abundance. US studies have shown that it directly limits the abundance of small birds – including sparrows.[16]

Nettles are attacked as if they are enemy combatants, invading our country. South Gloucestershire council is doing its best to make the county nectar-free. Honeybees and spotted flycatchers are long gone from the fringes of most villages. Groves of bramble, feeding butterflies and prime habitat for birds, are routinely ripped out without thought. Roadside flowers are a noteworthy event, except for stands of planted daffodils. You wonder, as you watch people paid by the council driving around, levelling hedges, scything flowers, killing the shared wildlife of our towns, how many birds have been lost, at a local level, through such petty yet expensive acts of ignorance. Worst of all, the benevolent intentions of many local people towards nature are also being mown to the ground.

In 2002, Cheshire resident Vera Shallcross planted an entire verge of the A534 near Sandbach with wild daisies and a variety of other flowers. Ten years of her stewardship later, in just two hours, the council had them mowed to the ground. The reason given was 'ensuring roads were visible to motorists'.[17] This has to be one of the strangest justifications of all time. Roadside accidents are a very serious matter. Aspiring daisies are not.

What hope do we have for restoring wildlife in our villages if even a promising verge, or a clump of caterpillar nettles, is a target for our taxes? There is only one glimmer of hope in this matter. Nobody, except councils, and a few vocal sufferers of ETD, think such cleansing is a good idea. You can read irate articles in the *Guardian* and *Telegraph*, *Mail* and *Times*, about the desecration of local flowers. You can find endless letters to local papers, deploring the vandalism of bees and lovely old trees, and petitions by local residents against the use of glyphosate. Yet tens of thousands of small campaigns have failed to halt the vandalism. None is more embarrassing to our country than the case of Sheffield's trees.

In 2017, two pensioners and a Green Party councillor were arrested. Sheffield City Council spent £250,000 of taxpayers" money on legal fees – in a time of austerity. Michael Gove, Secretary of State for the Environment, visited and described the whole thing as 'bonkers'. Nick Clegg, a former local MP and leader of the Liberal Democrats, said that the sight of peaceful activists being arrested was like something out of 'Putin's Russia'. What is happening? Sheffield's veteran urban trees are set to be cut down.[18]

The council, which is using a private contractor to resurface the roads, has decided that the ancient trees, some of which were planted in 1919 as war memorials to soldiers lost in the First World War, have to go. The town council is so determined that they have welcomed the convictions of local people who have dared to oppose them. Campaigner Calvin Payne, whose crime, it appears, was stepping inside a 'safety zone' and urging people on Facebook to 'save the trees' (this was noted as a call to protest), was given a suspended sentence, and ordered to pay £16,000 in costs.[19] Even after a series of court cases, nobody knows why, to pave a road, trees on a pavement need to be cut down.

This new manifestation of ecological tidiness disorder has incensed everybody. Some commentators have focused on the terrible precedent set by destroying wartime memorials. Others have emphasised the fact that councils can defy the will of local people, ban protest and strike at the roots of our democracy. 'Tree-gate' represents the controlling urges of many who menace our public spaces – an urge to cleanse the land at every turn.

No other European local government seems to destroy local heritage this way. In Berlin, local government, regardless of its politics, has for decades filled the city with native trees and wild areas. In the Netherlands, Belgium and France, villages are filled with flowers, not just for 'in bloom' competitions, but as standard practice. You do not see hedgerows given absurd haircuts to preserve national order.

Britain's squared hedges, the sprayed bases of trees, the lifeless lawns, in aggregate, wipe out wildlife on a massive scale. Garden warblers were first noticed in the bushy maze of our village gardens. How many villages are bushy enough to hold them now?

Yet in spite of all this, we are not, at all, a petty country. Suburban England is, in fact, one of the most involved areas in wildlife preservation. Giving any thought to this matter, we would deplore the fact our taxes are used to destroy the wildlife around us. Many people, of varied political leanings, want the buzz of bees back in their lives. And that should pave the way for a *new* campaign.

Keep Britain Messy

Anyone who has visited countries in eastern Europe for their wildlife is likely to have spent some time in its rural villages. Such places reveal a staggeringly different attitude to 'tidiness' and one we urgently need to copy if we're to get our urban and suburban wildlife back. In almost any space outside of a tended garden, nature is rampant. Scruffy willow stands characterise many shared spaces. So do open, earthy areas where swallows collect mud. These are modern rural villages – they are paved and hygienic. The houses have small gardens, similar to our own. This wonderful 'encroachment' of nature is a *cultural* difference, and not an economic one at all.

In Britain, curing ecological tidiness disorder has to happen first. Pushing for an incentivised ban on herbicides in our public spaces would be a start. Pushing to remove the 'Streetcare' budgets that councils use to tidy, senselessly, whilst claiming they never have enough money, would be a powerful step by government – but our nature charities must ask for it first.

A 'Keep Britain Messy' campaign, backed by our wildlife charities, was something I first ventured in *Birdwatching* magazine.[20] The aim of the campaign would be simple: to avoid expensive desecration, and allow the growth of 'scruffy' areas to bring birds back into our lives, providing a richer playground for Britain's nature-starved children in the process.

As a child, I remember finding teeming caterpillars among nettles, carefully extracting them and watching them turn into peacock butter-flies later in the summer. Yet today you read fearful articles in papers from parents, concerned about their children being stung. But nettles teach you a lesson you don't forget. Would you really want to raise a child afraid of nettles? What would happen to them later on – if something serious happened in their lives?

The richer the world around us – the scruffier, messier, the more full of life – the more that life will reward us in turn. The tidier our world, the more effectively we will drive hedgehogs, sparrows and honeybees to extinction. Not only will their charisma vanish from our lives, but the practical services they provide will be gone as well.

One hundred years ago, the villages of southern England held nightjars, wrynecks, nightingales and red-backed shrikes. Joined together, they would, in the future, have the scale to do so again. If the 'tidiers' are pushing for cleansed verges and consequent kestrel decimation, then we, the 'messy' camp, must push back. If the people of Poland can enjoy village nightingales, we should be able to as well. An RSPB 'Villages in Voice' campaign could see a collaboration between villages in southeast England, growing hawthorn scrublands to promote nightingales and other birds to recolonise villages and towns.

Indeed, it is only very recently that Britain has seen an epidemic of ETD. A look at any village photo from the first half of the twentieth century proves that people gardened, lived happily, played happily, in villages populated with the chaos of nature. Heritage-heavy newspapers like the *Telegraph* and the *Mail* could help play a role in urging a return to such flower-filled times. But how large has the change been in our villages, as they have been sterilised over time? For the evidence, look at Plates 27 and 28. These two contrasting images of a Hampshire home encapsulate many of the changes from our acceptance of mini-wilderness, to the enforcement of tidiness. The first image is taken in 1914, the second in 2017. The thoroughfare remains, widened a little for traffic, but otherwise the house, the wall, the structure, is all the same. The difference lies in the detail.

The 1914 picture is fuzzy not with age but with grasses, flowers and bits of twittering ivy. The 2017 image is cleansed, its smooth green lawn flower-free. This narrative, country-wide, removes billions of insects and millions of birds from our lives. And whilst many aspects of our lives have improved since 1914, our local wildlife has grown ever worse.

Plate 29 shows Zywkowo in Poland. This 'stork's village', now a tourist attraction, is in many respects very like a small village in southern England. It has a falling rural population, a church, a pub and a park. But in recent years, 8,000 tourists have turned the fortunes of the village around. The local Association of Agriculture and Tourism oversees a thriving B&B operation – and ensures the whole village is kept wonderful and wild.

The chemical imprint in such a village in eastern Europe must be close to zero. The garden shrikes attest to that, as do the hundreds of swallows and swifts, the spotted flycatchers and butterflies swinging through the streets. Nobody is spraying nature into oblivion, one impeccably lifeless lawn at a time. We could have our villages this way – as soon as we *want* them this way. Emotionally, our lives would also grow infinitely richer as a result. And a village tolerance of nature may, in time, give way to ever more respect of the natural world around us – and what it can provide. Such an attitude is best built from the bottom up. So here's what you and I can do.

Save your Sparrows

House sparrows are sown around the world by people. They thrive in mud palaces in the Sahara. They thrive on cattle ranches in the Amazon. They thrive in New York City, where they were introduced in the 1850s to eat rampant linden moths,[21] and have since become North America's commonest bird. It has been calculated that the range of the house sparrow can extend by as much as 225 kilometres in a year.[22]

It's quite an achievement, then, that in Britain we've been able to halve our number of house sparrows in four decades, from 12 million in 1970 to just over 6 million today.[23] No species better proves the degree of wildlife desert we are now able to create. House sparrows might be found around houses, but it's the conditions houses and gardens create – small invertebrates, dense bushes and nooks in which to wedge a nest – that ensure their survival. Across the country, sparrow declines have been strongest in the south and east, overall our wealthiest areas. But the correlation becomes more interesting – and compelling – when examined at a smaller scale.

In London, boroughs with lower income, like Hackney, have more stable sparrow populations. In London, and in Britain as a whole, suburban areas, however, show the strongest declines. Greater income leads to more spare cash, more home improvement and the power to tidy our surroundings. The overall outcome, wandering through a suburb in middle England, is that 'scruffiness' has gone. Get tidy, spray your lawn, pave your drive or clad your garden – and you deprive sparrows of invertebrates. Starvation, alone, now hammers many populations in suburban England, with fewer chicks leaving the nest, in ever poorer condition. Feeding sparrows seeds may seem kind, but chicks being fed vegetable seed matter are more likely to starve than those fed on invertebrates.[24] Instead, it is far better to plant native bushes, such as hawthorns, in which sparrows can feed themselves.

Cover, too, is just as important. Take exception to that large, untidy and very noisy bush – and you've deprived sparrows of the place where they hide from sparrowhawks and hold their daily coffee mornings. Dense bramble bushes, ivy creeping up walls, roofs with little gaps, nearby 'fallow' habitats like railway embankments – all of these sparrow habitats are vanishing. In their place, you have bare walls, paved driveways and no vegetation. House sparrows have a triple habitat that we need to protect: communal nesting nooks, dense communal bushes, and vegetated areas with aphids.

Income improvement makes people's lives better – but it doesn't have to signal the end for these noisy brown seeds. Nest boxes, put under one's roof, replace nest sites lost to cladded roofs. Just a couple of dense bushes provide shelter from predators. Bushy native plants like gooseberries can provide the aphids.

Take a careful look at your local suburb. It may be a well-to-do area, but for a house sparrow it's a world away from the far richer opportunities of Hackney. But you, me, our neighbours, can all make room for a little more scruffiness – and a few more sparrows. In return, sparrows reciprocate by eating aphid pests, cleansing our gardens without chemicals or cost. Many communities welcome sparrows as quiet forces of tidiness. Given a chance, they're on hand to help with the gardening too.

Feed your Flycatchers

Whilst a sparrow colony can spend its whole life in a couple of gardens, isolation is one of the most dangerous situations for many birds to face. For a number of declining species, restoring nature across whole villages is now needed, if we're to hang on to the birds that delight us in our gardens each summer.

Spotted flycatchers, as much an afterthought in the 1960s as honeybees, are almost gone – along with the clouds of aerial insects they once skewered in mid-air. Spread piecemeal across the countryside, their remaining 'island' populations are all extremely fragile. This is where rural gardens, joined up, can act as lifelines. Studies have shown that spotted flycatchers are declining less in rural gardens than in the wider countryside around us.[25] A large rural garden has a better chance of retaining bees and butterflies than most woods or farms. Rural gardens are often planted for the enjoyment of colour and insects – the countryside is not.

Our attitudes in rural gardens play a huge role in whether flycatchers live or die. Keep your flowers, bees, butterflies and the scruffiness of creepers and little nooks, and there's every chance your garden could still be graced by a flycatcher. Foster the same attitude in your neighbours, and suddenly your entire village can turn into a flycatcher landscape –more useful as a refuge than a single garden. Large areas of mid-Wales, Herefordshire and Worcestershire, in particular, have good concentrations of spotted flycatchers in rural villages – and could, at sufficient scale, keep them for future gardeners to enjoy. Tidiness, of any kind, removes flycatchers. Scruffy bee-filled gardens, free from chemicals, are their friend.

Mud, Glorious Mud

Fifteen thousand years ago, swallows were nesting in the Creswell caves in Derbyshire, sharing with cave bears, cave lions – and us.[26] Twenty-first century swallows have forgotten caves. They've moved on. There are no regular natural swallow colonies in Britain and they abandoned Creswell, at last, in the 1990s. House martins, like swallows, are 'high maintenance' in their relationship with humans, but it's amazing how recently that relationship was forged. At the start of the nineteenth century, many were still nesting on cliffs and under river banks, just beginning the process of colonising our towns.[27] Few birds remind us how swiftly our own lives have changed in the past century. London still has millions of houses, but the cart-horses that sowed manure and insects have come and gone. The sewers that sowed insects and disease have thankfully gone too. So have London's teeming house martins. The house martin's decline reflects, in part, the ever-decreasing volume of flying insects in the wider countryside. As we saw in Chapter 3, air pollution dramatically reduces the chances for colonies to survive.[28] But house martins are also threatened by the dangers of an ever-tidier world.

House martins need access to wet mud, in scruffy ditches or stream edges, within close reach of villages, if they are to successfully glue their home onto yours. They need flying insects, and the sources that generate them, close to their home, to raise a family. Britain's house martins now thrive best in areas where houses 'grow wild' – where our homes grow alongside insects, clean air – and glorious mud. Cultural attitudes, town by town, are so important in the survival of our house-dwelling birds, and the little muddy areas beside our streams, in our yards and along our tracks, which may seem just an untidy nuisance to us, are vital to these sprightly summer birds.

Lawns and Leatherjackets

Starlings, evolved in open grazing woodlands, nest in deep cavities, and grub for leatherjackets in pasture. Over time, they have shifted this adaptation from our wood-pastures into our towns and cities. Until the 1980s, it seems food-rich lawn and a deep nook weren't hard to find, either in the countryside or around our homes.

In a short space of time, as our roofs have become clad in PVC and our lawns subject to chemicals, starlings, like wrynecks before them, have relentlessly vanished from our villages. Overall, the BTO calculate that 25 *million* starlings have vanished from Britain since the late 1960s.[29] Now, most remaining starlings depend on a dual habitat created by us. They need our houses to have enough hollow entry points, or nest boxes, for them to hide a family each April. They require enough insect-rich lawn to feed that family.

If you've had your roof done but still have a soft spot for starlings, a nest box with a 45-millimetre entrance hole, 15 centimetres square and 45 centimetres deep, provides an excellent home.[30] Put it high on your house, ideally below the eaves, and not facing into direct sunlight. Starlings are adapted to seek out nooks. They'll find their new home in no time.

If you're spraying your garden, stop. You are wasting money. You are killing starlings. And the whole point of nature is that starlings deal with the pests in your lawn. They also provide comedic value, being one of few birds, as the comic poet Pam Ayres points out, to *walk*, not hop. Looked at in sunlight, the purple and green on a starling is the most ornate decoration in your garden.

Fruits of an Ancient Orchard

Some of us, however, may have larger gardens yet – and older ones. There are few more remarkable havens for wildlife left in Britain than some of our very last wood-pastures. When old orchards vanished from Kent and Herefordshire, many of Britain's last wrynecks did as well. When orchards vanished from Somerset, so did most of its lesser spotted woodpeckers. So rich are some ancient orchards in the Welsh Marches that a study of just three, in the Malverns, discovered the existence of 1,868 species within them.[31]

For the past seven years, I have never failed to be surprised by the layers of life within an ancient orchard. In the Malvern Hills, in Herefordshire, a small group of us have, over the past five years, tried to unravel its secrets. The orchard, it seems, works like a complex block of flats.

Every apple yields a secret. Mistle thrushes hide successful nests in mistletoe.[32] In a neighbouring cavity lie the eggs of a mandarin duck. In another, the squawking chicks of a jackdaw. In a deeper, older tree you sometimes find the snuggled ferocity of a tawny owlet. Below sloping branches lie the tiny drilled homes of lesser spotted woodpeckers; drilled into the heart of the oldest apples, the homes of their larger green cousins. Honeybees can still be found in abundance, covering pussy willows at the orchard's edge. Hornets thrive in the old trees. Dormice from adjacent woods creep through the maze of overlapping apple branches. Rare bracket fungi colonise the oldest trees. Goshawks and sparrowhawks hunt the orchard for birds.

If you own an ancient orchard, you're in possession of a place more reminiscent than you might think of Britain's wild wood. Much of an ancient orchard's life is hidden, encrypted: out of sight. But an orchard sprayed, its dead wood removed, is little more than a shell. It will hold just a fraction of the life we have found in our Herefordshire haven. In chemical orchards, birds like redstarts, spotted flycatchers and lesser spotted woodpeckers, all dependent on caterpillars, butterflies or bees, need not apply.

Whilst the obvious answer, if an orchard has exhausted its economic life producing fruit, is to remove it, the heritage lost extends far beyond the history. Britain's wood-pasture birds are in serious trouble. Sixty per cent of traditional orchards have been removed since 1960. Your orchard, growing ever richer over time, is the ageing nature reserve that no one else is growing. And that, all in all, makes it a very special place.[33]

Cliffs in the Sky

Urban gulls have engineered cities built on top of our own. This is because there are no foxes in the sky, scaling our buildings with cunning and crampons. So urban gull productivity, as any Bristol resident knows, is very high. The 'urban heat island' effect means that temperatures can be 4–6 degrees higher in our cities than in the surrounding countryside. This allows gull nesting to begin earlier, and streetlights also allow the gulls to forage through the night.[34] A long way north of Bristol, the arctic terns of Montrose, on the east coast of Scotland, are now thriving on flat factory roofs that resemble shingle beaches. Kittiwakes have bred on the rocky ledges of the Tyne Bridge and other man-made structures in Newcastle and Gateshead since at least the 1960s.

These gulls and terns may just be the beginning. Does it seem unfeasible that in future, guillemots, razorbills and shags might all form colonies on the ledges of buildings or bridges? If swifts can forsake trees for churches, for how much longer will Britain's seabirds nest exclusively on cliffs?

Peregrines on Derby Cathedral have amazed researchers by snatching migratory woodcock, at night – using the lights of the building itself.[35] Migrating so clumsily that they almost fall out of the sky, woodcock could, for millennia, bumble over their sleeping predators below. Now, should they pass a cathedral at night, there is every chance of getting nailed by a day-hunting predator. The bumbling night life of the woodcock just became a lot less fun.

Urban peregrines are on the rise – entirely because of human activity. A peregrine in a city is safe from persecution. Its chicks, falling early from the nest, are more likely to be rescued by concerned passers-by. An urban peregrine not only has a smattering of migrants it can catch by streetlight but a sushi-style procession of city pigeons – and in London, gaudy ring-necked parakeets – to snack on by day.

The movement of birds from natural habitats into our cities has, however, a long history. It is a wonderful but unnatural thing for a peregrine to make its home on a cathedral. These commitments reflect a moment in history when human habitats were as appealing to birds as those available in the wild.

Birds committing to humans, however, come to rely on us doing what we've always done. But we're the fastest-changing species on earth. One day, a few decades hence, during an epidemic of avian flu, we may choose to remove feral pigeons from our cities. And then, having committed to the human world, peregrines could be endangered once again. Indeed, Britain's endangered kittiwakes, at risk of global extinction, nesting on ledges on Newcastle's buildings, should be seen as something of a colonising triumph. But even here the disease of ETD has crept in. Each year, nets placed on buildings to discourage the birds end up senselessly entrapping the kittiwakes instead.[36] The urban life can bring safety – but it's a fragile existence. Some of our most iconic birds are already paying the price of change.

Saving Swifts

Swifts have been a part of Built Britain for a very long time. But it's only when you visit ramshackle old towns in France, Spain, Poland and so on that you realise swifts should be *really* common – because they are so well adapted to living in brick trees.

In Britain, we now have a very small number of ramshackle roofs, stone tiles and other swift-friendly nooks left in our ever-neater cities. The website swift-conservation.org sells and proposes a number of nest boxes that can be wedged under the eaves of almost any house, specifically designed for swifts.[37] The fun part, of course, is that by playing loud tape calls of swifts, in areas where they arrive each summer, we can, literally, call them down from the sky. The first year, swifts may simply inspect your nest box. The second year, they may stay to breed. Entire new colonies have now been established this way – with a nest box, a set of very loud speakers, and some rather puzzled neighbours.

Swifts make, each year, a journey that we made just once, over hundreds of thousands of years – from African woodlands to temperate houses. It will be a sign of our nature-friendly cities if they're still cleaning our skies in a hundred years to come.

Bomb-Site Birds

The Greenwich Docks, the Isle of Dogs, the rubble of Dagenham. Five or more degrees warmer than the surrounding countryside, on average, the industrial greenhouse of London, and the sheltered environment of its dockyards, provides the perfect combination of insects and dust for a special summer visitor.

If you can see rubble, gantries and 'keep out' signs, if there are Dobermann guard dogs watching you with silent hunger – you are in the land of Britain's industrial songbird. And if London were a country, the black redstart would surely be its national bird. Black redstarts famously established themselves on the bomb sites of London after the Second World War. They have a habit of falling in love with places that some might describe as hellish, blasted and forlorn. That flashing orange tail on a little black bird is hope: hope that even rubble can be claimed by nature over time.

From choughs in Welsh mine-shafts to peregrines on old grain silos, industrial structures across Britain teem with unexpected life. As primary industry fades with the decline of manufacturing, our industrial buildings are being taken away too – along with our swifts and black redstarts. A part of this is inevitable, of course. Most people wouldn't miss a crumbling building any more than they would want to live in one. We need a flourishing economy and people need new homes. But when all of our buildings are shiny and new, a weedy stone warehouse becomes an oddly precious thing.

There is huge urban tourism interest in areas where nature has taken back control. The popularity of films like *28 Days Later* has only shown how pervasive that idea can be. There are profitable ghost-towns in the USA with prairie-dogs living in the old wooden ranches. Perhaps, as we tidy away our industrial sites, we too could set a few aside – monuments to the past, refuges for the wildlife that claimed it. Listed buildings – for wildlife. These miniature urban jungles could, if properly marketed and managed, prove popular attractions for city-dwellers craving the intrigue of wilderness in their very urban lives.

Coal Wilderness

The post-industrial areas of northern England have a very different aspect to other places when it comes to saving birds. In these areas, vegetation freestyles in a way rarely permitted in any nature reserve or across much of the country. There are floodplain woodlands rich in willows and willow tits, and fallow grasslands filled with rare flowers and butterflies. Here, less land is managed – and more is simply left. Durham, in particular, is a fascinating county for birds declining elsewhere.

Many pairs of long-eared owls haunt Durham's moors and grasslands. You can find whinchats by the roadsides. As many as 300 pairs of willow tit are thriving, many in the scrubby embankments of old railway lines.[38] And almost every time a willow tit pops up, it's not doing so in pristine woodland – but on a brownfield site. The reason? An iron hold on landscape control has been loosened. Durham is scruffy. And in nature terms, that is the highest compliment.

Recognition that northern England's brownfields are vital areas for nature has been withheld by conservationists for a very long time. Ecological purity – the reedbed, the canopy woodland, habitats that are often much poorer for birds – has been prized over the dynamism of a brownfield river valley. Slowly, however, things are starting to change. Nature reserves like Potteric Carr, in Yorkshire, and the RSPB's Swillington and Fairburn Ings reserves, are beginning to popularise the idea of brownfield wild. Spoonbills have recently started to breed at Fairburn – a gem in Britain's brownfield crown.

It's now time for wastelands, scrublands, to become protected as the wilderness lacking in our fields. It's time we kept our brownfield wilderness, and willow tits, protected – for good. Brownfield Biospheres would be a good place to start: land purchased for the preservation of nature, thriving beside and enriching some of the most populated areas in our country, bringing nature to an entire generation of children otherwise sealed off from it.

Glorious Gravel

Little ringed plovers are to gravel pits what swallows are to barns. Flying from the Mediterranean each year, this is a bird deeply in love with small chunks of rock. The more dangerous, the more risk of being squashed by a forklift truck, the more little ringed plovers seem to thrive. But they're not alone. Britain's gravel pits are extraordinary magnets of life.

Each year, dragonflies emerging from flooded gravel-pit reedbeds draw droves of migrant hobbies, as they arrive from Africa to hunt them on the wing. Many of our ducks thrive in these places, alongside balletic great crested grebes. Sand martins build colonies in abandoned sand and gravel banks and hawk over the insect-rich water.

Flooded gravel pits are more than diverse ponds – they also protect habitats no longer seen in the wider countryside. The area around pits, water-retentive, naturally regrows with riverine scrub. The social importance of gravel pits, as areas to walk and enjoy nature close to cities, acts to protect this habitat, even if that protection is sometimes accidental. Nightingales share this rich scrub world with bullfinches, cuckoos, and a range of other vanishing birds.

Given that our birds evolved in places where scrublands played a vital role, our gravel pits come as a strange rescue formula in a countryside now free from nuance. A visit to Paxton Pits, in Cambridgeshire, or the Cotswold Water Park, in Gloucestershire, will reveal a richer spring chorus than most of our woodlands. Cuckoos watch reed warblers from scruffy willow stands. Grasshopper warblers reel beside patient fishermen in the brambles. At Cheshire's Woolston Eyes each May, you can watch black-necked grebes in all their finery, escorting their chicks on their backs, protected by noisy colonies of black-headed gulls. Over 110,000 individual birds of all species have been ringed at the Eyes – an amazing testament to the power of gravel. Gravel pits are truly the wetland hay meadows of the twenty-first century.[39]

For brownfield rewilding to work, however, it's not enough to buy the pits. Nature charities, local financiers or local government must buy the *corridors*. The Dearne valley, the Durham coalfields, the scrublands of Liverpool and Manchester, are thriving as collective landscapes. But there is another danger to these industrial jungles – and it's not just being built on. These places could one day be managed to extinction. The magic of brownfield wilderness is freestyling and decay. In areas too small for full ecosystem dynamics, the brownfield wild showcases the last of our freestyled scrublands. Scrub clearance, the 'conservation' tool that decimates so much wildlife, is best kept well away.

Britain's brownfield wilderness is special and unique. It preserves processes not seen elsewhere in Britain for hundreds, maybe thousands of years, like the growth of wild floodplain thickets. Management would kill it outright. Far better to buy brownfield – then sit back, relax, and let it rot. That is how brownfield nature works – and it's a rare and special thing.

The Opening Ape

Each day, forest elephants and lowland gorillas make an enormous mess of the Congo rainforest. Trees are cleared. Soils are disturbed. Dung is dropped. Dung beetles aerate the soils. New trees grow. The gorillas play their part by unwittingly scattering thousands of seeds. Indeed, without elephants and gorillas, each of which disperses invaluable species of trees, the Congo jungle would grow infinitely poorer.

If humans have any ecosystem role left, which I believe we do, we act, like our gorilla cousins, as the Opening Ape. We are, after all, a species too – one with a sustained habit of clearing and planting. Small allotments and hand-cut hay meadows are the results of our acting in a manner not too different to our primate ancestors.

Robins, for example, are so familiar that it's worth reconsidering their gamble – from following pigs with predictable snouts to unpredictable humans with hoes. With a 45% increase in the British population since 1970, robins are thriving, not just on the earthworms we dig up but on the mealworms we put out. Even wild boars can't compete with such a service.

Creating open spaces, disturbance and tree chaos can be a role that's very good for wildlife, provided this doesn't come to replace the restoration of our landscapes. In our cities, there won't be many places large enough to accommodate the stewardship of wild cattle, horses or lynx. There will always, however, be areas for the Opening Ape to thrive.

Iron Age farming has been employed on the islands of Tiree and Coll, in Scotland, for so long that there is little record of what was there before. Tiree has become, under human stewardship, a paradise of flowers, seeds and insects, a network of crofted meadows that dovetails with the 'machair' of the coast: the natural grassland kept low by the wind, and fertilised by the calcium-rich remains of seashells. The same can be found in the rich crofting settlements of North and South Uist, in the Outer Hebrides, where the fields are alive with lapwings and redshanks, corncrakes and skylarks – all in glorious abundance. These Hebridean farmlands remind us how well we have, in the past, integrated with nature. How small-scale gardening and planting, writ large, can add up to thriving populations of birds. Each summer, most of Britain's corncrakes, hundreds of breeding waders, starlings and cuckoos, as unaffected as the island's thriving bumblebees, carry on here much as they did a century before.

There will always be smaller areas of Britain – little islands, islands within cities and our gardens – where the Opening Ape can sow life. In a nation of 65 million people, the earthy disturbance of humans still has a vital role to play. And as gardeners, in our scruffiest, wildest and most wildlife-friendly capacity – we can still act as the animal that makes the rest of nature proud.

CHAPTER 14

Conservation Begins

Expanding our Minds, our Ambitions, and our Wildlife

> *Combinations have always been the most intriguing*
> *aspect of chess … They represent the triumph of mind*
> *over matter.*
> —Reuben Fine[1]

Birdwatching is often perceived as a male pursuit. But we males would not have many birds left to watch at all were it not for a small group of women. In 1889, Emily Williamson got really fed up. People were turning grebes into hats.

Emily founded the Plumage League to lobby against Britain's thriving trade in plumes, based on the slaughter of a staggering range of birds, from our own native kingfishers to birds of paradise imported from New Guinea. In 1891, Eliza Philips, head of the Fur and Feather League in Croydon, decided she wanted in. And so the Society for the Protection of Birds was formed. In 1904, sixteen years before women got the vote, the society got the royal seal of approval – and a royal charter.

In the first half of the twentieth century, the RSPB worked under the legacy started by Emily, growing to 20,000 members by 1900. In 1921, after over twenty years of graft, the Importation of Plumage (Prohibition) Act was passed – banning the killing of birds overseas for their plumes. Emily had won her battle.[2] To this day, the RSPB and other charities have become adept at opposing attacks against the natural world. Egg-collecting, for example, is now an illegal pastime, furtively practised by perhaps a dozen living fossils in our country. Most birds, most of the time, are valued. And nobody, anywhere, is turning grebes into hats.

By 1930, the RSPB had entered a new period in its history – the protection of habitat – and in that year it bought its first nature reserve at Cheyne Court in Kent. Its now famous Minsmere reserve followed in

1947.[3] In Britain, the purchase of lands such as Minsmere pre-dated any British national park, but set the tone for the very small areas of land that our country has, so far, been able to claim back for nature. By contrast, as early as 1872, Yellowstone National Park had been formed in the USA, and was being policed against poaching before 1900.

Whilst the buying of land would seem to be the absolute imperative for conservation, in 2017, of an available income of £104 million, the RSPB spent only £1.7 million of its money on land, adding a further 1,200 hectares to its reserves, spread across the country.[4] This represents just 1.6% of the charity's expenditure, and, to this day, Britain's areas set aside for nature remain some of the smallest and most isolated in Europe.

In 1948, the RSPB bought the island of Grassholm, an epic gannetry off the Pembrokeshire coast. This marked the start of a vital realisation that one pristine jewel was left in our wildlife crown. If Britain's bitterns vanished overnight, global bitterns would do just fine. If our seabird cities vanished, however, the entire world would be robbed. To this day, the RSPB has an outstanding record of protecting our seabird cities, and lobbying to protect the fishing grounds that keep them alive.

In 1959, the power of ecotourism was first explored by the charity. The Loch Garten watch-point for ospreys was opened. In an age long before *Springwatch*, over just six weeks, 14,000 visitors arrived. Had the government of the time noticed the significance of this, perhaps our national future would now be very different, with large areas set aside for ecotourism – and an increase in rural jobs. Organisations, from the RSPB to Facebook, are often defined by their pioneers. In the early days of the RSPB, we see a charity driven by two main goals. First, the prevention of wrong-doing and crimes against nature. Second, the recognition that key islands of life are worth saving. Had the great auk hung on into the 1920s, there is little doubt the RSPB would have intervened – and perhaps, through advocacy, saved it in time. To this day, the protection of habitat islands, and crime-prevention, remain at the core of the RSPB's ethos.

Even by 1960, however, the nature network of the RSPB was tiny compared to the catastrophic collapses under way in the countryside itself. By this stage in world history, in most developed countries, national governments, with enormous spending power, had stepped in to buy whole landscapes, not just reserves, for the nation. In Britain, however, one of the things that makes our wildlife so poor is that nature charities alone, not governments, considered this the right thing to do. British government of either party has, to date, taken little interest in the economic potential of nature, or considered our nation's right to landscape restoration. Whilst other countries considered it self-evidently in their interests to protect large areas for wildlife, the British government did not. All the RSPB could do was buy up ecosystem scraps, and turn them into gems as best it could. Given the odds, it has done a truly remarkable job.

British people, as a whole, however, love nature. They do so in their millions, on every side of politics. In spite of wildlife that is far more meagre than that of a country like Germany, we have many more people interested in it – and this is surely one of the greatest contradictions in our country.[5]

By the 1970s, the RSPB had undergone an explosion in numbers. In 1972, it had 100,000 members – eight years later, this had tripled. By 1981, it had shaped the creation of the Wildlife and Countryside Act, which, at the time, was an innovative way of protecting our rarest species and their habitats. In 1988, a new milestone saw the purchase of Abernethy Forest. The largest European land acquisition by a nature charity, this was one of the first moves towards landscape-level oases: areas home to not tens but hundreds of birds like capercaillies. In 1995, the purchase of Forsinard, in the Flow Country of Sutherland, secured another area large enough for landscape restoration. What is interesting is that Abernethy cost the charity only £1.8 million. This exceptionally large reserve did not bankrupt the RSPB at all, and many more such places could, it seems, be purchased, were the acquisition of land for protection to become the driving force of the charity in the future.

In 1997, membership passed the one-million mark, but more significant was a demonstration, in the year 2000, of the people power the charity could mobilise. Petitioning to secure bird protection laws in the European Union, the RSPB collected more than 500,000 of two million signatures.

By the 2000s, reintroductions of birds such as the corncrake revealed the scientific edge the charity could bring to rewilding when it chose. The resurgence of red kites and cranes has shown its amazing ability to repopulate landscapes with birds not seen there in centuries.

In the past eighteen years, with over 200 nature reserves across the country, the RSPB has become the largest nature charity in Europe. Yet Britain has experienced the worst bird declines, faces the most prospects of bird extinction, and suffers the most degraded landscapes and smallest natural areas on the continent. In the words of the RSPB's current director of conservation, Martin Harper, 'although we're winning some battles, we're losing the war.'[6] But with a million voices for nature, how *might* we win the war to save our wildlife? Here are eight rules that might help conservation leave the future better off.

1. Create New Outcomes

By the second move in a game of chess, there are over 72,000 possible games. By the third, there are 9 million. So far, only the first few moves have been made on the conservation chessboard. Other pieces have been cautiously prodded, a few times, then quietly left alone.

Born in an age of opposition, nature charities have largely defined themselves *against* wrong-doing, which in itself does little to rebuild what has been already lost. In recent decades, charities have purchased the remnants of ecosystems, often after the ingredients that make those ecosystems robust have already disappeared, though in fairness there was nothing they could have done otherwise.

Where species nest at high densities (avocets), thrive in single habitat types (bitterns), fill small areas of land (seabirds) or require protection from persecution (ospreys), British conservation has largely succeeded. The birds of permanent wetlands, some of our seabirds and most of our birds of prey have a solid future. The majority of our birds – evolved in wooded grasslands, river valleys and uplands, dependent on insect food chains and landscape dynamism – have little future at all.

If the first move on the chessboard has been to buy up the gems, the second move, decades under way elsewhere in Europe, has been the restoration of ecosystems. Before pointing to the 'lack of space' in Britain, we must remember, as we saw earlier in the book, that we have all the space we need, both in our national parks and through reform of our hunting estates and forestry parks. In moving towards restoration, however, our nature charities have faced not one but six industries covering almost all the land: the grouse, deer, sheep, forestry, dairy and cereal farms that uniquely dominate our island. Of these, the farms of grouse and deer are optional – and better hunting models exist. Dairy and sheep farms exist beyond the capacity required. Forestry has chosen to leave us sterile deserts not seen in Europe. Only cereal farms are self-evidently essential to our food production needs.

Rather than challenge the proportionality of these Big Six, charities have mostly accepted them, believing that 'natural' land must lie elsewhere, or that a few more birds must be put back into one or other of these crops. Unfortunately, almost all of Britain lies under the Big Six. And the 'elsewhere' has provided just tiny pockets of land. This has left our wildlife uniquely isolated and vulnerable.

Charities have ended up having to fight ecosystem rules, instead of working with them. Rather than pushing for extensive areas governed by nature, they've advocated instead the modification of crops and intensive human management. Few other countries have seen this to be a workable idea.

The problem with our nature-reserve network is that, as we discussed in Chapter 5, birds have evolved in landscapes. British nature is unsuited to hundreds of tiny reserves, bought in isolation from one another. Population dynamics do not confirm to human rules. Islands of farms, forestry clearings or tiny wetlands are what writer David Quammen describes as 'ragged fragments'. Whatever you do with the threads, the carpet is already ripped up.

Most landscape conservation in Britain, as in developing countries like Malaysia, has therefore happened by chance – not design. 'Expedient' conservation is where ecosystems are protected by accident. In Malaysia, this has been in mountainous areas, where logging cannot reach. In Britain, the New Forest was set aside by William the Conqueror for hunting. The large grasslands of Salisbury Plain were saved by the military. Conservation was not, however, the intention. The reason large wild areas have not been achieved *on purpose* is that charities do not have the power to buy areas the size of the New Forest, and invest only small percentages of their income in land. And government has so far overlooked the benefits of restoring wilderness in our national parks.

At present, there appear to be only two solutions available to solving this space issue, without which we have no viable future for wildlife at all. One – charities grow infinitely richer, invest more of their money in land, or collaborate with the private sector to buy huge areas of unproductive farmland, after it has ceased to be viable. Alongside this, hunting estates and the Forestry Commission could adapt their roles, acting as patrons of restoration in some large areas. Two – charities lobby government relentlessly for the subsidised restoration of our national parks. Then, government pays farmers, very well, to wind down, or extensively farm the land, with native animals, within large areas of our national parks. At the moment, neither of these options, of which one or the other is followed to some degree in most countries, is being followed here.

In other countries, national parks are the answer to solving the space issue. In Britain, nature charities have never seen national parks as the answer to single-handedly restoring ecosystems, and therefore arresting the decline not of separate birds but of wildlife as a whole. Indeed, ecosystem restoration has, uniquely, not truly been sought at all.

In some cases, British conservation has seen reserves as genuine solutions to nature decline, even though this approach is opposed to how nature works. This has been complemented by grants to farming landowners. Whilst most countries work with farmers to help nature, no other country has entrusted its 'core' nature to farmers. British farmers are excellent at farming. Nature is best at looking after itself. Even in strongly pro-farming economies such as Alaska, the land labels of 'wilderness' and 'farming' are rarely mixed up.

Nature reserve publicity often emphasises the difference of nature reserves from the surrounding land – and how much life they have. This is in fact the greatest failure of all. By this stage, isolation has won: ecosystems have lost and the countryside around is a desert, leaving species within those reserves isolated and doomed to expensive life support. A lapwing behind an electric fence, for example, is the most profound conservation failure of all – a zoo exhibit in a sterile wider world. Such manicured management, of course, drains charities of the money that could, instead, be used to buy and restore larger landscapes.

2. Let Nature Do the Work

The RSPB, at present, is hindered by the enormous cost of management, and the idea that birds cannot do without it. Replacing intensive management, a scheme of conservation less than a hundred years old, with near-original management, or rewilding, a scheme developed over millions of years, would be extremely cost-effective. Rewilding saves an enormous amount of cash, which could then be spent elsewhere on the acquisition of more land.

Once you start to think about natural agency, life-support systems for birds look as redundant as they are boring. Why worry about coppicing for nightingales, if free-roaming cattle can break trees and create that coppice for you free of charge? Why create a skylark plot at great expense when skylarks can thrive on a floodplain or beside grazing animals? Why maintain a piece of very expensive soil for stone-curlews, a bird that evolved in the wake of soil-disrupting mammals, as can be seen from its thick-knee cousins in the grasslands of southern Africa?

Extensive grazing, with native configurations of animals, makes your money go a very long way. Instead of desperately controlling each inch of a reserve, ignoring population dynamics and ensuring trophic isolation, your main investment is the animals themselves. If the RSPB and other nature charities followed the Knepp model described in Chapter 4, their money would go so much further. Far fewer purchases, of larger areas of land, would be the 'heavy hit' on the purse – instead of small pockets that then need to be managed at great cost. The 'restoration' part would cost much less – because nature would be doing the work.

Once the initial investment in animal architects has been made, the mosaic perpetuates itself. Successional scrub-cutting? Gone. Animals and trees compete. Skylark plots? Gone. Grasslands have their own innate diversity into which our native species fit. Rich grassland giving way to scrub? Not an issue. Herbivores maintain open grasslands but plenty of scrublands too. Birds, returning to their long lost niches, may fluctuate, but provided the landscape is large enough, extinction is unnatural. Extinction happens all the time, however, on nature reserves too small to support dynamic change. Look back through your bird notes. How many protected sites near you have lost their nightingales? Is that cuckoo you logged in 1995 still singing on your local nature reserve?

Almost every example of successful management in Britain amounts to creating permanent habitats, or a few habitats side by side, for birds. Almost every vanishing bird, however, belongs to our lost wooded grassland mosaic. There is not, anywhere I can find, a place where management has saved these birds in their true configuration. This is because management cannot create dynamic wooded grasslands. Only natural processes can.

Birds from wood warblers to cuckoos, lesser spotted woodpeckers to red-backed shrikes, have evolved in mosaics with so many rules, written over so many thousands of years, that seeking to recreate them with management is expensive and presumptuous. Managed conservation, in short, has been as successful as it can be.

The restoration of original processes and stewards, played out at scale, is now the only way to break the management deadlock. With the right animals, or their closest proxies, in the original natural herd sizes, you can, instead, rebuild habitats that cater for 'farmland' birds, 'moorland' birds and 'woodland' birds in one place, because these habitats once freestyled side by side. You save millions of pounds, and a great deal of time, by no longer separating the components of nature. You stop having to worry about in which order to repatriate the flowers, the bees and the birds. You move beyond putting down seeds for turtle doves, and towards real landscapes governed by wild actions – where weeds plant themselves.

If wildlife charities embraced extensive grazing mosaics as their *lead* policy for reversing bird decline, what a richer and more exciting country we would live in. Reversals in bird decline would be profound. Instead of buying land based on what was there already, the default position of a wildlife charity would be to allow nature to reveal her potential. The most crucial thing to remember is that the Knepp Estate started from almost nothing. Now, it has more life than most reserves that started with a 'key' species in mind.

One of the greatest leaps of faith is to accept that nature has its own rhythms. Surprise should be continual. Nobody guessed that Knepp's turtle doves would forage in ground disturbed by pigs. This now seems obvious: there were no tractors in the Pleistocene. Fluctuations, too, could be better accepted if areas were larger and more robust, allowing birds to move to avoid predators.

The complex maze of management schemes that is British conservation may contain good science, but the details don't add up to a vision that's viable or exciting. Whilst our neighbours in the Netherlands continue to expand areas under natural processes, seeing huge resurgences in wildlife at low financial cost, British conservation is stuck at the stage of damage limitation and continues to buy postage stamps of land.

Because farmland has existed for so long, the vision for farmland birds is to make their lives a little better. But imagine, for a second, we were standing in a cereal field in the Amazon basin. Would we be having a conversation about helping the sparrows, twittering amid the ghosts of the macaws? Britain's nature charities needs to decide whether all they can see is a field that must be modified and made less awful, or a wildland waiting to happen. Whether habitats should be static, or dynamic. Birds evolved beside other animals, not under conservation policies. But until charities believe this, they are unlikely to push for ecological restoration with conviction. Expanding minds, therefore, is just as important as exploring the economic potential of nature.

3. Tear up the Targets

> *Marsh fritillary larval webs will be found at a density of at least 200 per hectare … The optimal breeding habitat comprises abundant Molinia, where the vegetation height is largely within the range of 10 to 20 cm … Molinia meadows will cover at least 4 ha … scrub and saplings will be of no more than scattered occurrence …*
>
> —Strict instructions to nature: part of a 25-page site plan for a Site of Special Scientific Interest in Wales[7]

There is no doubt, to be fair, that nature reserves are safer, more comfortable, and obey written targets, when compared to the unwieldy surprise of wilder ecosystems. If a reserve sets itself a target and then meets it, its investors can see that a thing was 'well done'. Yet if we carry on the same way, we will get the same results. Our wider countryside will continue to starve and fragment. Most of its birds will continue to vanish as the last of the food chain collapses and tiny islands of birds die out one by one.

For wildlife to have any future in our country, reserve managers must create broader, expansive targets that allow nature to express itself. They could select broad-scale ambitions, and, like the 'managers' at Knepp, remain broad-minded about how nature shows her hand. Yet if one thing is standing in the way of a true natural resurgence in Britain, it's not pesticides. It's paperwork.

Knepp's rewilded lands have seen the largest turtle dove resurgence in Britain. Against a backdrop of 96% national decline, Knepp has gone from three pairs in 1999 to sixteen by 2017, with most acceleration in the past few years. But it isn't just rewilding that's made this possible. It's freedom. When the Burrells brought nature back to Knepp, they did not have to contend with the 'recognised' interests of a conventional reserve on large areas of their land. Ironically, because the land was barren, the Burrells were far freer than any reserve manager to rekindle the area to richness. It's a provocative thought, but the landowners most successfully restoring turtle doves in Britain are doing so *because*, not in spite of the fact that, they never set the target of conserving turtle doves at all.

Knepp's cattle did not have to nibble vegetation to a certain height. Their pigs did not use their snouts in a prescribed manner. Nobody, anywhere, planted weed strips for turtle doves. And turtle doves are loving things at Knepp, more so than on any farm now managed for their needs.

The resurgence of wildlife at Knepp exceeds the triumph of any Site of Special Scientific Interest (SSSI). What might be harder to accept is that in its present form, the iron hold of an SSSI actively hinders, as much as contributing to, the full possibilities of nature. Not only are 'targets' for nature at odds with freestyling populations of wildlife, they are at odds with nature itself. If the natural world can only express itself by meeting targets, like a school curriculum, Britain is doomed to a terrible future for wildlife. To illustrate better what we are talking about, I picked out a SSSI statement at random – and have quoted some of it above.

The SSSI consists of eleven small fragments of wet grassland, all isolated by farmland, managed for the marsh fritillary butterfly. It is the kind of place where the writing was already on the wall. Extinction debt (see Chapters 2 and 5) necessitates that by this stage of isolation, it's probably best to protect a larger area somewhere else. Instead, across 25 pages, the devastating level of prescription for our wildlife becomes clear. Marsh fritillaries, the statement demands, require extensive areas of *Molinia* grass. *Molinia* is a moor grass. Marsh fritillaries, however, can be found across Europe in a range of habitats, damp and dry, where devil's-bit scabious, their food plant, is common. Crucially, marsh fritillary colonies range widely across a landscape. In this country, only connected grasslands such as Salisbury Plain still enjoy thriving populations today.

Across 25 pages, however, grass height, the proportions of willow and larval webs per square metre are mapped out as if planning for the Chelsea Flower Show. In spite of all of this, marsh fritillaries, the statement says, are now in 'unfavourable' condition. The SSSI, it turns out, was too small to meet their needs in the first place. If the aim of all this bureaucracy is to save a beautiful butterfly, then surely that's worthwhile? In fact, quite the opposite. Once a target is set, that target becomes the goal, however stunted it was in aspiration to begin with. Many SSSI statements date back to the 1980s, yet still the targets remain. Hit the target, and you've won, no matter how many other species, or ecosystem rules, are overlooked in the process.

In a recent visit to an area bought for rewilding in Wales, a colleague and I were shown an SSSI within the site, which was being managed for a particular species of geranium. As we discussed the full possibilities of nature on the site with the warden, including the return of beavers, we stopped, in bemusement, as he stooped down and started picking up rocks. They were, he explained, blocking the emergence of a single geranium. We watched, with increasing bafflement, as he continued to tenderly prune the site as if it were his garden.

So divorced are the targets of many an SSSI from the true exuberance of nature, that their destructive aspirations would thwart many a more ambitious scheme for restoration. But how did such a decidedly weird situation come about?

In twentieth-century Britain, with life battered into tiny remnants, the creation of SSSIs and SPAs (Special Protection Areas) aimed to protect some of these remnants from vanishing completely. There was recognition that some areas of Britain were richer than others. In Europe's larger nature reserves or national parks, there are, of course, overall features of interest – and targets. In Africa, too, ecologists do not simply shrug their shoulders if rhinos disappear. But the smallness, the prescription, of British targets, the tenacious attention to the prevention of nature, appears to be unique.

For millions of years, the world enjoyed considerably more life – without targets. Very rarely have elephants or bison been proven to write them. Paper targets sit at odds with the dynamism of nature. Charities like Buglife have argued that rewilding could threaten fine-scale management for smaller groups such as butterflies.[8] Yet butterfly colonies, too, are intrinsically evolved to shift across mosaic landscapes over time. As seen from the Krefeld study (see Chapter 5), landscapes are organisms that live or die collectively, *regardless* of the size of their inhabitants.

Imagine, now, that a large rewilding scheme, such as Knepp, had to contend with 25 pages of marsh fritillary protection. In place of the return of butterfly clouds flying in abundance, paperwork would have required they continue to cosset and manage a few grasses, freezing 'marsh fritillary' habitat separately from nature itself. A fritillary monoculture would have stood in place of nature. Right now, this is what happens across many reserves where nature, instead, could be in charge.

So how can we keep the worthwhile target of 'special' places for nature, whilst removing such crushing restraints? In place of 25-page plans prescribing how nature should be, how about a one-paragraph statement, which simply protects, in the strongest terms, the legal *right* of nature to have the run of the place. As a draft, how about this:

> *Minsmere forms part of the low-lying Suffolk coast. Its marshlands are of enormous value to the natural world, whilst wooded grasslands surrounding the site are also full of life. In decades to come, the full expansion of these marshes along the coast could allow for a higher degree of naturalness, including but not limited to beavers creating wetlands, and wild native animals creating mosaic habitats of their own. This might enhance existing management with a range of dynamic processes.*

Such a recognition would enshrine the conservation work done on a site so far, but free up conservation for the future. Whatever breeds within Minsmere would continue to receive de-facto protection, but the designation would be open-ended, rather like nature itself.

At the moment, using 4,700 words, Natural England monitors 62 SSSI 'units' of Minsmere's wetlands, although the land is managed by the RSPB.[9] And it is this fearfully controlled management that thwarts any exciting future. Let's take Unit 001. In 2013, it had a problem. A ditch was aspiring beyond its station. Too many ditches in Unit 001 were in 'late succession to meet target, and water depth was less than target'. A late successional ditch, rich in longer sward, could be home to grasshopper or marsh warblers, a developing habitat for willow tits or a suitable place for a colonising bluethroat. But nature was being naughty. The target was there to put it right.

Over time, with targets, you can forget the wild completely. If even a ditch can be seen as a threat, the ecological excitement of an original site manager, such as a beaver, would play glorious havoc with the unit system. To this day, beavers have yet to be released into prime conservation wetlands by nature charities. There are still no officially-sanctioned beavers shaping the wetland landscapes of the East Anglian coast.

In 2017, the results of a twelve-year study of beavers habitat transformation was published, funded by the Natural Environment Research Council. It studied what happened when beavers were released onto drained pasture, reclaimed for conservation. As the beavers dammed, creating small wetlands and felled trees, promoting decay, overall biodiversity increased by 148%. Landscape variety, the key ingredient for nature to thrive, by 71%. The study summarised that:

> Our study illustrates that a well-known ecosystem engineer, the beaver, can with time transform agricultural land into a comparatively species-rich and heterogeneous wetland environment, thus meeting common restoration objectives.[10]

It is interesting that such studies are required at all. Nobody in Canada or most of Europe feels the need to prove that ecosystem engineers shape ecosystems to maximal advantage. The two images in Plate 30 show exactly how, in twelve years, beavers can transform a landscape – without a single digger or costly conservation plan in sight.

Succession, herbivory, flooding, local colonisation and desertion – these are universal patterns. We have failed to realise how strange it is to create a new system from scratch, replacing animal-managed landscapes with human ones, often without this being necessary at all. Each SSSI unit functions exactly as nature didn't intend: *alone*. Any site managers of real vision are currently thwarted by such destructive constraints.

Under the current system, people have mistaken gardening for conservation. This gives us the most tamed areas of nature in the world. It contributes towards, rather than preventing, the wider collapse of our landscapes. Only with the bureaucracy of conservation shredded, small-mindedness torn up and the broadest of targets put in place, can nature return in the way that she sees fit – a way in which she has evolved over millions of years.

4. Be Messy

> Wet scrub occurs across a range of damp soils. If unmanaged, however, scrub encroachment into these habitats will damage their wildlife interest.
>
> —A nature charity's unintended advice on removing willow tit habitat[11]

Scruffy habitats are problematic for British conservationists, far more so than elsewhere in Europe. The precise target system that must be followed by site managers prizes tidy ecological order – such as pure deciduous woodland – in place of freestyling ecosystems that are more chaotic, more varied and therefore most diverse.

Chaos, dereliction and decay are the last words you'll hear in British conservation. And the failure to understand and halt the decline of birds such as lesser spotted woodpecker, willow tit and long-eared owl encapsulates this problem. All are landscape specialists with a love of neglect.

Long-eared owls prosper where voles infest long grass. Even in original habitats, these conditions would have required areas where grazers were sparse or had been decimated by disease, as concentrated herbivory reduces rodent abundance.[12] Grassy moors with pines, old airfields or fallow grasslands near plantations are, therefore, the ideal habitats for long-eared owls, as well as fallow brownfield sites.

If you're standing in a place where the oaks are ragged, the alders fungal, and old apple trees bend under the weight of time, you might still glimpse a lesser spotted woodpecker clinging below a branch. Only the actions of time in trees create birds like lesser spotted woodpeckers, which both feed and excavate nests in dead wood. Managing for such birds is impossible. Interference is opposed to their needs.

If you're surrounded by a jumble of rotting scrub trees, and you can see mushrooms and mosses on the branches, you have reached willow tit standards of decay. I have watched, first-hand, disastrous efforts to 'manage' sites for willow tits. Management risks removing, or ordering, the rotting tree chaos in which these birds thrive. Preserving decay, however, means doing nothing. And whilst it reflects well on a charity to show they have dug a reedbed for bitterns, it takes more courage to admit that you left a coalfield to its own devices, and by stepping away, nature moved in and thrived best by itself. Protecting areas for birds like willow tits is vital. What is needed, then, is restraint – as decay and dereliction take hold.

The inability for many nature reserves to embrace scruffiness is why many of our counties already have more avocets than willow tits or spotted flycatchers. This is because a crucial ecosystem process, *freestyling*, has been put second to ecological gardening.

Whilst nature reserves may have to meet targets, the longer-term legacy of doing so is now becoming apparent. Whilst a few target species thrive in locked-down habitats, almost every bird evolved in margins, from cuckoos to turtle doves, continues to vanish. Almost every British bird haunts two or three micro-habitats. These must dovetail into one another for that species to survive. So habitat segregation is one of the most dangerous things in conservation.

One solution to these problems is for larger reserves to be freed up for dynamic processes to take hold. Areas like Abernethy Forest are big enough that you find plenty of freestyling, scruffy areas, outgrowing the need for manicure. The more tiny nature reserves are bought, however, the more they must be managed one square centimetre at a time. And the poorer, over time, our wildlife will become.

5. See the Bigger Picture

Life is really simple, but we insist on making it complicated.
—Confucius

If we can recognise fundamental truths, see common patterns such as insect loss, see that ecosystem restoration is the only solution, then get to work fast, British wildlife has some chance of a future. To do this, we need to massively simplify the battle plan, treating species as vanishing collectives and spot the right patterns in time.

In the 1990s and 2000s, the RSPB led the way. Bitterns were scarce because reedbeds were too, but it was realised there were far more reedbeds than bitterns. Soon after, it was found that a shortage of fish like rudd was starving bitterns in the reeds – so rudd were put back and the food chain was partly restored. It was realised that landscape scale was essential – so reedbeds were joined up.[13] Bitterns no longer face any prospect of extinction: key ecological lessons were acted on in time. Corncrakes, it was realised, were missing from everywhere except the Hebrides not because of some complex migratory strategy, but because crofted meadow habitats persisted here and nowhere else. The RSPB staved off damage to these last hay meadows – and expanded them in time.

Tackling landscape stewardship, food supplies and habitat connection is what works best for birds. So it is baffling when organisations disregard the fundamentals of ecology. Nowhere does this become stranger than in the case of our vanishing insectivores. Insectivorous birds are the living expressions of insects. The disappearance of an insectivore is, the world over, the quickest way to spot the decline of the insects that it eats. So universal is this rule that we are taught it at school – in the form of a food chain, or pyramid.

One of our fastest vanishing insectivores, the spotted flycatcher, for example, is the trophic expression of butterflies and bees. Yet virtually no scientific paper written in Britain correlates insect loss to flycatcher decline, or even tries to do so.[14] When you get to a situation where the clue to the bird's decline is so obvious as to be included in the *name* of the species, yet still overlooked, it appears some basic ecological common sense is rapidly required in the fight against bird decline.

The biggest problem in failing to see the big picture is that each bird decline is considered as a separate entity. Birds are often assigned 'species officers', which, from the start, helps to isolate them from the wider rules governing ecosystems. Birds become cherished individuals – rather than imperilled canaries in the same collapsing mine.

As long as we ignore universal rules, think about birds as isolated units and see Britain as a country with its own unique set of bird declines, our wildlife will continue to vanish – as more papers are cautiously written on why.

In burying ourselves deep in the ecology of a single species, there is also the risk that entire landscapes become sacrificed to the need of one species. This is arguably one of the greatest dangers in British conservation, where this practice can be taken to extremes.

Across the world, conservation has chosen 'flagship' species, like the tiger, to raise cash and attention for protecting valuable ecosystems. This is a common and understandable way of allowing the public to connect with a place, through the medium of a species. But you need to pick those species with care. Attribute a bird too much importance, or render it sacrosanct through misguided targets, and whole reserves, or even landscapes, can be written off for a wide range of other species. Indeed, one peculiar trademark of British conservation is the curious degree of investment in birds that require landscape degradation.

Stone-curlews are boggle-eyed wading birds with a penchant for bare earth, and have the guilty expression of a bird that's taken magic mushrooms. If you travel to the sheep-grazed plains of Extremedura, in Spain, you can see stone-curlews in abundance, in habitats that aren't truly natural, but recall the 'sheep-wastes' of seventeenth-century Britain. If you visit southern Europe, however, you will also find grazing woodlands or olive groves with earthy areas. Here, you find stone-curlews nesting, quite happily, under the shade of trees. Stone-curlews are adapted to places where things get very hot. Trees, it seems, provide them with shade. Their 'core' requirement is simply the significant disturbance of soil.

For hundreds of years, stone-curlews thrived in treeless Britain – but that doesn't mean this is their natural habitat. As with any 'farmland' bird, they evolved in a contest between trees and animals; a contest very much alive in areas like the Serengeti, where thick-knees, of the same family, nest below trees in earthy wallows disrupted by wild herbivores. Yet rather than accept stone-curlews as an earthy oddity, British conservation has turned these birds into a species that costs the earth.

Stone-curlew plots, strange rectangles of weedy soil, have been created across downland farms in Britain, often at considerable labour and expense. Forgetting how ecosystems work, these plots achieve what pigs or grazing herds, stopping to wallow, could achieve for free. No other country is creating stone-curlew plots. This may be because stone-curlews represent just a fraction of an ecosystem: they emblemise areas of the highest disruption by digging or wallowing animals – nothing more, nothing less. Conservation now removes many other birds by sustaining such a habitat as a rectangular museum, isolated from the dynamics that once created it. The weed-rich earth, separated from the danger of trees, benefits just a fraction of our grassland birds. Here, most sources of plants, insects and birds never get a look in. Woodlarks feed here – but need trees to sing from. Lapwings and curlews can do well. And that is pretty much it. These expensive allotments preclude trees, marginal birds like red-backed shrikes, scrub-grassland birds like cuckoos and long-grassland birds like quail. Yet sacrificing diversity for an ecological oddity is often regarded as success.

Nobody has really asked why stone-curlews, using an atypically barren earth habitat, are worth so much investment. Worse, no one seems to have wondered what this species once did in the absence of stone-curlew plots. Stone-curlews can, across Europe, be found in a range of habitats where not a conservation penny is spent. They are common on the edge of Spanish salt-pans, in disrupted industrial habitats, in Italian olive-groves, and even on Suffolk pig-farms – they thrive in any environment with plenty of earthy disturbance.

Another specialist with a toehold in Britain is the Dartford warbler, noted for its fussy preference for low heather, interspersed with stands of gorse. A warbler originating from the Mediterranean region, the Dartford warbler has, most probably, evolved from the *maquis*, a form of warm, coastal shrubland not native to our country. As a result, Dartford warblers share their world of gorse and mid-height heather with few other birds. And to maintain their habitat, endless forms of scrubland complexity, from aspiring young trees to dense bramble and thornlands, all need to be ripped out.

Being adapted to warmer climes, the British winter hits Dartford warblers hard. Every decade or so, enormous population crashes take place. Yet whilst British weather subtly hints to us that Dartfords are supposed to be rare, by wiping them out, conservationists diligently burn the land, and remove invaluable scrub, in order to put them back in. The destruction wrought to save one Mediterranean overshoot has become impressive. Heathlands managed for Dartford warblers are cleared of that life-filled menace, scrub, through cutting or burning. Yet whilst a heather-gorse mosaic accounts for just one specialist bird – the Dartford warbler – full, varied scrublands account for twenty; scrub grasslands, fifteen more. All in all, therefore, over thirty species are written off – for the future of just one.

By burning for Dartford warblers, we remove the interface of dense hawthorns and disturbed soils that sow turtle doves. We destroy the wet elder-lands of the willow tit. We remove the herby bushes of the garden warbler, the thorn blanket of the nightingale, and the basis for the next generation of oak – and the richest of all biomes: wood-pasture. Critical of the gamekeepers who burn for red grouse, conservation is doing just the same. Ironically, on closer inspection, Dartford warblers don't even require this at all. Whilst in a management textbook called *Bird Life of Woodland and Forest*, Robert Fuller wrote that 'scrub-cutting ... has been successfully employed by the RSPB on its Suffolk heathland, where it has been invaded by birch', in the 1970s Colin Bibby found that Dartford warblers often fed in birches, which helped them through the lean times by providing food.[15] Farmers do not create gold-plated pillows for their cows to sleep on, but managed conservation is doing just this for birds that evolved long before these methods existed. Dartford warblers and stone-curlews aren't that fussy: we are.

This folly of manicure, across hundreds of nature reserves, thousands of times, happens because conservationists, managing land for under a century, are trying at great expense to replicate one part of one ecosystem in one place, at one time. This is profoundly unnatural, and extremely damaging for the recovery of the natural world. Site managers have watched trees growing, and concluded that the best thing to do is to cut them down. They have forgotten there are many animals designed to best administer such a balance. And the results are damaging indeed. There are now four times more burned heather-loving Dartford warblers than willow tits: a scrubland bird that often makes its home in 'invading' birch, our third most diverse tree of all.

Micro-management is unnatural, extraordinarily expensive per hectare, and extracts only a fraction of an ecosystem, and its wildlife, from the ashes. Whilst initial moves to prevent extinction are understandable, these have reached dogma levels and become self-serving ends in themselves. New flagship species, indicative of diverse mosaic landscapes and thriving insect populations, urgently need to be moved to the fore. This may help to move conservation beyond freezing time, burning, and excluding the many for the few. Without seeing the bigger picture, managed conservation, taken to extremes, may ultimately kill more species than it saves.

6. Invest Only in Scale

Rushpole Wood, Denny Wood, Frame Wood – these are perhaps the last places where you can enjoy, in one morning's walk, the true richness of our ancient pasture woods. The New Forest's core woodlands are a national treasure. They are also an anomaly.

Ongoing studies are revealing over 100 pairs of lesser spotted woodpecker, 500 pairs of hawfinch, 500 pairs of redstart, and hundreds of tree pipits, woodlarks, nightjars and spotted flycatchers still existing, in apparently stable populations, across the New Forest.[16] Cuckoos are far more common than elsewhere, and perhaps as many as 100 birds still sing across the forest. The New Forest is perhaps the greatest living proof that large, food-rich areas of prime habitat are often enough to save our birds.

When buying land, two things stand out as overdue priorities for nature charities. First of all, virtually every British bird vanishing today is connected to our lost wooded grasslands. Yet if you look at our nature reserves, there is a huge bias towards permanent wetlands. So seeing parts of the New Forest transition into conservation ownership could be interesting. In Scotland, reserves like the RSPB's Corrimony and Abernethy offer better blueprints for the future, stepping in an exciting direction in terms of scale.

In decades to come, other landscapes may, for the first time in over a century, come onto the market. Salisbury Plain, light on roads, devoid of populations or farming and 28,000 hectares in size, is another huge area awaiting conservation vision, if its military priorities change. Historically, the fossil record shows the plain may have been the closest area Britain had to a Serengeti, even after the mega-herbivores had vanished.[17] This is shown, in part, by the density of herbivore fossils found around Stonehenge. There seems no better area in Britain where 'plains' dynamics could once more play out. Politically, less stands in the way here than in any other area. Thousands of butterflies can still be seen on the wing each summer. The bottom of the food chain is, in places, intact.

Full rewilding of Salisbury Plain would see a more diverse landscape emerge. The stone-curlews, great bustards and corn buntings would be returned to their original stewards. But the sight of horses, wild cattle, and even bison wandering the plain could become iconic and draw hundreds of thousands of tourists.

In a country we wrongly believe to be crowded, the possibilities of scale are sitting there, waiting to be explored. Even now, areas like Salisbury Plain, northern Dartmoor and the New Forest are gifts. For our birds to have any future, such landscapes are the only currency that nature understands.

7. Ask for More

Vision without action is a daydream. Action without vision is a nightmare.
—Japanese proverb

If you don't ask, you don't get. And think, for a moment, what kind of country we would live in, if people had settled for what was most readily achieved, not for what was best.

In 1940, appeasement was most readily achieved. Whilst Churchill did not stand alone in opposing it, large parts of the press, large sections of liberal and conservative society alike, were prepared to give Hitler a chance. In retrospect, it seems inevitable that Britain stood against Fascism, but even after entering the war, Italian mediation was a real prospect.[18] Only the soaring conviction of a few, Churchill most of all, won the day.

In 1946, doctors voted ten to one against the creation of the NHS. Against such opposition, many would simply have backed down. 'Elements of free care' might have been added to a paying system, covering the worst of illnesses, for some people. But the Attlee government held its nerve – and asked for more. Now, whatever your stance on the NHS, you enjoy it free at the point of entry. No ifs, no buts.

Great achievements in Britain, on both sides of politics, have only ever been achieved through tenacity, vision and asking for more. Often, at the time, far more people opposed than supported what we now know to be right. Whilst the battle to restore Britain's wildlife may not seem to most as epic as the Second World War or the foundation of the NHS, we can never hope to enjoy true national parks, amazing wildlife or secure rural jobs without a similar vision to that of our ancestors.

Right now, 1.1 million members of the RSPB both support and receive advice from Britain's leading nature charity. And unlike the previous struggles, a people-powered movement of enormous power already exists to fight for the natural world. At the moment, however, even mobilising small acts of change can appear a step too far.

In a recent press release, a well-respected lead scientist at the RSPB announced the government's recent ban on neonicotinoid pesticides. He reacted to the news, as you might expect an expert scientist to do, with 'delightful surprise'. But why did this come as a surprise? Mobilising a one-million-member petition against a pesticide proven to be of immense damage to the natural world – robbing us of bees and flycatchers, wagtails and starlings – would seem to be a basic activity for a large wildlife charity. Individual scientists are not expected to lobby, but nature charities are. Yet in this case the motion came from government – a surprise to those who barely pushed for it at all.

With such extreme caution, the vision and boldness to ask for wild parks, wild processes and wild animals will never come to pass – nor will the return of Britain's natural heritage. Understanding such failures is important if British wildlife is to have any future. Perhaps the underlying reason is that, as charities grow to a certain scale, they are prone to become more cautious. Emily Williamson, campaigning against the plume trade, could be absolute of purpose. But the larger an organisation grows, the more spread out it becomes, the more people it has to employ, the more the plurality of voices within it pull in a number of directions. Today it is mostly farmers' unions, not nature charities, that lead the way in absolute conviction and pure determination. In their present form, nature charities trade in doubt, nuance, uncertainty and deferment. To the victors go the spoils. At the same time, fund-raising risks becoming a circular activity, whereby more and more money is ploughed back into existing projects, maintaining an ever-growing number of jobs and products. If, by contrast, we were seeing 50% of charity money invested in land per year, Britain's wildlife gains might be quite remarkable. Yet for this to happen, there would need to be a radical restructuring in our largest nature charities, and a difficult cutting back of many non-essential operations and jobs.

Visionary ideas for change now often come from individuals outside our nature charities, often without the means to implement them fully. Yet our charities are full of world-class scientists, bright minds and dedicated volunteers. Somewhere, the brake is being applied. Small triumphs, damage limitation, are being mistaken for the goal. With big possibilities written off from the start, a meaningful future for our wildlife is being wiped out before it has the chance to begin.

Alongside inherent caution, charities seem to fear, at all costs, causing offence, negative headlines and vanishing members. Ironically, hundreds of thousands of new, younger members await our nature charities – the *moment* that an audacious vision of how the future should look is finally revealed. Calls for epic restoration of our landscapes would yield epic support. Calls for a few more birds of prey to drift over burned moorland, or for mild changes to crops, are not very appealing to a species, ourselves, intrinsically drawn to positivity and hope.

Whilst John Clare could wander out into a countryside filled with nightingales and haunted by wildcats, most people today live among the last dregs of nature. Why, in fairness, should they engage at all? The excitement of larger herbivores or lynx, the kinds of animals that people pay to see on safari in other countries, is no more than a dream. But daring to imagine profitable pelicans in Somerset is how real ventures like the NHS have come into being. In mapping out vision of this scale, it's important to remember that you won't become everybody's friend. But this is not the purpose of charities, any more than it is the purpose of politicians. The aim is to leave the next generation better off, not worse, than the last. The stronger the vision, the more principled the fight, history shows us that more, not less, will rally to the cause. In a century's time, if that vision was strong enough, those people will have a wonderful legacy to enjoy.

In this regard, concentrating energies on key areas would go a long way. A clear plan for two or three areas to become reliant on ecotourism, a clear plan for landscape restoration, generating rural jobs, is a positive prospect. Areas like the Somerset Levels, Snowdonia or the Peak District could be three regions where nature charities could focus their firepower. Push hard on the 75% of Snowdonia, making no profit on sheep, push hard for golden eagles, lynx, a whole new rural future for Wales and billions for the economy – and politicians might, at least, begin to listen. Mumble for a few more trees – and nobody will care.

Job creation is the reason governments listen. Costed plans for hundred-year futures, in new wild areas, are more likely to gain traction than laments on the fortunes of the skylark. Looking at the absurd economies of our failing uplands, it is clear that the economy is on the side of nature. The economy, in truth, is the biggest friend of conservation.

8. Remember Emily

In 1904, on receipt of the royal charter, Emily Williamson spoke proudly of the RSPB as 'a very small fledgling', which 'had no dreams of soaring to the heights which it had reached'.[19] The chances of such a flight had indeed been truly minute. Emily, a lone female voice without the right to vote, had the odds stacked against her in a way no wildlife charity need worry about today. In the 'plume boom' of the late nineteenth century, hundreds of millions of birds were harvested across the world. In London, the centre of trade in exotic feathers, a single order placed in 1892 consisted of 6,000 birds of paradise and 40,000 hummingbirds. In 1897, a representative of the Society for the Protection of Birds, W.H. Hudson, described the sight of 80,000 parrot skins:

> *Spread out in Trafalgar Square they would have*
> *covered a large proportion of that space with a*
> *grass-green carpet, flecked with vivid purple, rose and*
> *scarlet.*[20]

By the turn of the century, the plume trade was big business. Worth £2 million annually, plume-traders, as influential in government as the grouse shooting lobby is today, assured the public that the majority of their trade was limited to plumes 'shed' from live birds. They carried on with impunity. In Britain, plume-hunters would visit the cliffs of

Flamborough, cutting the wings off live kittiwakes and 'flinging their victims into the sea'.[21]

All this must have seemed hopeless, overwhelming, to the SPB members of the time. Yet by the turn of the century membership had grown to 20,000 as people rallied around their clear convictions. Following a successful 'Murderous Millinery' campaign, the Plumage Act of 1921 heralded the beginning of the end. Now, we take for granted the great crested grebes that adorn not our hats, but the canals beside our towns. This, in truth, is a miracle.

Since that time, much has been achieved, often against all the odds. This chapter was not written to criticise what has happened, or to belittle the work of many inspirational people along the way. It was written to make the case that things cannot continue this way.

If we want to have turtle doves, let alone pelicans, in the centuries to come, our charities must ask for more. They have to replace small ambition, incompatible with the natural world, with a costed plan for large wild areas, economies and jobs.

The British lobby for nature is mighty – decency and the economy are on its side. With legitimate democratic power, it is time for Britain's nature-loving voices to put forward a true vision for what our grand-children deserve to enjoy. Then, for the first time in our history, the natural heritage of Britain, and all the riches that go with it, may finally be given back.

Notes

Introduction

1. Alan Watson Featherstone, cited in Monbiot, G. *Feral: Searching for Enchantment on the Frontiers of Rewilding*. Penguin, London, 2013, p. 122.
2. Ceballos G., Ehrlich, P.R. & Dirzo, R. 2017. Biological annihilation via the ongoing sixth mass extinction signaled by vertebrate population losses and declines. *Proceedings of the National Academy of Sciences* 114: E6089–E6096. doi: 10.1073/pnas.1704949114.
3. RSPB and partners. 2012. *The State of the UK's Birds 2012*. RSPB, Sandy. Available via www.rspb.org.uk.
4. Urban population (% of total) in the United Kingdom was reported at 82.84% in 2016, according to the World Bank collection of development indicators, compiled from officially recognised sources.
5. Rae A. 2017. A land coverage atlas of the United Kingdom. https://doi.org/10.15131/shef.data.5266495. The citation for the underlying data is given as: Cole, B., King, S., Ogutu, B. *et al.* 2015. *Corine Land Cover 2012 for the UK, Jersey and Guernsey*. NERC Environmental Information Data Centre.
6. A comparative calculation made by the author, outlined in more detail in Chapter 7. Snowdonia National Park is 2,141 square kilometres. The Maasai Mara National Park is 1,821 square kilometres.
7. Yellowstone National Park is around 9,100 square kilometres. The Cairngorms, Britain's largest park, is still an enormous 4,528 square kilometres and virtually depopulated, yet huge areas are written off to burning for driven grouse shooting.

1. Taming Britain

1. Monbiot, G. 2015. Attacks on the last elephants and rhinos threaten entire ecosystems. *The Guardian*, 22 May 2015.
2. Ashton, N., Lewis, S.G., De Groote, I. *et al.* 2014. Hominid footprints from Early Pleistocene deposits at Happisburgh, UK. *PLoS ONE* 9(2): e88329. doi: 10.1371/journal.pone.0088329.
3. Yalden, D.W. & Albarella, U. 2009. *The History of British Birds*. Oxford University Press, Oxford, p. 33. This chapter draws heavily on the outstanding scholarship in this book. Unless given a separate citation, further references to the avian fossil record all come from this source.
4. Connor, S. 2010. Traces found of the earliest Britons from 900,000 years ago. *The Independent*, 7 July 2010. Simon Parfitt, of University College London,

explains how the Norfolk coastline would have looked during the time of early human colonisation, with grasslands dominated by giant elk, sabre-toothed cats and mammoths.

5. Alleyne, R. 2008. Remains of a sabre-toothed tiger the size of a horse found off British coast. *Daily Telegraph*, 19 November 2008. The findings detailed are those of Dick Mol, a palaeontologist at the Natural History Museum in Rotterdam, referring to a 'huge male' sabre-toothed cat he estimated to weigh about 400 kg.

6. Burger, J., Rosendahl, W., Loreille, O. *et al.* 2004. Molecular phylogeny of the extinct cave lion *Panthera leo spelaea*. *Molecular Phylogenetics and Evolution* 30: 841–849. doi: 10.1016/j.ympev.2003.07.020.

7. Curry, A. 2011. Dissecting the cave lion diet. *Science Now.* www.sciencemag.org/news/2011/11/dissecting-cave-lion-diet.

8. Ashton *et al.* 2014.

9. Owen, J. 2006. Stone-age elephant found at ancient UK hunt site. *National Geographic*, 7 July 2006. Details the bone remains of *Palaeoloxodon antiquus* found around Ebbsfleet, Kent. The primary research was led by Francis Wenban-Smith, principal research fellow in ecology at Southampton University.

10. Yalden & Albarella 2009.

11. Harrison (1980), cited in Yalden & Albarella 2009.

12. Diedrich, C.G. & Žak, K. 2006. Prey deposits and den sites of the Upper Pleistocene hyena *Crocuta crocuta spelaea* (Goldfuss, 1823) in horizontal and vertical caves of the Bohemian Karst (Czech Republic). *Bulletin of Geosciences* 81: 237–276. Czech Geological Survey, Prague.

13. Hunting for the first humans in Britain – British Archaeology Features – accessed 6 September 2015 but article since archived or taken offline.

14. Lister, A.M. 1987. Late-glacial mammoth skeletons (*Mammuthus primigenius*) from Condover (Shropshire, UK): anatomy, pathology, taphonomy and chronological significance. *Geological Journal* 44: 447–479. doi: 10.1002/gj.1162.

15. Cooper, A., Turney, C., Hughen, K.A. *et al.* 2015. Abrupt warming events drove Late Pleistocene Holarctic megafaunal turnover. *Science* 349: 602–606. doi: 10.1126/science.aac4315.

16. Jacobi, R.M., Rose, J., MacLeod, A. *et al.* 2009. Revised radiocarbon ages on woolly rhinoceros (*Coelodonta antiquitatis*) from western central Scotland: significance for timing the extinction of woolly rhinoceros in Britain and the onset of the LGM in central Scotland. *Quaternary Science Reviews* 28: 2551–2556.

17. Stiner, C.M. 2004. Comparative ecology and taphonomy of spotted hyenas, humans, and wolves in Pleistocene Italy. *Revue de Paléobiologie, Genève* 23: 771–785.

18. Cardoso, J.L. 1993. *Contribuição para o conhecimento dos grandes mamíferos do Plistocénico Superior de Portugal.* Camara Municipal de Oeiras.

19. Herring, D. 2005. Time on the Shelf. NASA Earth Observatory. https://earthobservatory.nasa.gov/Features/TimeShelf (accessed October 2018).

20. Pettitt, P. & White, M. 2012. *The British Palaeolithic: Human Societies at the Edge of the Pleistocene World*. Routledge, London, p. 133.

21. Lane, M. 2011. The moment Britain became an island. *BBC News Magazine*, 15 February 2011. Quotes the evidence of Professor David. E. Smith at Oxford University on the tsunamis that isolated Britain at the end of the last glacial period. www.bbc.co.uk/news/magazine-12244964 (accessed October 2018).

22. Bondevik, S., Lovholt, F., Harbitz, C., Stormo, S. & Skjerdal, G. 2006. The Storegga Slide tsunami: deposits, run-up heights and radiocarbon dating of the 8000-year-old tsunami in the North Atlantic. American Geophysical Union fall meeting 2006, abstract id. OS34C-01.

23. Lane 2011.

24. Weninger, B., Schulting, R., Bradtmöller, M. *et al.* 2008. The catastrophic final flooding of Doggerland by the Storegga Slide tsunami. *Documenta Praehistorica* XXXV.

25. Cromsigt, J.P.G.M. & Beest, M. 2014. Restoration of a mega-herbivore: landscape-level impacts of white rhinoceros in Kruger National Park, South Africa.' *Journal of Ecology* 102: 566–575. doi: 10.1111/1365-2745.12218.

26. Vera, F.W.M. 2013. Can't see the trees for the forest. In Rotherham, D. (ed.). *Trees, Forested Landscapes and Grazing Animals: a European Perspective on Woodlands and Grazed Treescapes*. Routledge, London, pp. 99–126.

27. Godwin (1956), cited in Kennedy, C.E.J. & Southwood, T.R.E. 1984. The number of species of insects associated with British trees: a re-analysis. *Journal of Animal Ecology* 53: 455–478. doi: 10.2307/4528. The subsequent 'trees argument' in this chapter draws several times on Southwood's study of insect abundance in our native trees.

28. Alexander, K.N.A. 1998. The links between forest history and biodiversity: the invertebrate fauna of ancient pasture-woodlands in Britain and its conservation. In Kirby, K.J & Watkins, C. (eds). *The Ecological History of European Forests*. CAB International, Wallingford, pp. 73–80. Alexander has argued widely for a broken wood-pasture landscape based on beetle evidence. See also: (a) Alexander, K. *et al.* 2006. The value of different tree and shrub species to wildlife. *British Wildlife* 18: 18–28; and (b) Alexander, K.N.A. 2014. Non-intervention v intervention – but balanced? I think not. *British Ecological Society Bulletin* 45: 36–37.

29. Allen, M.J. & Gardiner, J. 2009. If you go down to the woods today; a re-evaluation of the chalkland postglacial woodland: implications for prehistoric communities. In Allen, M.J., Sharples, N. & O'Connor, T (eds)., *Land and People: Papers in Memory of John G. Evans*. Prehistoric Society Research Paper No. 2, Oxbow Books, pp. 49–66.

30. (a) Rose, F. 1974. The epiphytes of oak. In Morris, M.G. & Perring, E.H. (eds). *The British Oak: its History and Natural History*. Botanical Society of the British

Isles, E.W. Classey, Berkshire, pp. 250–273; (b) Rose, F. 1992. Temperate forest management: its effects on bryophyte and lichen floras and habitats'. In Bates, J.W. & Farmar, A.M. (eds). *Bryophytes and Lichens in a Changing Environment.* Clarendon Press, Oxford, pp. 211–233.

31. Hall, S.J.G. 2008. A comparative analysis of the habitat of the extinct aurochs and other prehistoric mammals in Britain. *Ecography* 31: 187–190. doi: 10.1111/j.0906-7590.2008.5193.x.

32. The discussion about Eurasian bison as native animals is ongoing, although, as the Knepp team point out, it seems unlikely that, being able to cross into Britain until 8,200 years ago, bison failed to do so. Doggerland fossils, in the North Sea, reveal bison. A point of view on bison rewilding can be seen at https://knepp.co.uk/bison.

33. Kaagan, L.M. 2000. The horse in Late Pleistocene and Holocene Britain. PhD thesis, University College London.

34. Thieme, H. 1997. Lower Palaeolithic hunting spears from Germany. *Nature* 385: 807–810. doi: 10.1038/385807a0.

35. Kerkdijk, H. 2012. Where did the wild horse go? This is a fascinating article available to view at the website Rewilding Europe (www.rewildingeurope.com). The author explores the ecologically logical idea that horses are an adaptable species, and you cannot simply judge that tarpan were steppe animals from their last known haunt. The idea of 'encrypted' wild animals is in the author's view, more convincing, and born out by zebra phenotypes in Africa.

36. The extraordinary actions of beavers in a landscape are diverse enough to fill a book, and indeed will feature more heavily in Macdonald, B., *Cornerstones,* Bloomsbury, 2021. The Derek Gow Consultancy, Scottish Wild Beaver Group, Cornwall Beaver Project and a number of other advisory groups can be contacted for further advice about, and demonstrations of, the extraordinary biodiversity gains brought about by beavers. The following paper provides the most complete summary of scientific studies into the actions of beavers, although it fails to acknowledge that beaver ponds have two temperature strata: the top sunlit layer is amenable to amphibians whilst the bottom, cool-water layer is ideal for the growth of salmonids, refuting arguments from some, such as the Angling Trust, that beavers are bad news for salmon. Janiszewski, P., Hanzal, V. & Misiukiewicz, W. 2014. The Eurasian beaver (*Castor fiber*) as a keystone species: a literature review. *Baltic Forestry* 20: 277–286.

37. Kitchener, A. 2010. The elk. In O'Connor, T. & Sykes, N. (eds). *Extinctions and Invasions: a Social History of British Fauna.* Windgather Press, Oxford.

38. The Orkneyinga Saga describes both red deer and reindeer, making this distinction, being hunted in Caithness by the Earls ('Jarls') of Orkney. This was brought to my attention by Isabella Tree at Knepp. However, it is clear this was, at this stage, a relict population: few other references are made to reindeer at so late a time in our history. Some have suggested this population was feral.

39. The impact of wolf reintroduction on the wooded landscape is well known and best studied in northern Yellowstone, USA: Beschta, R.L. & Ripple, W.J. 2010. Recovering riparian plant communities with wolves in northern Yellowstone, U.S.A. *Restoration Ecology* 18: 380–389. doi: 10.1111/j.1526-100X.2008.00450.x.

40. The fascinating *British Tree Guide* by Owen Johnson and David Moore (Collins, 2015) draws attention to the site-specificity of a number of 'refugee' tree species of the genus *Sorbus*. Some of these are now restricted to just single geographical localities. Evidently these trees would once have been far more widespread within their region, and their significance for Britain's fauna is now almost impossible to unravel, although many, like the familiar Rowan, are fruit-bearers. In spite of early deforestation, however, it seems these trees may always have been specific to far smaller areas of the UK than, for example, the oaks. This leads me to believe that some of the faunal communities of our steep valleys may, like the flora and fauna of New Guinea, have developed over time in isolation – which, in turn, would suggest that our steep woodland valleys may have been considerably more cut off, in ecological terms, than our lowland habitats.

41. Rackham, O. 2006. *Woodlands*. New Naturalist 100. Collins, London.

42. Birds favouring predominantly open grassland habitats (20 species): skylark, quail, great bustard, Montagu's harrier, stone-curlew (predominantly dry grasslands); hen harrier, short-eared owl, meadow pipit, twite (rough grassland); chough, wheatear (coastal or upland pasture), curlew, lapwing, greylag goose, redshank, black-tailed godwit, ruff, snipe, yellow wagtail, corncrake (damp grassland).

43. Birds favouring scrub-grassland, i.e. requiring both habitats to thrive (16 species): corn bunting, yellowhammer, whinchat, red-backed shrike, cuckoo, magpie (thorn stands in grassland); grey partridge, turtle dove, tree sparrow, cirl bunting, linnet (scrub near disturbed soils); blackbird, song thrush, dunnock (scrub adjoining pasture); long-eared owl, merlin (hawthorn or pine stands in rough grassland).

44. True scrubland birds (20 species), in approximate order of vegetation succession: stonechat, Dartford warbler (heather, gorse); whitethroat, grasshopper warbler, chiffchaff (low scrubland); marsh warbler, sedge warbler, reed bunting (herbaceous scrub near water); nightingale, lesser whitethroat, bullfinch (dense thornlands aged 5+ years); garden warbler, willow warbler, blackcap (maturing shrub stands aged 8+ years); marsh tit (maturing hazel scrub); willow tit (maturing elder, willow or birch scrub aged 15+ years); greenfinch, chaffinch, redpoll, long-tailed tit (all thriving best where young trees meet scrub).

45. Birds of wooded grasslands, with mature trees but plenty of open space (18 species): rook, stock dove (seeds and insects); jackdaw, carrion crow (generalists); white-tailed eagle, red kite, raven (scavengers); hobby, buzzard, sparrowhawk, kestrel, barn owl (predators); black grouse, woodlark, tree pipit,

nightjar, ring ouzel, red grouse (upland edge with some willow, juniper and rowan).

46. Many authors, including Oliver Rackham and Isabella Tree, have argued against the role of fire in our uplands and deciduous wood-pastures. They have perhaps underestimated the important role of fire in the estimated 20% of the UK that would have been scrub-grassland, or indeed, in the marshlands that would have dried out in the hot summers. Recent fires in the Biebrza Marshes of Poland, a truly natural temperate landscape, are a reminder that fire can even ravage reedbeds if they are dry enough. Such processes are usually devastating in the immediate term but beneficial for wildlife in the longer (2–10-year) term.

47. Birds of wood-pasture, spacious mature woodlands characterised by standing dead wood, glades and open pasture (30 species): wryneck, mistle thrush, redstart, starling, green woodpecker, robin (pasture); blue tit, great tit, nuthatch, jay (oak); treecreeper; goldfinch; spotted flycatcher; tawny owl; goldcrest (yew); coal tit (pine); great spotted woodpecker, lesser spotted woodpecker, swift (standing deadwood and old growth); goshawk, honey-buzzard, hawfinch, woodcock (large, broken wooded landscapes); siskin, redwing, common crossbill (preference for pine-led pastures with birch); capercaillie, parrot crossbill, green sandpiper, crested tit (old pine forest with glades).

48. Yalden & Albarella 2009.

49. We often see just a fraction of wetland habitats in one place, but our wetland birds are adapted to exploit a myriad of changing, dynamic habitats. These include ephemeral floodplains with shallow standing water (black tern, shoveler, garganey, Baillon's and spotted crakes, black-necked and great crested grebes), eroding river banks (kingfisher, sand martin); gravel spits or meanders (avocet, little ringed plover, common sandpiper, common tern); mature trees in wetlands (osprey, goosander, cormorant, grey heron); scrublands in marshes (bluethroat, little egret, spoonbill, reed bunting); and deeper wetlands with reedbeds (bittern, water rail, bearded tit, reed warbler, Savi's warbler, pochard, little grebe). In addition, there are landscape-level species reliant on varied wetland ecosystems such as the familiar mute swan, the increasingly familiar crane, and the long-forgotten Dalmatian pelican, a specialist of very extensive deltas rich in fish.

50. (a) Holloway, S. 1996. *The Historical Atlas of Breeding Birds in Britain and Ireland 1875–1900*. Poyser, London (this outstanding book is used extremely often across this chapter); (b) Yalden & Albarella 2009, p. 82.

51. Yalden & Albarella 2009. Page 88 includes a 1957 map (Darby & Versey) of pre-Bronze-Age fenland extent (Fig. 4.4).

52. Adkins, R., Adkins, L. & Leitch, V. 2008. *The Handbook of British Archaeology*, 2nd edition. Constable, London.

53. Pearson 2005, pp. 16–17, cited in Hunter, J. & Ralston, I. 2009. *The Archaeology of Britain: an Introduction from Earliest Times to the Twenty-first Century*, 2nd edition. Routledge, London.

54. Claris, P., Quartermaine, J. & Woolley, A.R. 1989. The Neolithic quarries and axe factory sites of Great Langdale and Scafell Pike: a new field survey. *Proceedings of the Prehistoric Society* 55: 1–25. doi: 10.1017/S0079497X00005326.

55. Rackham, O. 2000. *The History of the Countryside*, new edition. Phoenix Press, London.

56. Kennedy, M. 2014. 4,000-year-old Dartmoor burial find rewrites British bronze age history. *The Guardian*, 9 March 2014. Details the findings of chief archaeologist at the Dartmoor National Park Authority, Jane Marchand.

57. Pyne, S.J. 1997. *Vestal Fire: an Environmental History, Told through Fire, of Europe and Europe's Encounter with the World*. University of Washington Press, Seattle, pp. 348–369.

58. Hetherington, D. 2006. The lynx in Britain's past, present and future. *ECOS* 27(1): 67–74.

59. Northcote (1979), cited in Stewart, J.R. 2004. Wetland birds in recent fossil record of Britain and northwest Europe. *British Birds* 97: 33–43. This is an excellent paper, now available for free online, which summarises the fossil record of vanished birds in our wetlands. This doesn't reveal as many lost species as thought. Evidence of night-heron is scarce; a possible extinct night-heron fossil record is still being looked into; one record of pygmy cormorant is hard to analyse. Dalmatian pelican, common crane and white-tailed eagle are, however, all robustly recorded as the native denizens of Fenland.

60. The details of this and the previous two paragraphs, describing the early human colonisation of the Fens, draw on historical information carefully researched, and made available by, the Great Fen Project, viewable online at www.greatfen. org.uk.

61. Stanbury, A. 2011. The changing status of the Common Crane in the UK. *British Birds* 104: 432–447.

62. Ekwall (1960), cited in Gow, D., Campbell-Palmer, R., Edgcumbe, C. *et al.* Feasibility report for the reintroduction of the white stork (*Ciconia ciconia*) to England. Available via https://knepp.co.uk/white-storks.

63. The role of religion and politics in the eradication of Britain's birdlife is neither well studied nor ever likely to be extrapolated in full by ecological historians. However, the ecologist Derek Gow argues that storks may, after the Restoration, have been regarded as symbols of Republicanism and wiped out for this reason. A similar fate, for different ideological beliefs, befell the Cornish Chough. In addition, wetland loss after the disappearance of beavers, and storks being an easy supply of food, would have accelerated their demise. An excellent compilation of firm and valid documentation of the white stork in Britain is well expressed in: Fair, J. Pole position. *BBC Wildlife*, June 2016.

64. Yalden, D.W. 2007. The older history of the White-tailed Eagle in Britain. *British Birds* 100: 471–480.

65. Evans, R.J., O'Toole, L. & Whitfield, D.P. 2012. The history of eagles in Britain and Ireland: an ecological review of place-name and documentary evidence from the last 1500 years. *Bird Study* 59: 335–349. doi: 10.1080/00063657.2012.683388.

66. The historical range of the golden eagle will be examined and revised, in light of a number of factors, in Macdonald, B. *Cornerstones*. Bloomsbury, 2021.

67. Lovegrove, R. 2008. *Silent Fields: the Long Decline of a Nation's Wildlife*. Oxford University Press, Oxford. This superb book documents, in particular, the killing of our national heritage across the Tudor period, and I am indebted to the research of its diligent author for many facts in the section *Hunted out*. Lovegrove explores in intricate detail the scale of killings, and reasons behind the fact we have almost none of our national heritage left today, and indeed had lost much of it before Victorian times.

68. Fryer, G. 1987. Evidence for former breeding of the Golden Eagle in Yorkshire. *Naturalist* 112: 3–7.

69. Lovegrove 2008.

70. Paoletti, M.G. 1999. *Invertebrate Biodiversity as Bioindicators of Sustainable Landscapes*, Elsevier, Amsterdam.

2. The Anthropocene

1. Schwägerl, C. 2014. *The Anthropocene: the Human Era and How it Shapes our Planet*. Synergetic Press, London.

2. The poems of John Clare are all available online, describing with great astuteness the nests of birds, and in harrowing detail the destruction of the countryside, as the 'cleansing' model of modern farming gets under way. The definitive biography of someone now regarded as the finest never-recognised poet is Bate, J. 2014. *John Clare*, Picador, London.

3. The concept of an 'Anthropocene' epoch is increasingly but not universally accepted, and those who do subscribe to it differ in their opinions as to when exactly it began. Some, such as George Monbiot, see it as dating back to the time of megafaunal extinctions. Others map it to as late as 1950, with the formation of plastic rocks in the sea. The most widely agreed consensus appears to be that man-made climate change, dating from the early industrial period, initiated the time when our control of the planet's systems was truly consolidated. In ornithological terms, the narrative of bird decline we face today, one of landscape simplification and 'cleansing', can, in my view, be most clearly mapped to 1760.

4. Golawski, A. 2006. Changes in number of some bird species in the agricultural landscape of eastern Poland. *The Ring* 28: 127–133.

5. Holloway, S. 1996. *The Historical Atlas of Breeding Birds in Britain and Ireland 1875–1900*. Poyser, London, p. 66. This superb and definitive account of

Victorian bird decline is used extensively throughout this chapter. I am very grateful to have had such an amazing text to rely on, not only because it shows that the industrial narrative of bird decline is consistent – and not new at all. Approximately 70 historical references to bird population change, unless specified otherwise, draw from this text.

6. Holloway 1996, pp. 77–78.
7. Paintings such as 'The Hay Wain' by Constable may appear, at first glance, to show a current English landscape, but on closer examination what is depicted actually resembles the earthy wooded landscape of eastern Polish farmland, with its huge outgrown trees and small grazing herds. This is actually one of the best depictions of lost English wryneck habitat that we have.
8. Shrubb, M. 2011. Some thoughts on the historical status of the Great Bustard in Britain. *British Birds* 104: 180–191.
9. Galasso, S. 2014. When the last great auk died, it was by the crush of a fisherman's boot. *Smithsonian Magazine*, 10 July 2014. This excellent article summarises the extinction of the great auk, both in Britain and globally, though various other accounts are available.
10. Bengston, S. 1984. Breeding ecology and extinction of the Great Auk (*Pinguinus impennis*): anecdotal evidence and conjectures. *The Auk* 101: 1–12.
11. Revive & Restore. https://reviverestore.org (accessed October 2018).
12. Lovegrove, R. 2008. *Silent Fields: the Long Decline of a Nation's Wildlife*. Oxford University Press, Oxford.
13. Lovegrove 2008.
14. Cenian, Z., Lontkowski, J. & Mizera, T. 2006. Wzrost liczebności i ekspansja terytorialna bielika *Haliaeetus albicilla* jako przykład skutecznej ochrony gatunku. *Studia i Materiały Centrum Edukacji Przyrodniczo-Leśnej* 2 (12): 55–63.
15. Yalden, D.W. & Albarella, U. 2009. *The History of British Birds*. Oxford University Press, Oxford.
16. Perry (1978), cited in Evans D. 2002. *A History of Nature Conservation in Britain*, 2nd edition. Routledge, London. Whilst it can be easy to blame gamekeepers for all ills, Glengarry's record really does epitomise the degree of national robbery that has taken place over more than two centuries in Britain, with gamekeepers certainly playing a key role in the removal of our national heritage, a heritage they now claim to protect when almost none of it is left.
17. Ashmole, P. 2006. The lost mountain woodland of Scotland and its restoration. *Scottish Forestry* 60 (1): 9–22.
18. Hart-Davis, D. 1978. *Monarchs of the Glen: a History of Deer-Stalking in the Scottish Highlands*. Jonathan Cape, London, p.139.
19. Hart-Davis 1978.
20. Avery, M. 2015. *Inglorious: Conflict in the Uplands*. Bloomsbury, London.
21. Holloway 1996.

22. Barkham, P. 2015. 97% of hay meadows have gone: here's why it matters. *The Guardian*, 18 May 2015.

23. Department for Environment, Food and Rural Affairs (Defra). 2013. Farming statistics: livestock populations at 1 December 2012, UK and England. www.gov.uk/government/statistics/farming-statistics-livestock-populations-at-1-december-2012-uk-and-england (accessed October 2018).

24. Collins, E.J.T. & Thirsk, J. 2000. *The Agrarian History of England and Wales*. Cambridge University Press, Cambridge.

25. Wrigley, E.A. & Schofield, R.S. 1981. *The Population History of England, 1541–1871: a Reconstruction*. Edward Arnold, London, Table 7.8, pp. 208–209. The figures are corroborated by official census records, which can be viewed at www.visionofbritain.org.uk.

26. Snow, D. 2014. Viewpoint: 10 big myths about World War One debunked. *BBC News*, 25 February 2014. www.bbc.co.uk/news/magazine-25776836 (accessed October 2018). I have used the article for one statistic of mortalities in the war only.

27. Pym, H. [2014?] Did World War One nearly bankrupt Britain? BBC *I-Wonder*. www.bbc.co.uk/guides/zqhxvcw (accessed October 2018). I have used this article for the state of Britain's post-war finances.

28. The Swift Conservation website (www.swift-conservation.org) offers some examples of natural swift nest sites remaining in Europe, and I have witnessed this in Abernethy Forest, Highland. As far as can be discovered, this is the last haunt of tree-nesting swifts in Britain – but it would be lovely to be proven wrong or discover that others still existed.

29. Holloway 1996, pp. 541–542.

30. UK Parliament. Living heritage: improving towns. www.parliament.uk/about/living-heritage/transformingsociety/towncountry/towns (accessed October 2018). This website details the housing schemes after the First World War, but these statistics are also widely available elsewhere.

31. The Swift Conservation website (www.swift-conservation.org) – last accessed October 2018 – outlines the changing suitability of our houses for swifts in the twentieth century.

32. Forestry Commission. 2017. History of the Forestry Commission. www.forestry.gov.uk/forestry/cmon-4uum6r (accessed October 2018). The text outlines the changes made to British woodlands, especially in relation to planting after the First World War.

33. Atkinson, S. & Townsend, M. 2011. *The State of the UK's Forests, Woods and Trees: Perspectives from the Sector*. Woodland Trust, Grantham. www.woodlandtrust.org.uk/mediafile/100229275/stake-of-uk-forest-report.pdf?cb=58d97f320c (accessed October 2018).

34. For elucidation of these statistics please see the notes to Chapter 9.

35. This and the quotation from the *Journal of Forestry and Estates Management* are both cited in Norton, P. 2013. Old Sloden Wood in the New Forest: a survey of the yews. Available via www.ancient-yew.org (accessed October 2018).
36. Henkel, M. 2015. *21st Century Homestead: Sustainable Agriculture III*. Lulu.
37. RSPB. A history of hedgerows. www.rspb.org.uk/our-work/conservation/ conservation-and-sustainability/advice/conservation-land-management -advice/farm-hedges/history-of-hedgerows (accessed October 2018).
38. As oak is the cornerstone of our ecosystem, and an ecosystem in itself, its removal can single-handedly wipe entire species off the map. A compelling description of post-war oak removal can be found in Tree, I. 2018. *Wilding: the Return of Nature to a British Farm*. Picador, London.
39. Kirby, A. 2004. Scarce insects duck UK splat test. *BBC News Online*, 1 September 2004. news.bbc.co.uk/1/hi/sci/tech/3618332.stm (accessed October 2018). Since 2004, insect populations have continued to decline, according to all available measures.
40. Krebs J.R., Wilson, J.D., Bradbury, R.B. & Siriwardena, G.M. 1999. The second Silent Spring? *Nature* 400: 611–612.
41. Defra. 2016. British food at a glance (published 21 March 2016, cited in Proagrica Online. 2017. Self-sufficiency in UK agriculture, published 22 February 2017). It states facts from the Office of National Statistics. Produced by the Food and Trade Statistics team, Defra. Further enquiries to David Lee: 020 8026 3006.
42. Balmer, D., Gillings, S., Caffrey, B. *et al.* 2013. *Bird Atlas 2007–11: the Breeding and Wintering Birds of Britain and Ireland*. BTO, Thetford.
43. Hole, D.G., Whittingham, M.J., Bradbury, R.B. *et al.* 2002. Widespread local house-sparrow extinctions. *Nature* 418: 931–932.
44. Walker, L.K. & Morris, A.J. 2016. *Evaluating Turtle-dove HLS Package*. Report to Natural England, Work Package Number ECM6924. Cited in RSPB. 2018. *International Single Species Action Plan for the Conservation of the European Turtle-dove* Streptopelia turtur *(2018 to 2028)*. www.trackingactionplans.org/ SAPTT/downloadDocuments/openDocument?idDocument=49 (accessed October 2018). The estimate of 4,300 pairs of turtle dove in the UK, even in 2016, seems extremely high, given that 'good' counties such as Sussex and Kent's annual reports suggest that each has just a few hundred pairs.
45. From here on, we are into the statistics of modern bird decline, universally available and agreed. The main sources for these include the BTO's three *Bird Atlases*, the BTO's 'Birds of the Wider Countryside' reports, and the RSPB-led *State of UK Birds* and *State of Nature* reports. To be clear on the facts, I have specified the interval of time over which a decline has taken place. No further citations refer to these facts as they are not generally held to be in question.
46. McGowan & Kirwan (2013), cited in BirdLife International. 2016. *Perdix perdix*. The IUCN Red List of Threatened Species 2016: e.T22678911A85929015. http://dx.doi.org/10.2305/IUCN.UK.2016-3.RLTS.T22678911A85929015.en.

47. AHDB Dairy. 2016. *Dairy Statistics: an Insider's Guide 2016*. https://dairy.ahdb.org.uk/resources-library/market-information/dairy-statistics (accessed October 2018). AHDB is a non-profit organisation advising dairy farms. The source statistics that it uses come from Defra, DARD, SEERAD and the Welsh government.

48. This statistic is available from the Polish Federation of Cattle Breeders and Dairy Farmers, drawing on the country's Central Statistic Office 2011 figures. The information is available in a presentation given to Enterprise Ireland. www.enterprise-ireland.com/en/Events/OurEvents/Poland's-Agricultural -Transformation/Technical-Equipment-on-Polish-Farms.pdf (accessed October 2018).

49. The decline of the wood warbler is a complex matter of climate change, caterpillar starvation and woodland fragmentation, leading in turn to nest predation having a greater impact than in large forest areas, such as those in Poland. The decline will be fully explored in a chapter in the author's next book, *Too Good to Lose*.

50. Lenton, T.M., Held, H., Kriegler, E. *et al.* 2008. Tipping elements in the earth's climate system. *Proceedings of the National Academy of Sciences* 105 (6): 1786–1793. doi: 10.1073/pnas.0705414105.

51. Somerset Ornithological Society. *The Bittern*, issue 20 (August 2017). https://somersetbirding.org.uk/uploads/Bittern/bittern20.pdf (accessed October 2018).

52. Birdlife International / IUCN. 2015. *European Turtle Dove: Supplementary Material*. Available within BirdLife International *European Red List of Birds*. Office for Official Publications of the European Communities, Luxembourg. Looking at country-by-country farmland bird declines, relative to the intensity of their agricultural practices, is key to seeing the bigger picture. In 2010–2013, there were still between 120,000 and 300,000 pairs of turtle doves in Romania, comparable to our population 40 years ago. In Hungary, between 2000 and 2012, there were 64,000–150,000 pairs. These hay-meadow-rich countries are not showing obvious declines. Germany has seen reductions of a magnitude of 38–58%. The continent-wide map is more revealing. Most eastern European countries are seeing stability or small reductions in turtle doves, with western European countries seeing large declines.

53. Estimates for both of these vanished species taken from the *Red List of Breeding Birds in Germany* (Südbeck, P. *et al.* 2007. Rote Liste der Brutvögel Deutschlands, 4. Fassung. *Berichte zum Vogelschutz* 44: 23–81).

54. Stevens, D.K. 2008. The breeding of the Spotted Flycatcher *Muscicapa striata* in lowland England. PhD thesis, University of Reading.

55. Kentie, R., Senner, N.R., Hooijmeijer, J.C.E.W. *et al.* 2016. Estimating the size of the Dutch breeding population of continental black-tailed godwits from 2007–2015 using resighting data from spring staging sites. *Ardea* 114: 213–225. doi: 10.5253/arde.v104i3.a7.

3. The First Imperative

1. Salmon M.A. 1997. *The Aurelian Legacy: British Butterflies and Their Collectors*. Harley, Colchester, p. 193.

2. Rothamsted Research. 2013. Declines in UK moth species could signal a 'potentially catastrophic loss of biodiversity' in the British countryside. *Rothamsted Research*, 1 February 2013. For more information, Rothamsted's press office can be contacted on 01582 938 855 (www.rothamsted.ac.uk).

3. McKenzie, S. 2015. 'Alarming trend' of decline amongst UK's dung beetles. *BBC News*, 17 November 2015. www.bbc.co.uk/news/uk-scotland-highlands-islands-34831400 (accessed October 2018). The British Beetles website (www.coleoptera.co.uk) offers the latest statistics and surveys on the UK's declining beetles, including dung beetles.

4. Brooks, D.R., Bater, J.E., Clark, S.J. *et al.* 2012. Large carabid beetle declines in a United Kingdom monitoring network increases evidence for a widespread loss in insect biodiversity. *Journal of Applied Ecology* 49: 1009–1019. doi: 10.1111/j.1365-2664.2012.02194.x.

5. According to Declan Butler of *Nature*, the international journal of science, in total around 1,500 scientific papers now correlate pesticides, specifically neonicotinoids, with aspects of bee decline. Two recent UK ones include: (a) Baines, D. *et al.* 2017. Neonicotinoids act like endocrine disrupting chemicals in newly-emerged bees and winter bees. *Scientific Reports* 7: 10979. doi: 10.1038/s41598-017-10489-6; (b) Woodcock, B.A. *et al.* 2016. Impacts of neonicotinoid use on long-term population changes in wild bees in England. *Nature Communications* 7: 12459. doi: 10.1038/ncomms12459.

6. Hallmann C.A., Sorg, M., Jongejans, E. *et al.* 2017. More than 75 percent decline over 27 years in total flying insect biomass in protected areas. *PLoS ONE* 12(10): e0185809. doi: 10.1371/journal.pone.0185809.

7. The comparison is based first on existing populations of key large insectivores between the UK and Germany (wryneck, red-backed shrike, spotted flycatcher densities), then, on the fact that cuckoos in Germany have, in the last 10 years, fallen by 20–30%. This fact comes from NABU, a German nature preservation organisation, cited in *Spiegel Online* on 13 March 2008. In the UK, by contrast, cuckoo numbers fell by 21% between 2008 and 2009 alone.

8. Wood, T.J. & Goulson, D. 2017. The environmental risks of neonicotinoid pesticides: a review of the evidence post-2013. *Environmental Science and Pollution Research* 24: 17285–17325. doi: 10.1007/s11356-017-9240-x.

9. The calculation that 1,500 scientific papers now correlate neonicotinoids with aspects of bee decline has been made by Declan Butler of *Nature*, the international journal of science. Hundreds of peer-reviewed papers, each reaching similar conclusions, can be found online, with two referenced in Chapter 2 as well as the wider studies above. Note that although few birds test positive for neonicotinoids in the UK, this is because this country does not usually test

for the presence of this chemical, whereas other countries, including France and the Netherlands, do.

10. Hallmann, C.A., Foppen, R.P.B., van Turnhout, C.A.M., de Kroon, H. & Jongejans, E. 2014. Declines in insectivorous birds are associated with high neonicotinoid concentrations. *Nature* 511: 341–343. doi: 10.1038/nature13531.

11. Hokkanen, H., Menzler-Hokkanen, I. & Keva, M. 2017. Long-term yield trends of insect-pollinated crops vary regionally and are linked to neonicotinoid use, landscape complexity, and availability of pollinators. *Arthropod–Plant Interactions* 11: 449–461.

12. Breeze, T.D., Bailey, A.P., Balcombe, K.G. & Potts, S.G. 2011. Pollination services in the UK: how important are honeybees? *Agriculture, Ecosystems and Environment* 142: 137–143. doi: 10.1016/j.agee.2011.03.020.

13. Swiss wryneck expert Anna Freitag (Museum of Technology, Swiss Federal Institute of Technology) to author, personal communications, 2016. An excellent paper on wrynecks, which elucidates why Britain can no longer support populations of them, is: Freitag, A., Martinoli, A. & Urzelai, J. 2001. Monitoring the feeding activity of nesting birds with an autonomous system: case study of the endangered Wryneck *Jynx torquilla*. *Bird Study* 48: 102–109. doi: 10.1080/00063650109461207.

14. Tryjanowski, P., Karg, M.K. & Karg, J. 2003. Food of the Red-backed Shrike *Lanius collurio*: a comparison of three methods of diet analysis. *Acta Ornithologica* 38: 59–64. A number of Polish studies have analysed shrike diet in the realisation that it's critical to their future. In all studies, beetles are followed by grasshoppers, flies, bees, wasps and other bugs, with vertebrates caught in poorer weather.

15. The information on red-backed shrike food requirements is based on a variety of sources, from peer-reviewed papers to my own observations of nesting birds in eastern Europe. My foraging calculations are based around a mean of 14 daylight hours. All are designed to err on the side of caution. These calculations do not purport to be peer-reviewed science; they are designed to produce a rough indication of the number of large invertebrates required for a one pair of birds to successfully raise a family. (a) **PRE-INCUBATION**. A pair of shrikes occupies territory prior to nesting for roughly 21 days. Birds will be feeding up after a long journey from Africa, the female building calcium reserves to lay up to six eggs. A successful catch of one invertebrate, every 8 minutes, by two birds, provides a catch of 210 invertebrates per day per pair, which means **4,410** invertebrates over 21 days prior to egg-laying. (b) **INCUBATION**. During incubation, a male will be feeding a female at the nest, as well as himself, for 15 days. Allowing for 14 foraging hours in a day and one catch fed to the female every half hour equates to 28 catches per day. Nest-camera footage shows that a 'catch' consists of multiple large invertebrates. Allowing for four invertebrates per catch, that is 112 insects per day. Over 15 days, therefore, 1,680 insects are provided for the female, and over two broods that is 3,360

insects. In addition, the male feeds himself. If we allow him half the female's intake, because his energy demands are lower than the hers during incubation, that is an additional 1,680. In total, therefore, our 'incubation period' invertebrate intake (for 30 days) comes to **5,040**. (c) **CHICKS IN NEST**. Shrike chicks remain in the nest for around 15 days. Nest-cam footage suggests that, between them, the two parents deliver food six times per hour. This results in 84 visits per day, and we allow that each visit averages a delivery of four insects, hence 336 insects per day. Across two broods in the nest, totalling 30 days, a minimum of **10,080** invertebrates may be brought to the chicks, excluding adult demand. (d) **POST-FLEDGING**. Two broods of shrikes will typically fledge, each spending around 21 days on territory before migrating south. Allowing for some mortality from an average clutch size of five, we allow for 2 sets of 3 chicks out of the nest. Across the post-fledging period, we allow these 6 chicks to consume an insect every 15 minutes for 21 days each. One 14-hour day provides 56 feeds, and 21 days provide 1,176 feeds. For six fledglings spread across the season, this comes to **7,056** invertebrates. It should be noted this last calculation is probably a large underestimate. We also disregard adult feeding requirements after the fledging period. (e) **TOTAL**. The sum total of these probable underestimates is that for just one pair of shrikes to survive, at least **26,586** invertebrates will be required. But for a landscape-level population of 50 shrike pairs, at least 1.33 million large invertebrates are required.

16. E.H. Forbush, cited by Bent (1940), in Fitzgerald, T.D. 1995. *The Tent Caterpillars*. Cornell University Press, Ithaca, NY, p.180.

17. Fox, R., Parsons, M.S., Chapman, J.W. *et al.* 2013. *The State of Britain's Larger Moths, 2013*. Butterfly Conservation and Rothamsted Research, Wareham. https://butterfly-conservation.org/files/1.state-of-britains-larger-moths-2013 -report.pdf (accessed October 2018).

18. The larval food-plant diet of 'hairy caterpillar' moth species is available online via www.butterfly-conservation.org.

19. Stevens, D.K. 2008. The breeding of the Spotted Flycatcher *Muscicapa striata* in lowland England. PhD thesis, University of Reading.

20. These calculations are more detailed, and likely to more accurate, than the underestimates provided for the shrikes, because more aspects of spotted flycatcher breeding biology are available. Indeed, the more information we have about dietary demands, the more 'demanding' our insectivorous birds can appear. (a) **ADULT INTAKE**. The demands of two adult flycatchers are considered over an average of 120 days spent in Britain each summer. We factor in adults feeding themselves to regain body mass, and the female feeding up for two periods of laying an average of five eggs. There was no peer-reviewed paper on spotted flycatcher capture rate found, but capture rates in the comparably sized verditer flycatcher (a Himalayan species) can be up to 100 insects per hour. If we allow each bird, conservatively, one spell of 100 insects caught, plus 25 per hour over 10 more hours' average daylight from April to August, that is

350 insects per bird per day. Excluding the 30 days of chick-feeding, to avoid duplication risk, the adults are feeding themselves for 90 days. An average of 350 insects per bird per day results in a total of **63,000** insects caught by both birds. (b) **CHICKS IN NEST**. Two broods of flycatchers are raised, with each in the nest for about 15 days. Nest-cams at Pensthorpe, Norfolk, suggest an average of 14 feeds per hour. Over an average of 14 daylight hours each day in June and July, this means 196 visits per day. Five nest-cams observed by the author, as well as nests watched in Herefordshire (2012–2018), suggest that at least 3 insects are brought in most deliveries, which equates to 588 insects per day, and **17,640** insects fed to two broods over two sets of 15 days. (c) **POST-FLEDGING**. Chicks out of the nest increase their capture rate from 30% to 80% in the first week (BTO). On average, four chicks fledge from each of two broods, remaining for at least three weeks. If one chick catches 80 insects per day in the first week, then 220 in each of the next two weeks, that makes 3,640 per chick across the three weeks. This results in **29,120** insects caught by eight chicks. (d) **TOTAL**. In total, therefore, the minimum intake for one flycatcher family may sit, very roughly of course, around **109,760** insects. And a self-sustaining population of 50 pairs would require close to 5.5 million flying insects to survive.

21. Walker, L.A., Chaplow, J.S., Moeckel, C. *et al.* 2013. Anticoagulant rodenticides in predatory birds 2011: a Predatory Bird Monitoring Scheme (PBMS) report. Centre for Ecology & Hydrology, Lancaster. 'The most prevalent rodenticides detected in kestrel livers were difenacoum and bromadiolone. The co-occurrence of multiple residues was also prevalent with 19 out of 20 kestrels having more than one SGAR present in their liver.'

22. Newton, I., Wyllie, I. & Freestone, P. 1990. Rodenticides in British Barn Owls. *Environmental Pollution* 68: 101–117. doi: 10.1016/0269-7491(90)90015-5.

23. Dowding, C.V., Shore, R.F., Worgan, A., Baker, P.J. & Harris, S. 2010. Accumulation of anticoagulant rodenticides in a non-target insectivore, the European hedgehog (*Erinaceus europaeus*). *Environmental Pollution* 158: 161–166. doi: 10.1016/j.envpol.2009.07.017.

24. Hallmann, C.A. Foppen, R.P.B., van Turnhout, C.A.M., de Kroon, H. & Jongejans, E. 2014. Declines in insectivorous birds are associated with high neonicotinoid concentrations. *Nature* 511: 341–343. doi: 10.1038/nature13531.

25. Bean, T.G., Boxall, A.B.A., Lane, J. *et al.* 2014. Behavioural and physiological responses of birds to environmentally relevant concentrations of an antidepressant. *Philosophical Transactions of the Royal Society B* 369: 20130575. doi: 10.1098/rstb.2013.0575.

26. Schroeder, J. 1976. Individual fitness correlates in the Black-tailed Godwit. Thesis, Animal Ecology Group at the University of Groningen.

27. Oates, M. 2015. *In Pursuit of Butterflies: a Fifty-year Affair*. Bloomsbury, London.

28. Charman, E.C., Smith, K.W., Dillon, I.A. *et al.* 2012. Drivers of low breeding success in the Lesser Spotted Woodpecker *Dendrocopos minor* in England: testing hypotheses for the decline. *Bird Study* 59: 255–265.

29. Kennedy, C.E.J. & Southwood, T.R.E. 1984. The number of species of insects associated with British trees: a re-analysis. *Journal of Animal Ecology* 53: 455–478. doi: 10.2307/4528. The figures for insect abundance on native trees are all taken from this source.

30. Tree, I. 2018. *Wilding: the Return of Nature to a British Farm.* Picador, London.

31. Rackham, O. 2006. *Woodlands.* New Naturalist 100. Collins, London.

32. Personal observations across the UK, 2012–2018.

33. Mason P. & Allsop J. 2010. *The Golden Oriole.* Poyser Monographs, Bloomsbury, London, p. 147.

34. Osborne, L. & Krebs, J. 1981. Replanting after Dutch elm disease. *New Scientist* 90: 212–215.

35. Author's field observations, 2009–2018, based on observing willow tits in a variety of plantation, deciduous woodland and scrubland/river valley habitats (including 14 active nests) in the Forest of Dean and Windrush valley, Gloucestershire; Shropshire Hills in Herefordshire/Shropshire; Thetford and Swaffham Forests in Norfolk; Dearne valley in Yorkshire; Hampshire/Berkshire downs. The only shared factor is rotting elder. Most sites, but not all, contain blackthorn, hawthorn and birch, whilst prime woodlands hold willow types.

36. Monbiot, G. *Feral: Searching for Enchantment on the Frontiers of Rewilding.* Penguin, London, 2013, p. 242.

37. Holloway, S. 1996. *The Historical Atlas of Breeding Birds in Britain and Ireland 1875–1900.* Poyser, London.

38. Perrins, C. 1971. Age of first breeding and adult survival rate in the Swift. *Bird Study* 18: 61–70. doi: 10.1080/00063657109476297.

39. Newman, J.R., Novakova, E. & McClave, J.T. 1985. The influence of industrial air emissions on the nesting ecology of the house martin *Delichon urbica* in Czechoslovakia. *Biological Conservation* 31: 229–248.

40. Zatonski, J. 2016. Population of common swift in Poznan (Poland) and ecosystem services provided by it. *Ekonomia i Środowiski* 4: 263–273.

41. Zatonski 2016.

4. The Lost Stewards

1. Muir, J. 1911. *My First Summer in the Sierra.* Houghton Mifflin, Boston; 1988 reprint, p. 110.

2. Thomas, J.A. 1995. The ecology and conservation of *Maculinea arion* and other European species of large blue butterfly. In Pullin A.S. (ed.). *Ecology and Conservation of Butterflies.* Springer, Dordrecht, pp. 180–197. The secret life of the Alcon blue is one of the true meadow fairy tales, and one I had

the good fortune to help film in 2017. One version can be watched in David Attenborough's BBC *Life in the Undergrowth* series. A series of papers are written on the life cycle by Jeremy Thomas at Oxford University.

3. Kunz, W. 2016. *Species Conservation in Managed Habitats: the Myth of a Pristine Nature.* Wiley-VCH, Weinheim. All further references to Kunz are from this book.

4. Stoate, C., Leake, A.R., Jarvis, P.E., Szczur, J. & Moreby, S.J. 2015. The Allerton project: twenty-three years of agricultural and environmental data collection on a commercial farm. *Aspects of Applied Biology* 128: 27–33, ref.9.

5. Tokarska, M., Pertoldi, C., Kowalczyk, R. & Perzanowski, K. 2011. Genetic status of the European bison *Bison bonasus* after extinction in the wild and subsequent recovery. *Mammal Review* 41: 151–162. doi: 10.1111/j.1365-2907.2010.00178.x.

6. Burton, M. & Burton, R. 2002. *International Wildlife Encyclopedia*, 3rd edition. Marshall Cavendish, New York, pp. 936–938.

7. Begle, G.G. 1900. Caesar's account of the animals in the Hercynian forest (De Bello Gallico, VI, 25–28). *The School Review* 8 (8): 457–465. *JSTOR* www.jstor.org/stable/1074182 (accessed October 2018).

8. Tree, I. 2018. *Wilding: the Return of Nature to a British Farm.* Picador, London.

9. Aykroyd, T.N.B. 2005. Holland goes wild: a message for developed landscapes. *Wild Europe*, 14 September 2005. www.wildeurope.org/index.php/restoration/national-strategies/rewilding-holland (accessed October 2018). An excellent and thoughtful summary of the Oostvaardersplassen rewilding.

10. Vera, F.W.M. 2000. *Grazing Ecology and Forest History.* CABI Publishing, Wallingford.

11. Vera, F.W.M. 2009. Large-scale nature development: the Oostvaardersplassen. *British Wildlife*, 20 (5): 28–36. This detailed article by Frans Vera, citing over 30 peer-reviewed sources, explains the concept of the Oostvaardersplassen and its challenge to 'canopy forest' ecological thinking, and explains the return of birds through the prism of rewilding.

12. The Oostvaardersplassen has 57 resident bird species, and an additional 43 regular summer visitors. This brings its breeding bird total to 100 species. This information is available from tour operator FlevoBirdwatching, not associated with the reserve (www.birdsnetherlands.nl). The only comparable UK reserve, and it's the exception to the rule, is Minsmere, with a blend of carefully managed habitats. I was unable to find a definitive list, but it is generally acknowledged to hold 90–100 breeding bird species. However, it might be argued that the diversity and robust populations of the Oostvaardersplassen, with white-tailed eagle, bluethroat, honey-buzzard, willow tit, lesser spotted woodpecker and many more, proves the rewards of a more integrated environment. No other UK reserve has this number of breeding birds.

13. In spring and early summer, 150 species are reliably seen on birding tours within the reserve, including resident woodland birds that are rare in the

UK and not present in most woods. This information comes from Birding Holland, a tour operator not associated with the reserve. See more at www. birdingholland.com.

14. Lorimer, J. & Driessen, C. 2014. Experiments with the wild at the Oostvaardersplassen. *ECOS* 35 (3/4). Available via www.geog.ox.ac.uk. This paper explores the dynamic shifting of the bird populations and, with due fairness, requests more peer-reviewed science to promote the rewilding scheme in the wider scientific community.

15. Vera 2000.

16. Bijlsma, R.G. 2008. Broedvogels van de buitenkaadse Oostvaardersplassen in 1997, 2002 en 2007. A & W-rapport 1051. Altenburg & Wymenga, Veenwouden.

17. Tree 2018. Frans Vera is quoted in the chapter on 'The secret of grazing animals'.

18. Burrell, C. 2016. Guest blog – what we do at Knepp. https://markavery. info/2016/11/25/guest-blog-knepp-charlie-burrell (accessed October 2018).

19. Tree 2018. The data from Knepp in this chapter come from this source, and also from the estate's surveys, which are published online at https://knepp.co.uk/ yearly-surveys. In addition to the testament of the estate's owners, the Knepp wildlife recovery has been independently documented by a range of organisations including Butterfly Conservation, the RSPB and a range of independent experts. Most increases are seen in the block where the range of herbivores is present; it must be remembered that this area is by no means all of the estate at this stage, and so bird numbers should be interpreted with that in mind.

20. Some information is available at https://knepp.co.uk on the roles reprised by herbivores at Knepp and their effect on the landscape, whilst a more detailed appraisal of the species' returns and recoveries can be found in Tree 2018.

21. Tree 2018.

5. A Question of Scale

1. Quammen D. 1997. *The Song of the Dodo*. Scribner, New York, p. 11.

2. The fundamental theory that animal populations exist best in connected landscapes, subject to fluctuation and dynamic shifts over time (and that isolation, conversely, disrupts this and can bring about local extinction), is not 'new' or challenged science. The general consensus is that some bird populations consist of 'sink' and 'source'. In source, birth rates exceed death rates, leading to emigration. Sink populations have higher mortality than birth rates, and cannot exist alone. Meta-populations, however, comprise separate populations, which can complement one another, allowing interchange of individuals between them. What this chapter aims to convey is the practical outcomes of population dynamics and how they work. It does not aim to fully expound each facet of these theories. For more, see: (a) Andrén, H. 1994. Effects of habitat fragmentation on birds and mammals in landscapes with different proportions of suitable

habitat: a review. *Oikos* 71: 355–366. www.jstor.org/stable/3545823; (b) Opdam, P. 1991. Metapopulation theory and habitat fragmentation: a review of Holarctic breeding bird studies. *Landscape Ecology* 5: 93–106. doi: 10.1007/BF00124663; (c) Esler, D. 2000. Applying metapopulation theory to conservation of migratory birds. *Conservation Biology* 14: 366–372. doi: 10.1046/j.1523-1739.2000.98147.x; (d) Simberloff, D. 1995. Habitat fragmentation and population extinction of birds. *Ibis* 137: S105–S111. doi: 10.1111/j.1474-919X.1995.tb08430.x.

3. Professor Ian Newton, personal communication, April 2018.

4. Taylor, J. 2015. Determinants of variation in productivity, adult survival and recruitment in a declining migrant bird: the Whinchat (*Saxicola rubetra*). PhD thesis, Lancaster University.

5. RSPB. 2011. New Forest is hotspot for vanishing bird. ww2.rspb.org.uk/our-work/rspb-news/news/289380-new-forest-is-hotspot-for-vanishing-woodland-bird (accessed October 2018).

6. Loch Lomond and Trossachs National Park. 2013. Isolated capercaillie population in southern Scotland may no longer be viable. Formerly available on website of the national park at the following URL: www.lochlomond-trossachs.org/looking-after/isolated-capercaillie-population-in-southern-scotland-may-no-longer-be-viable/menu-id-522.html.

7. Durham Bird Club. 2011. *Lek* magazine, winter 2011, p. 48. No longer available online but available in print from the Durham Bird Club: www.durhambirdclub.org.

8. Cotswold Water Park Trust. 2012. Newsletter, summer 2012. PDF available at the time of writing via www.waterpark.org.

9. Balmer, D., Gillings, S., Caffrey, B. *et al.* 2013. *Bird Atlas 2007–11: the Breeding and Wintering Birds of Britain and Ireland*. BTO, Thetford. This definitive account is used for all statistics of decline or increase in this chapter unless specified otherwise.

10. Cotswold Water Park Trust 2012.

11. BBC. 2016. *Planet Earth 2*. Broadcast November to December 2016.

12. Morrison, P. & Gurney, M. 2007. Nest boxes for roseate terns *Sterna dougallii* on Coquet Island RSPB reserve, Northumberland, England. *Conservation Evidence* 4: 1–3.

13. Peakall, D.B. 1962. The past and present status of the Red-backed Shrike in Great Britain. *Bird Study* 9: 198–216. doi: 10.1080/00063656209476030.

14. Takács, V., Kuzniak, S. & Tryjanowski, P. 2004. Predictions of changes in population size of the Red-backed Shrike (*Lanius collurio*) in Poland: population viability analysis. *Biological Letters* 41(2): 103–111. All further references to Viktoria's work are from this paper and correspondence with the author (2016).

15. Green, R.E. 1995. The decline of the Corncrake *Crex crex* in Britain continues. *Bird Study* 42: 66–75. doi: 10.1080/00063659509477150.

16. Green 1995.

17. Green, R.E. 2004. A new method for estimating the adult survival rate of the Corncrake *Crex crex* and comparison with estimates from ring-recovery and ring-recapture data. *Ibis* 146: 501–508.

18. Holloway, S. 1996. *The Historical Atlas of Breeding Birds in Britain and Ireland 1875–1900*. Poyser, London.

19. Gribble, F.C. 1983. Nightjars in Britain and Ireland in 1981. *Bird Study* 30: 165–176, doi: 10.1080/00063658309476794.

20. Balmer *et al.* 2013, p. 461.

21. Bright J.A., Langston, R.H.W. & Bierman, S. 2007. Habitat associations of nightjar *Caprimulgus europaeus* breeding on heathland in England. RSPB Research Report No. 25. ww2.rspb.org.uk/images/bright_langston_bierman_tcm9-192440.pdf (accessed October 2018).

22. The live sightings map is a screen-grab from the website Devon Birds: www.devonbirds.org/birdwatching/recording/cuckoos (accessed 30 June 2018).

23. Musgrove (2013) cited in Holling, M. *et al.* 2016. Rare breeding birds in the United Kingdom in 2014. *British Birds* 109: 491–545. In spite of predictions of 1,000–2,000 pairs of lesser spotted woodpecker in Britain, the only 'connected' populations being reported are from the New Forest (Rob Clements) and Herefordshire (Herefordshire Woodpecker Project 2010; Ben Macdonald & Ken Smith – ongoing). The RBBP (www.rbbp.org.uk) reported just 300 pairs in 2014, acknowledging that this is an underestimate.

24. New Forest naturalist and local expert Rob Clements has projected a figure of 200 pairs, based on the density in areas checked so far; the UK species expert Ken Smith puts this estimate slightly lower, but I am happy with Rob's methodology.

25. Courchamp, F., Berec, L. & Gascoigne, J. 2008. *Allee Effects in Ecology and Conservation*. Oxford University Press, Oxford. This work explains the shift towards the Allee effect, whilst hundreds of classic standard works cover competition theory, something we do not have time to explore in this book.

26. Allee effects are complex. I am not arguing that these are a universal rule of nature; instead, that they can sometimes override resource-dependent competition. They are one of the competing rules of nature, not the only one. For more see: (a) Courchamp *et al.* 2008; (b) Allee, W.C. *et al.* 1932. Studies in animal aggregations: mass protection against colloidal silver among goldfishes. *Journal of Experimental Zoology* 61: 185–207. doi:10.1002/jez.1400610202; (c) Stephens, P.A. *et al.* 1999. What is the Allee effect? *Oikos* 87: 185–190. doi:10.2307/3547011; (d) Kramer, A.M. *et al.* 2009. The evidence for Allee effects. *Population Ecology* 51: 341–354. doi: 10.1007/s10144-009-0152-6.

27. Rob Clements, personal communication, March 2017, regarding ongoing fieldwork in the New Forest, mapping hawfinch roosts, partly in relation to active goshawk nests known to licensed recorders.

28. Tomiałojć, L. & Neubauer, G. 2017. Song thrush *Turdus philomelos* and hawfinch *Coccothraustes coccothraustes* exhibit non-random nest orientation in dense temperate forest. *Acta Ornithologica* 52: 209–220.

29. Roy Dennis Wildlife Foundation. Nest building. www.roydennis.org/animals/raptors/osprey/nest-building (accessed October 2018).

30. Author's observations on the Volga Delta, 21 June 2016.

31. Soloviev, M.Y. & Tomkovich, P.S. 1998. The phenomenon of brood aggregations and their structure in waders in northern Taimyr. *International Wader Studies* 10: 201–206.

32. Author's satellite measurement made on Google's 'My Maps'.

33. Eriksson, M.O.G. & Götmark, F. 1982. Habitat selection: do passerines nest in association with lapwings *Vanellus vanellus* as defence against predators? *Ornis Scandinavica* 13: 189–192.

34. Balmer *et al.* 2013. The abundance maps for the Mosslands, for farmland birds, are interesting, often the most clustered in the UK for the named species.

35. This goes for summer migrant passerines, but not, as we'll explain later, long-lived birds of prey such as ospreys. This is not a definitive percentage; each year varies dependent on migratory conditions. The wider point is that most small, fragile migrants do not make it back. Ongoing BTO studies show rates of around 30–35% return in nightingales. For corncrake mortality see Green 1995 (see note 16, above). Turtle dove mortality can be around 64% for first-year birds: Calladine, J.R. *et al.* 1997. *The Summer Ecology and Habitat Use of the Turtle Dove* Streptopelia turtur: *a Pilot Study*. English Nature Research Report No. 219. English Nature, Peterborough.

36. The slaughter of migratory songbirds, many heading for Europe to breed, is well known in conservation circles, and the 'Malta Migration Massacre' campaign has raised awareness in and outside of Malta to a new level. A recent referendum in the country yielded a tight result, only marginally in favour of retaining hunting, and there are further signs that the 'sport' is becoming less popular in the country. For more, see birdlifemalta.org and www.chrispackham.co.uk.

37. (a) Holt, C.A., Fraser, K.H., Bull, A.J. & Dolman, P.M. 2012. Habitat use by nightingales in a scrub-woodland mosaic in central England. *Bird Study* 59: 416–425. doi: 10.1080/00063657.2012.722191; (b) Wilson, A.M., Fuller, R.J., Day, C. & Smith, G. 2005. Nightingales *Luscinia megarhynchos* in scrub habitats in the southern fens of East Anglia, England: associations with soil type and vegetation structure. *Ibis* 147: 498–511. doi: 10.1111/j.1474-919x.2005.00420.x.

38. British Trust for Ornithology. 2015. *Managing Scrub for Nightingales*. BTO, Thetford. Conservation Advice Notes no. 1. www.bto.org/research-data-services/publications/conservation-advice-notes/2015/managing-scrub-nightingales (accessed October 2018).

39. Norrdahl, K., Suhonen, J., Hemminki, O. & Korpimäki, K. 1995. Predator presence may benefit: kestrels protect curlew nests against predators. *Oecologia* 101: 105–109. doi: 10.1007/BF00328906.

40. Though used as an erroneous excuse for grouse moors and predator eradication, the susceptibility of the curlew to predators is a serious concern. Yet, clearly, this species managed fine for around 2 million years without dying out. Isolation, however, confines pairs in certain areas, making them 'sitting targets'. This paper charts the devastation wrought in Ireland: Grant, M.C., Orsman, C., Easton, J. *et al.* 1999. Breeding success and causes of breeding failure of curlew *Numenius arquata* in Northern Ireland. *Journal of Applied Ecology* 36: 59–74. doi: 10.1046/j.1365-2664.1999.00379.x.

41. Galbreath, R. & Brown, D. 2004. The tale of the lighthouse-keeper's cat: discovery and extinction of the Stephens Island wren (*Traversia lyalli*). *Notornis* 51 (4): 193–200. The story of the Lyall's or Stephens Island wren, first identified when it was brought to a lighthouse-keeper in the mouth of a pet cat, is one of the great 'warning' stories of isolation and the dangers it poses to bird populations.

6. Memory

1. Welshman, M. 2014. Never mess with a swallow who thinks your home belongs to him. *Daily Mail Online*, 26 May 2014. www.dailymail.co.uk/news/article-2639205/Never-mess-swallow-thinks-home-belongs-Malcolm-Welshman-tried-discovered-nature-beaten.html (accessed October 2018).

2. Horton, J. 2013. Corncrakes tracked to Congo in new study. *The Scotsman*, 28 April 2013. Cites the evidence of Rhys Green, Honorary Professor of Conservation Science at the RSPB (based at Department of Zoology, University of Cambridge). Even if the megafauna at one end of their journey has disappeared, corncrakes still enjoy bounty of elephants at the other.

3. With the exception of the Nene Washes reintroduction scheme, covered later in this chapter, corncrakes, like a whole range of migrants, may have increased in strongholds but never expanded back into Britain, where they once bred in every country. This isn't purely down to habitat loss; there are now plenty of suitable meadows spread around the country. See for example: (a) Wotton, S.R., Eaton, M., Ewing, S.R. & Green, R.E. 2015. The increase in the Corncrake *Crex crex* population of the United Kingdom has slowed. *Bird Study* 62: 486–497; (b) Balmer, D., Gillings, S., Caffrey, B. *et al.* 2013. *Bird Atlas 2007–11: the Breeding and Wintering Birds of Britain and Ireland*. BTO, Thetford.

4. Pevernagi, E. 1996. Walking down the memory lane 2. https://commons.wikimedia.org/wiki/File:Z-_Walking_down_the_memory_lane.jpg (accessed October 2018).

5. Metcalfe, K., Agamboué, P.D., Augowet, E. *et al.* 2015. Going the extra mile: Ground-based monitoring of Olive Ridley turtles reveals Gabon hosts the largest rookery in the Atlantic. *Biological Conservation* 190: 14–22.
6. Zim, H.S. & Smith, H.M. 1956. *Reptiles and Amphibians: a Guide to Familiar American Species*. Simon & Schuster, New York, p. 18.
7. Frazier, J.G. 2007. The oldest place where there is always something new. *Marine Turtle Newsletter* 116: 3–6. http://www.seaturtle.org/mtn/PDF/MTN116.pdf (accessed October 2018).
8. (a) Lohmann, K.J., Putman, N.F. & Lohmann, C.M.F. 2008. Geomagnetic imprinting: a unifying hypothesis of long-distance natal homing in salmon and sea turtles. *Proceedings of the National Academy of Sciences* 105(49): 19096–19101. doi: 10.1073/pnas.0801859105; (b) Lohmann, K., Lohmann, C.M.F., Ehrhart, L.M., Bagley, D.A. & Swing, T. 2004. Geomagnetic map used in sea-turtle navigation. *Nature* 428: 909–910. doi: 10.1038/428909a; (c) Lohmann, K. & Lohmann, C. 1996. Detection of magnetic field intensity by sea turtles. *Nature* 380: 59–61. doi: 10.1038/380059a0.
9. Weidensaul, S. 2003. *Living on the Wind: Across the Hemisphere with Migratory Birds*, Palgrave Macmillan, London. This fascinating book summarises the progress of science in uncovering the migrations across the North American continent, and the universal science of how birds find home.
10. Weidensaul 2003.
11. Wolfgang Wiltschko's is a lifetime of work on this subject. Three pertinent papers, all available online, are as follows: (a) Wiltschko, W. & Wiltschko, R. 2005. Magnetic orientation and magnetoreception in birds and other animals. *Journal of Comparative Physiology A* 191: 675–693. doi: 10.1007/s00359-005-0627-7; (b) Wiltschko, W, Traudt, J., Güntürkün, O., Prior, H. & Wiltschko, R. 2003. Lateralization of magnetic compass orientation in a migratory bird. *Nature* 419: 467–470. doi: 10.1038/nature00958; (c) Wiltschko, W. & Wiltschko, R. 1976. Magnetic compass of European robins. *Science* 176: 62–64. doi: 10.1126/science.176.4030.62.
12. Wiltschko & Wiltschko 1976.
13. Wiltschko, R. & Wiltschko, W. 1978. Relative importance of stars and the magnetic field for the accuracy of orientation in night-migrating birds. *Oikos* 30: 195–206.
14. Ritz, T. 2004. Resonance effects indicate a radical-pair mechanism for avian magnetic compass. *Nature* 429: 177–180. doi: 10.1038/nature02534.
15. The work of Schulten and others, such as Thorsten Ritz, is ongoing in this matter, and equally fascinating and difficult to understand. Some of the 'breakthrough' papers, from newer to older: (a) Solov'yov, I.A. & Schulten, K. 2012. Reaction kinetics and mechanism of magnetic field effects in cryptochrome. *Journal of Physical Chemistry B* 116 (3), 1089–1099. doi: 10.1021/jp209508y; (b) Ritz, T., Wiltschko, R., Hore, P.J. *et al.* 2009. Magnetic compass of birds is based on a molecule with optimal directional sensitivity. *Biophysical Journal* 96: 3451–3457. (c) Möller, A., Sagasser, S., Wiltschko, W. & Schierwater, B. 2004.

Retinal cryptochrome in a migratory passerine bird: a possible transducer for the avian magnetic compass. *Naturwissenschaften* 91: 585–588. doi: 10.1007/s00114-004-0578-9.

16. Professor Ian Newton, personal communication, April 2018.
17. Camacho, C., Canal, D. & Potti, J. 2016. Natal habitat imprinting counteracts the diversifying effects of phenotype-dependent dispersal in a spatially structured population. *BMC Evolutionary Biology* 16: 158. doi: 10.1186/s12862-016-0724-y.
18. Welshman 2014.
19. Dr Caroline Schuppli, personal communication, February 2017. For more information about saving Sumatra's orangutans, visit or donate to the Sumatran Orangutan Conservation Project: www.sumatranorangutan.org.
20. Ishida, Y., Van Coeverden de Groot, P.J., Leggett, K.E. *et al.* 2016. Genetic connectivity across marginal habitats: the elephants of the Namib Desert. *Ecology and Evolution* 6: 6189–6201. This story is filmed in detail for the upcoming Netflix series, *Our Planet* (2019).
21. British Trust for Ornithology. Why ring birds? www.bto.org/volunteer-surveys/ringing/about-ringing/why-ring-birds (accessed October 2018).
22. All of the data on the ringing recoveries described are available online: British Trust for Ornithology. Online ringing and nest recording report. www.bto.org/volunteer-surveys/ringing/publications/online-ringing-reports (accessed October 2018).
23. Stephen Roberts, personal communication, 2015.
24. Takács, V., Kuzniak, S. & Tryjanowski, P. 2004. Predictions of changes in population size of the Red-backed Shrike (*Lanius collurio*) in Poland: population viability analysis. *Biological Letters* 41(2): 103–111.
25. Carter, I., Newbery, P., Grice, P. & Hughes, P. 2008. The role of reintroductions in conserving British birds. *British Birds* 101: 2–25.
26. Brooks, M. 2014. *At the Edge of Uncertainty: 11 Discoveries Taking Science by Surprise.* Profile Books, London.
27. Stevenson, G.B. 2007. An historical account of the social and ecological causes of Capercaillie *Tetrao urogallus* extinction and reintroduction in Scotland. PhD thesis, University of Stirling. EthOS ID: 513634.
28. Carter *et al.* 2008.
29. Stevenson 2007.
30. Love, J. 2013. *A Saga of Sea Eagles.* Whittles Publishing, Dunbeath.
31. Balmer *et al.* 2013, p. 296.
32. Carter *et al.* 2008.
33. Forrester, R. & Andrews, I. 2007. *The Birds of Scotland*, Scottish Ornithologists' Club, Aberlady, p. 452.
34. Burnside, R.J., Carter, I., Dawes, A. & Waters, D. 2012. The UK great bustard *Otis tarda* reintroduction trial: a 5-year progress report. *Oryx* 46: 112–121. doi: 10.1017/S0030605311000627.

35. Great Crane Project. 2017. Where are we at? [September 2017 update.] www. thegreatcraneproject.org.uk/project/where-are-we-now (accessed October 2018).

36. Carter, I. & Newbery, P. 2004. Reintroduction as a tool for population recovery of farmland birds. *Ibis* 146: 221–229. doi: 10.1111/j.1474-919X.2004.00353.x.

37. Walker, R.H., Robinson, R.A., Leech, D.I. *et al.* 2017. Bird ringing and nest recording in Britain and Ireland in 2015. *Ringing & Migration* 31: 115–159. doi: 10.1080/03078698.2016.1298316.

38. Jeffs, C., Davies, M., Carter, I. *et al.* 2016. Reintroducing the Cirl Bunting to Cornwall. *British Birds* 109: 374–388.

39. RSPB. 2017. Hen harrier reintroduction in southern England: a view from the RSPB. ww2.rspb.org.uk/community/ourwork/b/southwest/archive/2017/07/25/hen-harrier-reintroduction-in-southern-england-a-view-from-the-rspb.aspx (accessed October 2018).

40. Baxter, A. 2016. Understanding the factors associated with declines of an alpine specialist bird species in Scotland. PhD thesis, University of Aberdeen. EthOS ID: 707500.

7. A Wild Economy

1. Nationalpark Bayerischer Wald (Bavarian Forest National Park). The founding sentiment of the national park, with an English version. www.nationalpark-bayerischer-wald.de/english (accessed October 2018).

2. Urban population (% of total) in the United Kingdom was reported at 83% in 2017, according to the World Bank collection of development indicators. https://data.worldbank.org/indicator/sp.urb.totl.in.zs (accessed October 2018).

3. Hepperle, E., Dixon-Gough, R., Mansberger, R. *et al.* (eds). 2017. *Land Ownership and Land Use Development: the Integration of Past, Present, and Future in Spatial Planning and Land Management Policies*. City Planning, Zurich, p. 47.

4. The Dark Sky designation is awarded to areas that are outstanding for stargazing. This also accentuates how avoidable it is that Northumberland National Park, and others, is so wildlife-depleted. In an area of this magnitude could be many golden eagles, elk, lynx and a thriving ecotourism industry. www.northumberlandnationalpark.org.uk, www.breconbeacons.org (both accessed October 2018).

5. Office for National Statistics. 2014. Annual mid-year population estimates for national parks, mid-2002 to mid-2012. ONS Population Estimates Unit, 20 March 2014. www.ons.gov.uk/peoplepopulationandcommunity/populationandmigration/populationestimates/bulletins/annualsmallarea populationestimates/2014-03-20 (accessed October 2018).

6. Augustin, B. & Kubena, J. 2009. *Yellowstone National Park*. Marshall Cavendish, New York. Wildlife numbers fluctuate constantly in Yellowstone; the figures were accurate as of the time of writing.

7. UNESCO. The English Lake District. https://whc.unesco.org/en/list/422 (accessed October 2018). 'Located in northwest England, the English Lake District is a mountainous area, whose valleys have been modelled by glaciers in the Ice Age and subsequently shaped by an agro-pastoral land-use system characterized by fields enclosed by walls.'

8. Lake District National Park Authority website. www.lakedistrict.gov.uk (accessed October 2018). No recent changes have been made to the boundaries of the park.

9. Culliford, A. 1999. National parks: the complete guide to Britain's national parks. *The Independent*, 24 July 1999. www.independent.co.uk/life-style/travel -national-parks-the-complete-guide-to-britains-national-parks-1108194.html (accessed October 2018).

10. Walpole, M., Karanga, G., Sitati, N. & Leader-Williams, N. 2003. *Wildlife and People: Conflict and Conservation in Masai Mara, Kenya*. Wildlife and Development Series. International Institute for Environment and Development, London.

11. Mioduszewski, W. 2004. Protection of water-related ecosystems and their role as water suppliers: Polish national report. Seminar on the role of ecosystems as water suppliers (Geneva, 13–14 December 2004). Convention on Protection and Use of Transboundary Watercourses and International Lakes. www.unece. org/fileadmin/DAM/env/water/meetings/ecosystem/Reports/Poland_en.pdf (accessed October 2018).

12. Dudley, N. (ed.). 2008. *Guidelines for Applying Protected Area Management Categories*. IUCN, Gland. https://cmsdata.iucn.org/downloads/guidelines_for_ applying_protected_area_management_categories.pdf (accessed October 2018).

13. The RSPB's Fairburn Ings and Swillington Ings reserves provide an excellent day's wildlife watching in northern England, and Potteric Carr, run by the Yorkshire Wildlife Trust, is an outstanding example of small-scale brownfield restoration.

14. The drive I am referring to was from Inveraray to Tarbet, although the aerial view shows that the combination of barren uplands, devoid of nuance, with dense forestry, creating a bipolar landscape, is not unique to this one area of the park.

15. These ancient woodlands have seen, among others, a 60% decline in wood warbler, pied flycatcher losses, and recent declines to just a few pairs of lesser spotted woodpecker (last Ben Hoare, personal communication, February 2018).

16. Boyce, D.C. 2012. *A Survey of Waders and Other Birds on Mires in Exmoor National Park, 2011–12*. RSPB South-west Region Report to South West Water. Curlews no longer breed on Exmoor. RSPB, Exeter. www.beachlive.

co.uk/media/pdf/r/e/Exmoor_Mires_Project_bird_survey_2012.pdf (accessed October 2018).

17. National Parks England. Farming in the English national parks. Paragraph 2.6. www.nationalparksengland.org.uk/__data/assets/pdf_file/0009/967905/ Farming-in-the-English-National-Parks.pdf (accessed October 2018).

18. Silcock, P., Rayment, M., Kieboom, E., White, A. & Brunyee, J. 2013. *Valuing England's National Parks: Final Report for National Parks England*. Cumulus Consultants Ltd, Broadway. www.nationalparksengland.org.uk/__data/ assets/pdf_file/0004/717637/Valuing-Englands-National-Parks-Final-Report -10-5-13.pdf (accessed October 2018).

8. The Wild Highlands

1. Shaw, G.B. *Back to Methuselah*, Act I. These words are spoken to Eve by the Serpent, the mouthpiece of the Devil – but, as shown by many centuries of English literature, the Devil always gets the best lines!

2. Public and Corporate Economic Consultants (PACEC). 2016. *The Contribution of Deer Management to the Scottish Economy: a report prepared by PACEC on behalf of the Association of Deer Management Groups*. PACEC, Cambridge. www.deer-management.co.uk/wp-content/uploads/2016/02/Final-25FEB.pdf (accessed October 2018).

3. Author's calculations as follows: deer estates cover 18,300 square kilometres; Yellowstone National Park is around 9,100 square kilometres; Greater Manchester is 1,276 square kilometres; Dartmoor National Park is 954 square kilometres.

4. Highland Council. Highland profile: key facts and figures. An analysis of the 2013 NRS mid-year estimates, showing human populations, and population densities, in the Highland region. Available via www.highland. gov.uk. Sutherland and Caithness, the most sparsely populated sub-counties, have population densities of 2.5 and 2.3 persons per square kilometre. This is extraordinarily low, even lower than the state of Montana. Highland as a whole has a density of 10.1 persons per square kilometre, still comparable to the state of Utah. The US information is from the 2013 census by the United States Census Bureau.

5. Author's calculations as follows: Scotland's land mass covers 80,077 square kilometres, of which deer estates cover 18,000, affording them 22.8% of the country.

6. PACEC 2016.

7. Office of National Statistics. 2018. Statistical Bulletin: UK labour market, September 2018. www.ons.gov.uk/employmentandlabourmarket/peopleinwork/ employmentandemployeetypes/bulletins/uklabourmarket/september2018 (accessed October 2018). States: '32.40 million people in work'. This means that 2,532 jobs represents 0.008% of the British economy.

8. Statista. Gross domestic product (GDP) of Scotland (excluding North Sea GDP) from 2011/2012 to 2016/2017 (in million GBP). www.statista.com/statistics/350717/scottish-gross-domestic-product-gdp (accessed October 2018).

9. Putman, R. 2012. *Scoping the Economic Benefits and Costs of Wild Deer and Their Management in Scotland*. Scottish Natural Heritage Commissioned Report no. 526.

10. Watson, A., Payne, S. & Rae, R. 1989. Golden Eagles *Aquila chrysaetos*: land use and food in northeast Scotland. *Ibis* 131: 336–348. doi: 10.1111/j.1474-919X.1989.tb02783.x.

11. Edwards, T. & Kenyon, W. 2013. *Wild Deer in Scotland*. SPICe Briefing, ref: 13/74. Scottish Parliament, Edinburgh. www.parliament.scot/ResearchBriefingsAndFactsheets/S4/SB_13-74.pdf (accessed October 2018).

12. Ewing, S.R., Eaton, M.A., Poole, T.F. *et al.* 2012. The size of the Scottish population of Capercaillie *Tetrao urogallus*: results of the fourth national survey. *Bird Study* 59: 126–138. doi: 10.1080/00063657.2011.652937.

13. Reforesting Scotland. 2013. The impact and management of deer in Scotland. A submission to the Scottish Parliament's Rural Affairs, Climate Change and Environment Committee. www.reforestingscotland.org/what-we-do/influencing-policy/the-impact-and-management-of-deer-in-scotland (accessed October 2018).

14. Reforesting Scotland 2013.

15. Rewilding Britain. Glenfeshie. www.rewildingbritain.org.uk/rewilding/rewilding-projects/glenfeshie (accessed October 2018). Many other articles online chart the recovery of the estate's tree-life following deer culls. Anders Holch Povlsen is a huge landowner in Scotland and appears set to play a very important role in its ecological restoration in years to come.

16. Putman, R.J., Duncan, P. & Scott, R. 2005. Demographic changes in a Scottish red deer population (*Cervus elaphus* L.) in response to sustained and heavy culling: an analysis of trends in deer populations of Creag Meagaidh National Nature Reserve 1986–2001. *Forest Ecology and Management* 206: 263–281.

17. Trees for Life. The Glen Affric Forest Landscape Project. https://treesforlife.org.uk/work/woodland-projects/the-glen-affric-forest-landscape-project (accessed October 2018).

18. Putman, R.J. 2011. A review of the legal and administrative systems governing management of large herbivores in Europe. In Putman, R.J., Apollonio, M. & Andersen, R. (eds). *Ungulate Management in Europe: Problems and Practices*. Cambridge University Press, Cambridge, pp. 54–79.

19. Orr-Ewing, D. 2016. Deer management in Scotland: report to the Scottish Government from Scottish Natural Heritage. Statement from Scottish Environment. www.scotlink.org/wp/files/documents/Deer-Statement-18-November-2016-Final.pdf (accessed October 2018). LINK is supported by:

RSPB Scotland; National Trust for Scotland; Scottish Wildlife Trust; Woodland Trust Scotland; Ramblers Scotland and Trees for Life.

20. You can only search for so long for an absence of statistics relating to a harmless animal. The Lynx Trust has found no recorded evidence of a Eurasian lynx attack on humans (regardless of severity) ever. No one in the sheep-farming sector, or others opposed to the reintroduction of lynx, has been able to find one instance either. Enquiries could be made with experts in the field of human–feline conflict, such as Professor David Macdonald at Oxford University, or indeed any other feline expert worldwide.

21. Office for National Statistics. Deaths from dog bites, England and Wales, 1981 to 2015. ONS, reference number 006077; released 5 September 2016. Available via www.ons.gov.uk.

22. (a) Red Sage Conservation. SCH-WTE Socio-economic Impact Report, Appendix 6: Case Study: Osprey; (b) Dickie, I., Hughes, J. & Esteban, A. 2006. Watched like never before … the local economic benefits of spectacular bird species. RSPB, Sandy. http://ww2.rspb.org.uk/Images/watchedlikeneverbefore_tcm9-133081.pdf (accessed October 2018); (c) Shiel, A., Rayment, M. & Burton, G. 2002. RSPB reserves and local economies. RSPB, Sandy. https://www.rspb.org.uk/globalassets/downloads/documents/positions/economics/rspb-reserves-and-local-economies.pdf (accessed October 2018).

9. New Forests

1. Rackham, O. 2000. *The History of the Countryside*, new edition. Phoenix Press, London, p. 29.

2. There is still no official 'extinction-imminent' index published for UK birds, although the *State of Nature* reports give a good idea. This statement is based on four calculations: (1) rates of decline in past 40, and past 10 years; (2) lack of remaining strongholds for these species; (3) size of these remaining populations; (4) degree of fragmentation in these populations. Readers can draw their own conclusions, from the BTO's data, on which species faces the fastest extinction risks.

3. State of Nature Partnership. 2013. *State of Nature Report 2013*. Available from the RSPB's website at http://ww2.rspb.org.uk/Images/stateofnature_tcm9-345839.pdf, or from many other sources.

4. Wäber, K., Spencer, J. & Dolman, P.M. 2013. Achieving landscape-scale deer management for biodiversity conservation: the need to consider sources and sinks. *Journal of Wildlife Management* 77: 726–736. doi: 10.1002/jwmg.530.

5. Deer Initiative. About wild deer. www.thedeerinitiative.co.uk/about_wild_deer (accessed October 2018). A wealth of information on current deer distribution, increases and associated issues can be explored via www.thedeerinitiative.co.uk.

6. Harris, S., Morris, P., Wray, S. & Yalden, D. 1995. *A Review of British Mammals: Population Estimates and Conservation Status of British Mammals Other*

than Cetaceans. JNCC, Peterborough. http://jncc.defra.gov.uk/pdf/pub03_areviewofbritishmammalsall.pdf (accessed October 2018).

7. Holt, C.A., Fuller, R.J. & Dolman, P.M. 2010. Experimental evidence that deer browsing reduces habitat suitability for breeding Common Nightingales *Luscinia megarhynchos. Ibis* 152: 335–346. doi: 10.1111/j.1474-919X.2010.01012.x.

8. Williams, I. 2013. Speech at the launch of the *State of Nature Report*, Cardiff, 22 May 2013. https://www.gwentwildlife.org/node/6762 (accessed October 2018).

9. Atkinson, S. & Townsend, M. 2011. *The State of the UK's Forests, Woods and Trees: Perspectives from the Sector.* Woodland Trust, Grantham. www.woodlandtrust.org.uk/mediafile/100229275/stake-of-uk-forest-report.pdf ?cb=58d97f320c (accessed October 2018).

10. Polley, H., *et al.* 2014. The forests in Germany. A paper published by the Federal Ministry of Food and Agriculture (BMEL) and available online (www.bmel.de).

11. Surrey's definitive tree-cover statistic comes from mapping by BlueSky, carried out by aerial surveys. Their National Tree Map, covering England and Wales, shows high degrees of woodland cover in Kent, Surrey, Sussex, Hampshire and southeast Wales, among other areas, and can be obtained at www.blueskymapshop.com.

12. Forest Research. 2017. Forestry statistics 2017. IFOS-Statistics, Forest Research, Edinburgh. Enquiries to statistics@forestry.gsi.gov.uk. www.forestresearch.gov.uk/tools-and-resources/statistics/forestry-statistics/forestry-statistics-2017 (accessed October 2018).

13. Atkinson & Townsend 2011.

14. Forestry Statistics 2017.

15. Forestry Commission. 2011. *National Forest Inventory Report: Standing Timber Volumes for Coniferous Trees in Britain.* Further enquiries on timber ratios can be addressed to nfi@forestry.gsi.gov.uk.

16. Woodland Investment Management Ltd. 2018. French forestry. www.woodlands.co.uk/blog/woodland-activities/french-forestry (accessed October 2018).

17. Conway & Henderson (2010), cited in Sharps, K., Henderson, I., Conway, G., Armour-Chelu, N. & Dolman, P.M. 2015. Home-range size and habitat use of European Nightjars *Caprimulgus europaeus* nesting in a complex plantation-forest landscape. *Ibis* 157: 260–272. doi: 10.1111/ibi.12251.

18. Conway, G., Wotton, S., Henderson, I. *et al.* 2007. Status and distribution of European Nightjars *Caprimulgus europaeus* in the UK in 2004. *Bird Study* 54: 98–111. doi: 10.1080/00063650709461461.

19. Atkinson & Townsend 2011.

20. The website http://taurosprogramme.com (accessed October 2018) offers full information on the project.

21. Oquiñena Valluerca, I. 2011. Analysis of vegetation changes induced by a European bison herd in the Kraansvlak area (2003–2009). Master's thesis, Faculty of Geosciences, University of Utrecht.

22. CJC Consulting. 2015. *The Economic Contribution of the Forestry Sector in Scotland*. CJC Consulting, Oxford. https://scotland.forestry.gov.uk/images/corporate/pdf/economic-contribution-forestry-2015.pdf (accessed October 2018).

23. Forestry statistics 2011: annual business survey, cited in Confor [2012]. *Forestry: 7,000 Green Jobs and Low-Carbon Growth*. www.confor.org.uk/media/79582/forestry7000greenjobsandlowcarbongrowthjune2012.pdf (accessed October 2018).

24. Public and Corporate Economic Consultants (PACEC). 2000. *English Forestry Contribution to Rural Economies Final Report: a report prepared by PACEC on behalf of the Forestry Commission*. PACEC, Cambridge. www.forestry.gov.uk/pdf/engmult.pdf/$FILE/engmult.pdf (accessed October 2018).

25. Forestry statistics 2011.

26. Bath, A., Olszanska, A. & Henryk Okarma, H. 2008. From a human dimensions perspective, the unknown large carnivore: public attitudes toward Eurasian lynx in Poland. *Human Dimensions of Wildlife* 13: 31–46, doi: 10.1080/10871200701812928.

10. The Golden Hills of Wales

1. Armstrong, E. 2016. *Research Briefing: the Farming Sector in Wales*. Paper number: 16-053. National Assembly for Wales, Cardiff. www.assembly.wales/research%20documents/16-053-farming-sector-in-wales/16-053-web-english2.pdf (accessed October 2018). An impartial research paper prepared by the Research Service of the National Assembly for Wales. The paper confirms the following: Welsh agricultural GVA (gross value added: national economic contribution) was 0.71% for all agriculture (including sheep farming); 58,300 people work in all sectors of agriculture in Wales (including sheep farming); total agricultural land coverage of Wales is 88%; 80% of all agricultural land in Wales is designated as 'less favourable' to farming; 29% of the UK's 33.34 million sheep (i.e. 9.66 million sheep) are found in Wales; Welsh agricultural land covers 17,530 square kilometres, and permanent pasture 13,260 square kilometres.' The latest population estimate for Wales is 3,113,200 (Office for National Statistics, mid-2016), so 58,300 people represent 1.87% of the Welsh population.

2. Glyn Roberts, president of the Farmers' Union of Wales, has, in many engagements with the press, stated that the average farmer earns £13,000 – but £2,600 without EU subsidy. This appears to be at odds with the statement of the official Welsh Assembly Research Briefing (note 1), which states that the average income is £29,400. However, no other figures are available on the amount

left after subsidy – which all parties seem to agree is very low and probably reflected by FUW's statements.

3. Author comparison: Welsh pasture farmland covers 13,260 square kilometres. Greater London covers an area of 1,583 square kilometres. The first is 11.9 times the size of the latter.

4. AHDB. 2013. Why the UK imports lamb from New Zealand. *Beef and Lamb Matters*, 17 July 2013. http://beefandlambmatters.blogspot.com/2013/07/why-uk-imports-lamb-from-new-zealand.html (accessed October 2018). Statistics provided by AHDB Beef & Lamb, a pro-industry organisation for beef and lamb levy-payers in the UK. Its blog, *Beef and Lamb Matters*, is available at http://beefandlambmatters.blogspot.com.

5. This calculation was first made by George Monbiot and subsequently checked by me. On average, each person consumes 5.0 kg of lamb per year. With 100 g of lamb containing around 294 calories, annual intake from lamb is therefore 14,700 kcal. Annual total calorie intake average is approximately 1,250,000 kcal. Lamb therefore provides 1.18% of calories. See: AHDB Beef and Lamb. 2016. *UK Yearbook 2016*. ADHB, Kenilworth. http://beefandlamb.ahdb.org.uk/wp/wp-content/uploads/2016/07/UK-Yearbook-2016-Sheep-050716.pdf (accessed October 2018).

6. International Agency for Research on Cancer. 2015. *Red Meat and Processed Meat*. IARC Monographs on the Evaluation of Carcinogenic Risks to Humans, vol. 114. IARC, Lyon. http://publications.iarc.fr/Book-And-Report-Series/Iarc-Monographs-On-The-Evaluation-Of-Carcinogenic-Risks-To-Humans/Red-Meat-And-Processed-Meat-2018 (accessed October 2018). An assessment of more than 800 epidemiological studies, summarised in: Bouvard, V., Loomis, D., Guyton, K.Z. *et al.* 2015. Carcinogenicity of consumption of red and processed meat. *Lancet Oncology* 16: 1599–1600. doi: 10.1016/S1470-2045(15)00444-1.

7. Office for National Statistics. 2001. *200 Years of the Census in … Wales*. ONS, London.

8. Davies, J. 2007. *A History of Wales*. Penguin, London.

9. Odih, P. 2007. *Gender and Work in Capitalist Economies*. McGraw-Hill, New York.

10. Office for National Statistics 2001.

11. Sim, I.M.W., Burfield, I.J., Grant, M.C., Pearce-Higgins, J.W. & Brooke, M. 2007. The role of habitat composition in determining breeding site occupancy in a declining Ring Ouzel *Turdus torquatus* population. *Ibis* 149: 374–385. doi: 10.1111/j.1474-919X.2007.00655.x.

12. Calladine, J., Baines, D. & Warren, P. 2002. Effects of reduced grazing on population density and breeding success of black grouse in northern England. *Journal of Applied Ecology* 39: 772–780. doi: 10.1046/j.1365-2664.2002.00753.x.

13. Furness, R.W. 1988. The predation of tern chicks by sheep. *Bird Study* 35: 199–202. doi: 10.1080/00063658809476989.

14. McCarthy, J. 2012. Curlew under threat of extinction in Wales. *Wales Online,* 20 July 2012. www.walesonline.co.uk/news/wales-news/curlew-under-threat-extinction-wales-2027839 (accessed October 2018). Dave Elliott (RSPB) estimated 576 pairs in 2012, with a 50–80% decline in the fifteen years before this time.

15. BBC News. 2012. Black grouse numbers rise in Wales, but only in RSBP-managed areas. *BBC News,* 24 March 2012. www.bbc.co.uk/news/uk-wales-17494491 (accessed October 2018). Cites RSPB Cymru figures.

16. Hayhow, D.B. Benn, S., Stevenson, A., Stirling-Aird, P.K. & Eaton, M.A. 2017. Status of Golden Eagle *Aquila chrysaetos* in Britain in 2015. *Bird Study* 64: 281–294. doi: 10.1080/00063657.2017.1366972.

17. State of Nature Partnership. 2013. *State of Nature Report 2013.* Available from the RSPB's website at http://ww2.rspb.org.uk/Images/stateofnature_tcm9-345839.pdf, or from many other sources. A collaborative report involving the 25 leading conservation organisations in the United Kingdom and includes information on overseas territories.

18. Armstrong 2016.

19. Bassi, E. 2017. Estimate of breeding pair's distribution and seasonal abundance patterns of floating Golden Eagle *Aquila chrysaetos* population in the Italian Central Alps through field surveys and contemporary censuses. *Avocetta* 41: 41–45. Author calculation. Uses an average of four mean home-range studies in the wooded Italian and French Alps, an upland habitat comparable to a rewilded, wooded Snowdonia (rather than the treeless and relatively prey-poor habitats of western Scotland, where eagle productivity is lower and home ranges larger). The paper showed the mean extent of an eagle territory is 97 square kilometres. However, territory boundaries can overlap. Eagle territories vary depending on prey abundance and competition, among other factors.

20. Schmidt, K., Jędrzejewski, W. & Okarma, H. 1997. Spatial organization and social relations in the Eurasian lynx population in Białowieża Primeval Forest, Poland. *Acta Theriologica* 42: 289–312. Author calculation. Estimates based on lynx male and female territories in the wooded Białowieża, Poland, where deer densities are reasonably high and the woodland temperate biome comparable to the native biome of Snowdonia. In Białowieża, densities of lynx average 2.9 animals per 100 square kilometres. I have allowed for a conservative 50% woodland or scrub cover in a restored Snowdonia, suitable for roe deer, and 50% of areas being unsuitable for deer and therefore lynx. This produces 1,066 square kilometres. At 2.9 individuals / 100 km², this produces 30.9 individuals within this area. Note that this is a carrying capacity for a fully restored ecosystem, not how many animals would initially be released – or could survive in Snowdonia in its present state.

21. National parks statistics available from Germany's Federal Agency for Nature Conservation, www.bfn.de.

22. National Parks Wales. 2013. *Valuing Wales' National Parks*. A PDF file was available via www.nationalparkswales.gov.uk. This website no longer exists (October 2018); for further enquiries, try www.nationalparks.gov.uk.

23. BBC News. 2018. Unemployment rate in Wales rises to 5%. *BBC News Online*, 21 February 2018. www.bbc.co.uk/news/uk-wales-43140756 (accessed October 2018). Figures used are from the Office for National Statistics, citing 76,000 people out of work in Wales as of Oct-December 2017.

24. Molloy, D. 2011. Wildlife at work. The economic impact of white-tailed eagles on the Isle of Mull. RSPB, Sandy. http://ww2.rspb.org.uk/Images/wildlifeatwork_tcm9-282134.pdf (accessed October 2018).

25. Snowdonia National Park. 2016. Visitor numbers. www.snowdonia.gov.wales/looking-after/state-of-the-park/tourism/visitor-numbers (accessed October 2018).

26. Bryden, D.M., Westbrook, S.R., Burns, B., Taylor, W.A. & Anderson, S. 2010. Assessing the economic impacts of nature based tourism in Scotland. Scottish Natural Heritage Commissioned Report no. 398. www.nature.scot/snh-commissioned-report-398-assessing-value-nature-based-tourism-scotland (accessed October 2018).

27. Natural England. 2017. Monitor of Engagement with the Natural Environment: 2015 to 2016. www.gov.uk/government/statistics/monitor-of-engagement-with-the-natural-environment-2015-to-2016 (accessed October 2018).

28. Snowdonia National Park. 2016. Employment: industry of employment. www.snowdonia.gov.wales/__data/assets/pdf_file/0003/340725/ParkData-Employment.pdf (accessed October 2018).

11. A Grouse Moor Wild

1. Roosevelt, T. 1905. *Outdoor Pastimes of an American Hunter*. C. Scribner's Sons, New York. Reprinted by Stackpole Books, 1990, p. 272.

2. Bennett, O. 2016. Grouse shooting. House of Commons Briefing Paper CBP-7709, 24 October 2016. https://researchbriefings.parliament.uk/ResearchBriefing/Summary/CBP-7709 (accessed October 2018).

3. Fraser of Allander Institute. 2010. An economic study of grouse moors. A report by the Fraser of Allander Institute to the Game & Wildlife Conservation Trust Scotland, July 2010. www.gwct.org.uk/media/350583/An-Economic-Study-of-Grouse-Moors.pdf (accessed October 2018).

4. Office of National Statistics. 2018. Statistical Bulletin: UK labour market, September 2018. www.ons.gov.uk/employmentandlabourmarket/peopleinwork/employmentandemployeetypes/bulletins/uklabourmarket/september2018 (accessed October 2018).

5. The England mid-year population estimate, released on 22 July 2017 by the Office of National Statistics, estimates 55,268,100 people in England. This renders 175 grouse-moor landowners 0.0003% of the population.

6. UK National Ecosystem Assessment: Technical Report Broad Habitats: Mountains, Moorlands and Heaths. UNEP-WCMC, June 2011 (cited in Bennett 2016: see note 1).

7. The Lake District National Park covers 2,362 square kilometres (Lake District National Park Authority website: www.lakedistrict.gov.uk, accessed October 2018).

8. Metropolis. Factsheet 027: London. Information from Metropolis, the World Association of the Major Metropolises. Greater London covers an area of 1,583 square kilometres.

9. Fraser of Allander Institute 2010.

10. World Bank national accounts data and OECD national accounts data files, available with regular updates from www.worldbank.org. These statistics were up to date as of 21 March 2018.

11. Conversion by author as follows: $2.647 trillion = £1.88 trillion. This figure divided by £23.3 million comes to 0.0012% of GDP.

12. Note that the Moorland Association is pro-shooting. It regularly uses the figure of £67 million as the English grouse moor contribution to the economy. Its website is at www.moorlandassociation.org.

13. Conversion by author as follows: $2.647 trillion = £1.88 trillion. This figure divided by £67.7 million comes to 0.0036% of GDP.

14. Friends of the Earth. 2016. Grouse moors in England: research methodology. https://friendsoftheearth.uk/sites/default/files/downloads/methodology -grouse-moors-research-101915.pdf (accessed October 2018). Contact FoE for more information, but a painstaking breakdown of each estate's payments is available through its website: friendsoftheearth.uk.

15. Silcock, P., Rayment, M., Kieboom, E., White, A. & Brunyee, J. 2013. *Valuing England's National Parks: Final Report for National Parks England.* Cumulus Consultants Ltd, Broadway. www.nationalparksengland.org.uk/__data/ assets/pdf_file/0004/717637/Valuing-Englands-National-Parks-Final-Report -10-5-13.pdf (accessed October 2018).

16. Silcock *et al.* 2013.

17. Moorland Association (2011), cited in British Association for Shooting and Conservation (BASC). 2015. Grouse shooting and management in the United Kingdom: its value and role in the provision of ecosystem services. BASC, Wrexham. https://basc.org.uk/wp-content/uploads/downloads/2015/03/ Research-White-Paper-Grouse-shooting-and-management.pdf (accessed October 2018).

18. BASC 2015.

19. Office for National Statistics population estimates for UK, 22 June 2017, state 65,648,054 people in the UK. This renders 40,000 grouse shooters 0.06% of the British population.

20. BASC 2015.

21. Davies, B.B., Pita, C., Lusseau, D. & Hunter, C. (2010), The value of tourism expenditure related to the East of Scotland bottlenose dolphin population:

final report. Aberdeen Centre for Environmental Sustainability and Moray Firth Partnership.

22. RSPB. Natural foundations: conservation and local employment in the UK. www.rspb.org.uk/globalassets/downloads/documents/positions/economics/natural-foundations---conservation-and-local-employment-in-the-uk.pdf (accessed October 2018).

23. Visit Scotland. 2017. Insight Department: wildlife tourism. https://www.visitscotland.org/binaries/content/assets/dot-org/pdf/research-papers-2/wildlife-topic-paper-2017.pdf (accessed October 2018).

24. Dickie, I., Hughes, J. & Esteban, A. 2006. Watched like never before ... the local economic benefits of spectacular bird species. RSPB, Sandy. http://ww2.rspb.org.uk/Images/watchedlikeneverbefore_tcm9-133081.pdf (accessed October 2018). For many more statistics that point out the worth of ecotourism, contact economics@rspb.org.uk.

25. Bryden, D.M., Westbrook, S.R., Burns, B., Taylor, W.A. & Anderson, S. 2010. Assessing the economic impacts of nature based tourism in Scotland. Scottish Natural Heritage Commissioned Report no. 398. www.nature.scot/snh-commissioned-report-398-assessing-value-nature-based-tourism-scotland (accessed October 2018).

26. Silcock *et al.* 2013.

27. Friends of the Earth. Map of grouse moors in England. https://friendsoftheearth.uk/page/map-grouse-moors-england (accessed October 2018). Note that, if anything, the estimates by FoE are conservative, in that the Moorland Association ascribes even larger areas to grouse shooting. The map also shows historic flooding in relation to grouse moors – a critical subject explored in detail in Avery, M. 2015. *Inglorious: Conflict in the Uplands*. Bloomsbury, London.

28. Fletcher, K., Aebischer, N.J., Baines, D., Foster, R. & Hoodless, A.N. 2010. Changes in breeding success and abundance of ground-nesting moorland birds in relation to the experimental deployment of legal predator control. *Journal of Applied Ecology* 47: 263–272. doi: 10.1111/j.1365-2664.2010.01793.x.

29. Harrison (1980), cited in Yalden, D.W. & Albarella, U. 2009. *The History of British Birds*. Oxford University Press, Oxford. Early fossils of whimbrels, recognisably evolved from curlews by this stage, date back 1.9 million years. This puts the latest date of curlew separation from the species around this time or most probably earlier.

30. Fletcher *et al.* 2010.

31. Etheridge, B., Summers, R.W. & Green, R.E. 1997. The effects of illegal killing and destruction of nests by humans on the population dynamics of the hen harrier *Circus cyaneus* in Scotland. *Journal of Applied Ecology* 34: 1081–1105.

32. Whitfield, D.P., McLeod, D.R.A., Watson, J., Fielding, A.H. & Haworth, P.F. 2003. The association of grouse moor in Scotland with the illegal use of poisons to control predators. *Biological Conservation* 114: 157–163.

33. Fielding, A., Haworth, P., Whitfield, P., McLeod, D. & Riley, H. 2011. A conservation framework for hen harriers in the United Kingdom. JNCC Report 441. Joint Nature Conservation Committee, Peterborough. http://jncc.defra.gov.uk/pdf/jncc441.pdf (accessed October 2018).

34. Hayhow, D.B., Eaton, M.A., Bladwell, S. *et al.* 2013. The status of the Hen Harrier, *Circus cyaneus*, in the UK and Isle of Man in 2010. *Bird Study* 60: 446–458, doi: 10.1080/00063657.2013.839621.

35. Annual hen harrier figures are available from the Rare Breeding Birds Panel (www.rbbp.org.uk). This is a cross-organisation panel, which compiles definitive statistics on our nation's rarer breeding birds.

36. The extensive studies of Langholm and summary thereof are best available via www.langholmproject.com/publications.html. Pertinent to this chapter is: Ludwig, S.C., Roos, S., Bubb, D. & Baines, D. 2017. Long-term trends in abundance and breeding success of red grouse and hen harriers in relation to changing management of a Scottish grouse moor. *Wildlife Biology*: wlb.00246. doi: 10.2981/wlb.00246.

37. Whitfield, D.P., Fielding, A.H., Mcleod, D.R.A. *et al.* 2007. Factors constraining the distribution of Golden Eagles *Aquila chrysaetos* in Scotland. *Bird Study* 54: 199–211. doi: 10.1080/00063650709461476. See also (a) Whitfield, D.P., Fielding, A.H., McLeod, D.R.A. & Haworth, P.F. 2004. Modelling the effects of persecution on the population dynamics of golden eagles in Scotland. *Biological Conservation* 119: 319–333; (b) Whitfield, D.P., Fielding, A.H., McLeod, D.R.A. & Haworth, P.F. 2004. The effects of persecution on age of breeding and territory occupation in golden eagles in Scotland. *Biological Conservation* 118: 249–259; (c) Watson, A., Payne, S. & Rae, R. 1989. Golden Eagles *Aquila chrysaetos*: land use and food in northeast Scotland. *Ibis* 131: 336–348. doi: 10.1111/j.1474-919X.1989.tb02783.x.

38. L. Waddell, cited in Beeston, R., Baines, D. & Richardson, M. 2005. Seasonal and between-sex differences in the diet of Black Grouse *Tetrao tetrix*. *Bird Study* 52: 276–281. doi: 10.1080/00063650509461400.

39. Little, B. & Davidson, M. 1992. Merlins *Falco columbarius* using crow nests in Kielder Forest, Northumberland. *Bird Study* 39: 13–16. doi: 10.1080/00063659209477093.

40. Heavisides, A., Barker, A. & Poxton, I. 2017. Population and breeding biology of merlins in the Lammermuir Hills. *British Birds* 110: 138–154.

41. Warren, P., & Baines, D. 2014. Changes in the abundance and distribution of upland breeding birds in the Berwyn Special Protection Area, North Wales 1983–2002. *Birds in Wales* 11: 32–42.

42. UK Government & Parliament. 2016. Petition: ban driven grouse shooting. https://petition.parliament.uk/archived/petitions/125003 (accessed October 2018).

43. The Fjelljakt website is fascinating and well worth a look: http://fjelljakt.se. It demonstrates how other countries are combining beautiful ecosystems with

commercial hunting – the normal state in a capitalist democracy. What we have in Britain's uplands is uniquely unprofitable for everyone.

44. OneKind. 2017. Mountain hare persecution in Scotland. www.onekind.scot/wp-content/uploads/Mountain-hare-persecution-in-Scotland-JULY-2017.pdf (accessed October 2018). This provides a damning and meticulously documented photographic account of the obscene slaughter of mountain hares in a national park, the Cairngorms. The relevant papers for this matter, which is being widely explored by conservationists at the moment, include: (a) Harrison, A., Newey, S., Gilbert, L., Haydon, D.T. & Thirgood, S. 2010. Culling wildlife hosts to control disease: mountain hares, red grouse and louping ill virus. *Journal of Applied Ecology* 47: 926–930; (b) Newey, S. 2008. The conservation status and management of mountain hares. Scottish Natural Heritage Commissioned Report no. 287 (ROAME No. F05AC316).

45. Watson, A. & Wilson, J.D. Seven decades of mountain hare counts show severe declines where high-yield recreational game bird hunting is practised. *Journal of Applied Ecology*, published online August 2018. doi: 10.1111/1365-2664.13235.

12. Pelican Possibility

1. Senator Frank Church, from Idaho, USA, arguing for passage of the Wild and Scenic Rivers Act (1968). The act was passed.

2. Baboianu, G. *Pelecanus crispus* Romania – Saving *Pelecanus crispus* in the Danube Delta. LIFE 05 NAT/RO/000169. A report outlining current conservation measures in Romania for the species, available via www.ec.europa.eu.

3. Author measurement using Google My Maps. I projected a similar area to the 'core' Danube onto the Somerset Levels, to see how much larger Somerset's reedbeds and wetlands would need to grow to accommodate such a scale of habitat.

4. Danube Delta Biosphere Reserve Authority. Distribution of the population in Danube Delta Biosphere Reserve. www.ddbra.ro/en/danube-delta-biosphere-reserve/danube-delta/population/distribution-of-the-population-in-danube-delta-biosphere-reserve-a909 (accessed October 2018).

5. Author measurement using Google My Maps. Takes in the full extent of the main Somerset reedbeds (Ham Wall, Shapwick Heath reserves etc.) at the present time.

6. Farm Business Survey. [2010?] Agriculture in the south west of England 2009/2010. http://farmbusinesssurvey.co.uk/regional/commentary/2009/southwest.pdf (accessed October 2018). The Farm Business Survey is conducted on behalf of and financed by the Department for Environment, Food and Rural Affairs (Defra).

7. Office for National Statistics. 2001. *200 Years of the Census in … Wales*. ONS, London.

8. Somerset County Council. 2016. State of the Somerset economy report, April 2016. Economy and Planning, Somerset County Council. Available via www.somerset.gov.uk/policies-and-plans/schemes-and-initiatives/

somerset-economic-assessment (accessed October 2018). See Chapter 3, Industry sectors. The report includes the following statistics on the Somerset agricultural sector: (a) agriculture, forestry and fishing combined provide just 600 jobs in Somerset (100 on Mendip, 100 on Sedgemoor, 200 in South Somerset, 100 in Taunton Deane and 100 in West Somerset); (b) percentage GVA for agriculture decreased by 8.9% between 2009 and 2014; (c) agriculture, forestry and fishing constitute the fourth least productive industry sector in Somerset by GVA by jobs, and in terms of overall GVA, not even in the Somerset top ten (£187 million).

9. Duchy College. 2016. The south west dairy industry: a vital cog in the economy. A document prepared for the National Farmers' Union by the Rural Business School, Duchy College, October 2016. www.nfuonline.com/assets/67942 (accessed October 2018).

10. Duchy College 2016.

11. Norton, S. 2015. Should we be drinking milk? Arguments for and against dairy. *The Independent*, 21 April 2015.

12. Matthews, S.B., Waud, J.P., Roberts, A.G. & Campbell, A.K. 2005. Systemic lactose intolerance: a new perspective on an old problem. *Postgraduate Medical Journal* 81: 167–173.

13. The Bug River is a large linear landscape but can be found on Google Maps by entering the following GPS: 52.684352, 22.238333. As can be seen, there is human activity in the landscape in the form of strip-farming, but this works by harvesting from the floodplain. The aerial map, taken in summer, reveals the river's natural course, with plenty of gravel barriers breaking its flow, and meanders to slow its course.

14. Carrington, D. 2014. Somerset Levels floodwater pumping operation costs £100,000 a week. *The Guardian*, 3 February 2014. www.theguardian.com/environment/2014/feb/03/somerset-levels-floodwater-pumping-environment (accessed October 2018).

15. Palmer, R.C. & Smith, R.P. 2013. Soil structural degradation in SW England and its impact on surface-water runoff generation. *Soil Use and Management* 29: 567–575. doi: 10.1111/sum.12068. The quote is as follows: 'Late-harvested crops such as maize had the most damaged soil where 75% of sites were found to have degraded structure generating enhanced surface-water runoff.'

16. Palmer & Smith 2013. The quote is as follows: 'The intensive use of well-drained, high-quality sandy and coarse loamy soils has led to soil structural damage resulting in enhanced surface-water runoff from fields that should naturally absorb winter rain. Surface-water pollution, localized flooding and reduced winter recharge rates to aquifers result from this damage.'

17. Roosevelt, F.D. 1937. Letter to all state governors on a uniform soil conservation law, February 26, 1937. *The American Presidency Project*. http://www.presidency.ucsb.edu/ws/?pid=15373 (accessed October 2018).

18. Rewilding Britain. 2015. Flooding, trees and rewilding. https://www.rewildingbritain.org.uk/blog/flooding-trees-and-rewilding (accessed October 2018).

19. Environment Agency. 2014. River dredging and flood defence: to dredge or not to dredge? A 2014 PDF presentation cited in Monbiot G. 2014. Dredged up. www.monbiot.com/2014/01/30/dredged-up (accessed October 2018).

20. Kucharska, A. & Znaniecka, M. [2004.] Extensive farming practices as a tool for active protection of Biebrza wetlands biodiversity on the example of model WWF projects. *wetHYDRO publications* – Workshop 4 (Warsaw, September 2004): 65–79. http://levis.sggw.waw.pl/wethydro/contents/monografie/ws4/065-079_MalgorzataZnaniecka_e.pdf (accessed October 2018).

21. Cross, M. 2017. Wallasea Island Wild Coast Project, UK: circular economy in the built environment. *Proceedings of the Institution of Civil Engineers – Waste and Resource Management* 170: 3–14. doi: 10.1680/jwarm.16.00006.

22. Boyle, M., Galpin, B. & Tinsley-Marshall, P. 2014. Great Fen Bird Report, 2005–2013. Great Fen Project, Ramsey Heights, Cambridgeshire. www.greatfen.org.uk/great-fen-bird-report-2005-2013 (accessed August 2018).

13. Our Birds

1. Thoreau, H.D. Letter to Harrison Blake (20 May 1860); published in Sanborn, F.B. (ed.). 1906. *Familiar Letters.* Houghton Mifflin, Boston, MA.

2. The abandoned airfield and its fauna is Tempelhofer Feld. The goshawks and boars are found across the city. The lake with black terns is Müggelsee. The park with wrynecks is Hobrechtsheide. The area with bison (enclosed) and common red-backed shrikes is Döberitzer Heide. The beaver rest platform is placed near Ostbahnhof railway station.

3. Witt, K. 1991. Rote Liste der Brutvoegel in Berlin, 1. Fassung. *Berliner Ornithologischer Bericht.*

4. Rae, A. 2017. A land cover atlas of the United Kingdom (document). https://figshare.com/articles/A_Land_Cover_Atlas_of_the_United_Kingdom_Document_/5266495/1 (accessed October 2018). Full set of maps: https://doi.org/10.15131/shef.data.5219956. The citation for the underlying data is given as: Cole, B., King, S., Ogutu, B. *et al.* 2015. Corine Land Cover 2012 for the UK, Jersey and Guernsey. NERC Environmental Information Data Centre.

5. Davies, Z.G., Fuller, R.A., Loram, A. *et al.* 2009. A national scale inventory of resource provision for biodiversity within domestic gardens. *Biological Conservation* 142: 761–771. doi: 10.1016/j.biocon.2008.12.016.

6. Plummer, K.E. 2011. The effects of over-winter dietary provisioning on health and productivity of garden birds. PhD thesis, University of Exeter. *Author comment*: the questionable effects of artificially feeding garden birds is something too long for this book to enter into at length, and still under investigation. However, a few things are worth bearing in mind. Firstly, blue and great

tits are competitors of a number of vanishing species, including marsh and willow tit (nesting resources), and wood warbler and pied flycatcher (lepidopteral resources). Artificially increasing their populations is very questionable. Secondly, if you are helping unhealthy blue or great tits through the winter, you are pushing otherwise unhealthy birds back into the natural selection process. We do not know how such a process will end. Thirdly, winter bird tables have been complicit in the contraction of diseases such as trichomonosis, devastating in the collapse of our greenfinch population, and chaffinch numbers too. Overall, I am fairly sceptical as to whether, warm fuzzy feelings aside, artificial feeding is a good idea for British birds as a whole. Instead, this chapter advocates nest sites and increasing natural feeding opportunities for birds, because this is in line with how they evolved in the first place. See: Lawson, B., Robinson, R.A., Colvile, K.M. *et al.* 2012. The emergence and spread of finch trichomonosis in the British Isles. *Philosophical Transactions of the Royal Society B* 367: 2852–2863. doi: 10.1098/rstb.2012.0130.

7. Daily Telegraph. 2012. Council workmen mow down tens of thousands of rare wildflowers in just two hours. *The Telegraph*, 29 June 2012. The same story was covered by a number of national papers including *The Daily Mail*.

8. Roos, S., Johnston, A. & Noble, D. 2012. UK hedgehog datasets and their potential for long-term monitoring. BTO Research Report No. 598. www.bto.org/sites/default/files/u12/hedgehogscopingreportfinal05042012ptes.pdf (accessed October 2018). A report on work carried out by the BTO, commissioned and funded by the People's Trust for Endangered Species and the British Hedgehog Preservation Society.

9. There are organic slug pellets available that do not kill or endanger hedgehogs – these are available online via the RSPB website. The dangerous slug pellet for the entire British food chain is that containing metaldehyde. These pellets are generally blue in colour. Metaldehyde has been proven to have negative effects on wild birds, hedgehogs and humans, causing liver damage if ingested in sufficient amounts, and can prove dangerous to dogs. It is the second most common cause of poisoning in dogs after chocolate. A long, impartial paper written by those outside of the industry (various papers turn out to be written by those who benefit from selling the chemical) can be found here: Castle G.D., Mills, G.A., Gravell, A. *et al.* 2017. Review of the molluscicide metaldehyde in the environment. *Environmental Science: Water Research and Technology* 3: 415–428. doi: 10.1039/C7EW00039A. See also: Reece, R.L., Scott, P.C., Forsyth, W.M., Gould, J.A. & Barr DA 1985. Toxicity episodes involving agricultural chemicals and other substances in birds in Victoria, Australia. *Veterinary Record* 117: 525–527.

10. Studies by Lisa Warnecke (2016) for the Society of Experimental Biology at the University of Hamburg, paper under preparation. For more information contact lisa.warnecke@uni-hamburg.de.

11. Walker, L.A. Chaplow, J.S., Llewellyn, N.R. *et al.* 2013. Anticoagulant rodenticides in predatory birds 2011: a Predatory Bird Monitoring Scheme (PBMS)

report. Centre for Ecology & Hydrology, Lancaster. https://pbms.ceh.ac.uk/sites/default/files/PBMS_Rodenticide_2011.pdf (accessed October 2018).

12. Plantlife. 2016. *The Good Verge Guide: a Different Approach to Managing our Waysides and Verges*. Plantlife, Salisbury. www.plantlife.org.uk/application/files/4614/8232/2916/Road_verge_guide_17_6.pdf (accessed October 2018). An excellent booklet on maintaining our roadside biodiversity.

13. Plantlife 2016.

14. Womack, A. 2016. South Gloucestershire Council urged to stop using glyphosate in weed killing programme by the Green Party. *The Gazette Series*, 22 April 2016. Various petitions have been levelled at the council to change its position.

15. Portier, C.J., Armstrong, B.K., Baguley, B.C. *et al.* 2016. Differences in the carcinogenic evaluation of glyphosate between the International Agency for Research on Cancer (IARC) and the European Food Safety Authority (EFSA). *Journal of Epidemiology and Community Health* 70: 741–745. doi: 10.1136/jech-2015-207005. Opinion on *human* impact of glyphosates is divided, as follows: (a) 'The IARC WG concluded that glyphosate is a "probable human carcinogen", putting it into IARC category 2A due to sufficient evidence of carcinogenicity in animals, limited evidence of carcinogenicity in humans and strong evidence for two carcinogenic mechanisms.' (b) 'The RAR concluded that "classification and labelling for carcinogenesis is not warranted" and "glyphosate is devoid of genotoxic potential".'

16. MacKinnon, D.S. & Freedman, B. 1993. Effects of silvicultural use of the herbicide glyphosate on breeding birds of regenerating clearcuts in Nova Scotia, Canada. *Journal of Applied Ecology* 30: 395–406. doi: 10.2307/2404181.

17. Daily Telegraph 2012.

18. Halliday, J. 2018. Sheffield council pauses tree-felling scheme after criticism. *The Guardian*, 26 March 2018. This story has been widely covered in a number of papers.

19. Perraudin, F. 2017. Sheffield tree protester given suspended jail sentence. *The Guardian*, 3 November 2017.

20. Macdonald, B. 2012. Lost world: how tidying up our countryside has created a fatal mess for our wildlife. *Bird Watching*, May 2012, 68–71.

21. Robbins, C.S. 1973. Introduction, spread, and present abundance of the house sparrow in North America.' *Ornithological Monographs* 14: 3–9. www.jstor.org/stable/40168051.

22. Blakers, M., Davies, S.J.J.F. & Reilly, P.N. 1984. *The Atlas of Australian Birds*. Melbourne University Press, Melbourne, p. 586.

23. Balmer, D., Gillings, S., Caffrey, B. *et al.* 2013. *Bird Atlas 2007–11: the Breeding and Wintering Birds of Britain and Ireland*. BTO, Thetford. This book is used for all statistics of bird population changes in this chapter unless specified otherwise.

24. Vincent, K.E. 2006. Investigating the causes of the decline of the urban House Sparrow *Passer domesticus* population in Britain. PhD thesis, De Montfort University.

25. Stoate, C. & Szczur, J. 2006. Potential influence of habitat and predation on local breeding success and population in Spotted Flycatchers *Muscicapa striata*. *Bird Study* 53: 328–330, doi: 10.1080/00063650609461450.

26. Derbyshire bird reports (R. Frost & R. Key) in Turner A. 2006. *The Barn Swallow*. Poyser Monographs. Bloomsbury, London, p. 100.

27. Holloway, S. 1996. *The Historical Atlas of Breeding Birds in Britain and Ireland 1875–1900*. Poyser, London.

28. Newman, J.R., Novakova, E. & McClave, J.T. 1985. The influence of industrial air emissions on the nesting ecology of the house martin *Delichon urbica* in Czechoslovakia. *Biological Conservation* 31: 229–248. doi: 10.1016/0006-3207(85)90069-2.

29. BTO. 2012. Wild bird populations in the UK 1970–2011. www.bto.org/news-events/news/2012-12/wild-bird-populations-uk-1970-2011 (accessed October 2018).

30. My own first-hand experience and design, based on siting successful starling nest boxes on houses and also in orchards (southwest England). Starlings are evolved to nest in deep woodpecker holes and seek out deep cavities in houses, which are not dissimilar; many designs on the internet are far too shallow for starlings.

31. Natural England. 2011. Traditional Orchard Project in England: the creation of an inventory to support the UK Habitat Action Plan. Natural England Commissioned Report NECR077. http://publications.naturalengland.org.uk/publication/47015 (accessed October 2018).

32. My own field observation of sixteen active nests in an orchard near Mathon, Herefordshire, 2012–2018.

33. For far more detail on the extraordinary biodiversity that even a single traditional organic orchard can contain, see Macdonald, B. & Gates, N. 2020. *Orchard: A Year in England's Eden*. HarperCollins, London.

34. Rock, P. 2005, Urban gulls: problems and solutions. *British Birds* 98: 338–355.

35. Kettel, E.F., Gentle, L.K. &. Yarnell, R.W. 2016. Evidence of an urban Peregrine Falcon (*Falco peregrinus*) feeding young at night. *Journal of Raptor Research* 50: 321–323. doi: 10.3356/JRR-16-13.1.

36. BBC News. 2018. Endangered birds' 'miserable deaths' in Newcastle netting. *BBC News*, 30 July 2018. www.bbc.co.uk/news/uk-england-tyne-45007207 (accessed October 2018).

37. See actionforswifts.blogspot.com and www.swift-conservation.org for all things related to how you can save swifts, and even start populations from scratch, around your home. Excellent nest boxes are available from the latter website, which also includes designs so that you can make your own.

38. For information about the current and ongoing status of willow tit research in Durham, visit the website of Durham Bird Club at www.durhambirdclub.org.

39. Data from ringing studies carried out by Professor David Norman. For more information, visit www.woolstoneyes.com.

14. Conservation Begins

1. Reuben Fine, quoted in Chernev, I. 1960. *Combinations: the Heart of Chess*. Crowell, New York.

2. Moore-Colyer, R.J. 2000. Feathered women and persecuted birds; the struggle against the plumage trade c. 1860–1922. *Rural History* 11: 57–73. doi: 10.1017/S0956793300001904.

3. RSPB. A history of the RSPB, from its humble beginnings, to the thriving far-reaching organisation it is today. www.rspb.org.uk/about-the-rspb/about-us/our-history (accessed October 2018).

4. RSPB. 2017. Trustees' report and accounts for the year ending 31 March 2017. www.rspb.org.uk/globalassets/downloads/about-us/rspb-annual-accounts-2017.pdf (accessed October 2018).

5. Author's comparison of available figures for 'nature lovers' between the UK and Germany is as follows. RSPB has 1.1 million members, and the National Trust has 4 million as of 2011. The largest nature conservation charity in Germany, NABU (Naturschutzbund Deutschland) had 530,000 members as of June 2016. Germany, in both its urban and rural rewilding schemes and scope of conservation, is far ahead of the UK, yet it appears that the number of people subscribing money to protect nature is perhaps half as great. This suggests much wiser national decision-making, and more ambition, as a whole.

6. Harper, M. 2013. One big thing for nature: a comment on George Monbiot's book *Feral*. Martin Harper's blog, RSPB, 16 June 2013. https://ww2.rspb.org.uk/community/ourwork/b/martinharper/archive/2013/06/16/one-big-thing-for-nature-a-comment-on-george-monbiot-s-book-feral.aspx (accessed October 2018).

7. Countryside Council for Wales. 2011. Core management plan including conservation objectives for Yerbeston Moors SSSI (including Yerbeston Tops SAC). This SSSI example was picked because it typifies a lot of the problems with letting nature be nature; I consider it pretty typical in doing so. Many SSSI statements are available via the websites of Natural England, Nature Resources Wales and Scottish Natural Heritage for readers to decide as to whether nature requires this level of management to survive.

8. Barkham, P. 2017. 'It is strange to see the British struggling with the beaver': why is rewilding so controversial? *The Guardian*, 3 July 2017. Quotes the views of Matt Shardlow, chief executive of Buglife, on the need for small-scale reserves for insect habitat, not to be left behind in the quest to rewild large areas of land.

9. Natural England. Condition of SSSI units for site Minsmere–Walberswick Heaths and Marshes SSSI. An ongoing assessment of the SSSI, available via https://designatedsites.naturalengland.org.uk (accessed October 2018).

10. Law, A., Gaywood, M.J., Jones, K.C., Ramsay, P. & Willby, N.J. 2017. Using ecosystem engineers as tools in habitat restoration and rewilding: beaver and

wetlands. *Science of the Total Environment* 605–606: 1021–1030. doi: 10.1016/j. scitotenv.2017.06.173.

11. RSPB. Managing wet scrub. www.rspb.org.uk/our-work/conservation/ conservation-and-sustainability/advice/conservation-land-management -advice/managing-wet-scrub (accessed October 2018).

12. Petty, S. & Avery, M.I. 1990. Forest bird communities: a review of the ecology and management of forest bird communities in relation to silvicultural practices in the British uplands. Cited in WallisDeVries M.F., Bakker, J.P. & van Wieren, S.E. (eds). 1998. *Grazing and Conservation Management*. Kluwer, Dordrecht.

13. Self, M. 2005. A review of management for fish and bitterns, *Botaurus stellaris*, in wetland reserves. *Fisheries Management and Ecology* 12: 387–394. doi: 10.1111/j.1365-2400.2005.00462.x.

14. It is interesting that in spite of the name 'flycatcher', almost every paper starts with a complex hypothesis for flycatcher decline, rather than the decline in flies, bees and butterflies – which is now known in most insect orders. These hypotheses have included (but are not limited to) increased predation (jays, grey squirrels), arrival date and habitat type used. Not one UK paper that I can find suggests trophic collapse as a main driver of decline.

15. (a) Fuller, R.J. 1995. *Bird Life of Woodland and Forest*. Cambridge University Press, Cambridge; (b) Bibby, C.J. 1977. Ecology of the Dartford Warbler *Sylvia undata* (Boddaert) in relation to its conservation in Britain. PhD thesis, CNAA.

16. Local expert Rob Clements continues to map breeding and roosting aggregations of hawfinches, and breeding lesser spotted woodpeckers, in the New Forest. Amazingly few baseline surveys exist for numbers of spotted flycatchers or cuckoos, but there are regular surveys of woodlarks and nightjars. The *overall* pattern for insectivorous birds is stasis. The wood warbler is declining, as is the woodlark, for different reasons. There seems an urgent need to develop a full baseline of insectivorous birds in the forest, something that would require serious funding and coordination.

17. Allen, M.J. & Gardiner, J. 2009. If you go down to the woods today: a re-evaluation of the chalkland postglacial woodland; implications for prehistoric communities. In Allen, M.J., Sharples, N. & O'Connor, T. (eds). *Land and People: Papers in Memory of John G. Evans*. Prehistoric Society Research Paper 2. Oxbow Books, Oxford, pp. 49–66.

18. Jones, N. 2015. The knives come out of the cabinet in Churchill's wartime government. *The Spectator*, 28 March 2015.

19. Kramer, M.B. [n.d.] Williamson [née Bateson], Emily (1855–1936). *Oxford Dictionary of National Biography*. doi: 10.1093/ref:odnb/54568.

20. Moore-Colyer 2000.

21. Moore-Colyer 2000.

Index